The Activities of Bacterial Pathogens *in Vivo*

**Based on contributions to a
Royal Society Discussion Meeting**

THE ROYAL
SOCIETY

The Activities of Bacterial Pathogens *in Vivo*

Based on contributions to a Royal Society Discussion Meeting

London, UK Meeting held on 20–21 October 1999

Editors

H. Smith
Birmingham University, UK

C. J. Dorman
University of Dublin, UK

G. Dougan
Imperial College, UK

D. W. Holden
Imperial College, UK

P. Williams
University of Nottingham, UK

Imperial College Press

ICP

Published by

Imperial College Press
57 Shelton Street
Covent Garden
London WC2H 9HE

Distributed by

World Scientific Publishing Co. Pte. Ltd.
P O Box 128, Farrer Road, Singapore 912805
USA office: Suite 1B, 1060 Main Street, River Edge, NJ 07661
UK office: 57 Shelton Street, Covent Garden, London WC2H 9HE

British Library Cataloguing-in-Publication Data
A catalogue record for this book is available from the British Library.

THE ACTIVITIES OF BACTERIAL PATHOGENS *IN VIVO*
First published in the *Philosophical Transactions of The Royal Society* series B in 2000
First published by Imperial College Press in 2001

Copyright © 2000 The Royal Society

The moral rights of the authors have been asserted.

ISBN   1-86094-272-5✓

Printed in Singapore.

PREFACE

Pathogenic bacteria, i.e. those that produce disease, have unique biological properties, which enable them to invade a host and produce sickness. The molecular bases of these biological properties are the determinants of pathogenicity and research objectives are to recognize them, identify them chemically and relate their structure to function. Most of our present knowledge comes from studies with cultures *in vitro*. However, there is a rising interest in bacterial behaviour in the infected host and new methods have been developed for studying it. The objective of the Discussion Meeting was to describe these methods and to show how they, and a recent surge in conventional studies, are shedding light on the activities of bacterial pathogens *in vivo*. Participants were asked to enquire about bacterial and host factors that operate *in vivo* to bring about sickness, to show how phenomena recognized *in vitro* relate to behaviour *in vivo* and, if evidence of relevance is not available now, to indicate how it might be obtained.

There are two introductory papers. The first, by Smith, outlines the new methods, poses questions about the behaviour of bacterial pathogens *in vivo* and indicates how answers may be obtained. Growth *in vivo* and the underpinning processes of nutrition and metabolism are given special emphasis because new methods are highlighting their importance. The second, Marshall *et al.*, describes how the cellular environment can affect gene expression. It deals with the expression of genes coding for determinants of DNA topology (DNA gyrase, integration host factor and the nucleoid-associated protein H-NS) during adaptation of *Salmonella typhimurium* to the intracellular environment of macrophages. These global systems influence the transcription of genes involved in virulence, e.g. the *spv* locus. Next, five papers describe the new methods and their use in understanding hostpathogen interactions. The first, by Philpott *et al.*, shows how a combination of studies with cell cultures and those with various animal models (infections of macaques, rabbit intestinal loops and murine lungs) have defined the molecular basis for mucosal invasion and the stimulation of inflammation by *Shigella flexneri*, which may apply to dysentry in man. In the second, Merrell & Camilli describe the use of recombinase-based *in vivo* expression technology (IVET) to detect genes that are transcriptionally induced during infection, including those expressed transiently or at low levels. Spatial and temporal expression of specific genes, e.g. for the toxin co-regulated pilus (*tcpA*) and cholera toxin (*cxtA*) can be monitored during the course of infection. Hautefort & Hinton discuss many techniques, other than IVET, for detecting gene expression *in vivo*, e.g. differential fluorescence induction and *in vivo* antigen technology (IVIAT). Some of these approaches can determine whether genes are expressed constitutively or in an organ-specific or cell-type-specific fashion. The paper by Unsworth & Holden describes signature-tagged mutagenesis and its use for *S. typhimurium* in a mouse model to identify many virulence genes required for growth *in vivo*, including several clustered on a chromosomal pathogenicity island. It also shows how the use of a temperature-sensitive, non-replicating plasmid and competitive index tests can demonstrate that virulence gene function *in vivo* may differ from that predicted from *in vitro* studies. Finlay & Brumell describe the interaction of *S. typhimurium* with relevant host cells both *in vitro* and in various animal models. Sophisticated imaging and molecular genetic tools are being used to monitor gene expression in both the pathogen and the host cell during infection. Tissue culture results have been confirmed and new questions evoked.

Three papers discuss the impact of the new methods. The first of these, by Heithoff *et al.*, describes identification by IVET of many housekeeping and virulence genes of *S. typhimurium*, which are induced only *in vivo*. Some of these genes are expressed *in vitro* if regulatory genes of the DNA adenine methylase system (Dam) are mutated. Dam-negative mutants illustrate how the loss of a single enzyme can completely block the ability of a pathogen to cause disease yet fully elicit a protective immune response. The paper by Moxon & Tang shows how a combination of genomics and methods for detecting gene expression *in vivo* are identifying genes that relate to virulence. It discusses practical and semantic difficulties in distinguishing between classical virulence factors and those that promote survival and growth in the host. It underlines the problem of obtaining animal models that reflect disease in the natural host. The paper by DiRita *et al.* relates knowledge of virulence gene regulation gained from studies *in vitro* to what occurs *in vivo* for two pathogens. For *Vibrio cholerae*, the ToxR regulon is active *in vivo* but the environmental factors that activate it are not clear. For *Streptococcus pyogenes*, capsule production occurs in an animal model of necrotizing skin infection. It is critical for virulence but dependent on mutation in a two-component regulatory system CsrR and CsrS, i.e. on the loss of the regulation that occurs *in vitro*.

The next three papers discuss important aspects of bacterial pathogenicity and the evidence for their operation *in vivo*. Williams *et al.* describes quorum sensing, i.e. the regulation of bacterial processes in a cell-density-dependent manner through cell-to-cell communication by signalling molecules. Relevant signalling molecules have been detected in animal models and human infections. These molecules not only control bacterial gene expression but can also modulate host-cell responses. The paper by Cornelis describes the Yop virulon of *Yersinia* species as an archetype for type III secretion systems, which are activated by contact with eukaryotic cells. They allow bacteria to inject their proteins across two bacterial membranes and the host-cell membrane to destroy or subvert target cells. Studies with macrophages are described, but proof that they operate during animal infections has not yet been obtained. Morschhauser *et al.* shows that point mutation, genetic rearrangements and horizontal gene transfer processes contribute to macroevolution, long-term processes leading to new species, and microevolution, short-term developments occurring in days or weeks. Microevolution occurs *in vivo*; genome variability of pathogenic microbes leads to new phenotypes, which are important in acute development of an infectious disease. Horizontal transfer *in vivo* of genes by plasmids, bacteriophages and pathogenicity islands is more important for macroevolution.

The final paper, by Dougan *et al.*, raises practical implications of the new knowledge. It deals with the handling of mucosally delivered antigens in attempts to design effective vaccines. Studies are needed of the mechanism of pathogenicity employed by microbial pathogens, of the combined mucosal and systemic immune response associated with infection and recovery, and of the mechanism of action of known good mucosal immunogens. The importance of studies in the natural host or whole animal systems is emphasized.

Some important aspects that emerged during the Meeting are summarized. Many of the new methods for studying bacterial behaviour *in vivo* require the pathogen to have robust genetics capable of easy manipulation. This is not always so, for example for *Campylobacter jejuni*, and it is fortunate that some methods, such as IVIAT, can be applied to such pathogens. The current surge in knowledge of bacterial behaviour *in vivo* comes as much from application of conventional methods (chemical and biological comparison of *in vivo*- and *in vitro*-grown organisms, mutation, virulence tests and complementation) as from new methods. A serious problem in applying both the new and conventional methods is the frequent lack of realistic animal models for human infections. This may severely limit our ability to get to the molecular basis of pathogenicity in humans. The problem may be mitigated in the future by the use of transgenic animals and the design of non-invasive methods for possible use in humans. Even when satisfactory animal models are available, better methods are needed for following the progress of infection spacially and in real time *in situ*. Because of its convenience, infection of macrophages in culture has been used as a halfway house between *in vitro* and *in vivo* conditions. Although these experiments may not reflect all the nuances of infection in animals, much useful information has been obtained from them, some of which has been confirmed by experiments with animals. At present, attention is largely concentrated on bacterial activities *in vivo* rather than the host factors that affect them and the interaction between the two. Most references to host factors are made in relation to the environment in macrophages or animals as a whole, rather than to specific factors and their changes during infection.

In the future, host DNA microarrays may be used to investigate global changes in eukaryotic gene expression in response to bacterial infection. Bacterial pathogens and their exoproducts are excellent probes for host cell biology. Some of the assumptions about the behaviour of pathogens *in vivo* based on research *in vitro* have been confirmed, particularly with regard to virulence determinants and regulatory systems. However, other assumptions have been shown to be too simplistic, e.g. operation of the ToxR regulatory system. Many of the genes expressed *in vivo* detected by the new methods are involved with nutrition, growth, metabolism and survival in the tissues of the host. Some well-known traditional virulence determinants — aggressins and toxins — have not been detected. Pathogens that are fully host adapted employ slip-strand mispairing to generate population diversity and have fewer transcription regulators than pathogens with both host and environmental lifestyles. We have moved from an era of the gene to the era of the genome and can now undertake 'top-down' approaches to problems of pathogenicity. Application of the new knowledge to the design of novel approaches to preventive therapeutic medicine has begun and will accelerate. Ways of inducing pathogen attenuation rather than death may be derived. Many of the genes detected by the new methods are not known to be involved in metabolism, stress response, regulation or virulence-determinant production by the pathogen under consideration, nor of any other bacterial pathogen. Thus, vast areas of the behaviour of pathogens *in vivo* remain unexplained. A major challenge for the future will be integration of the vast amount of information that will accumulate from genomics with equally voluminous data derived from intensive use of the new methodologies for studying bacterial behaviour *in vivo*.

March 2000

C. J. Dorman[1]
G. Dougan[2]
D. W. Holden[2]
P. Williams[3]
H. Smith[4]

[1]Department of Microbiology, Moyne Institute of Preventive Medicine, University of Dublin, Trinity College, Dublin 2, Republic of Ireland

[2]Department of Biochemistry, Wolfson Laboratories, Imperial College of Science, Technology and Medicine, London SW7 2AY, UK

[3]School of Pharmaceutical Sciences, University of Nottingham, University Park, Nottingham NG7 2RD, UK

[4]The Medical School, University of Birmingham, Birmingham B15 2TT, UK

CONTENTS

Introduction

QUESTIONS ABOUT THE BEHAVIOUR OF BACTERIAL PATHOGENS *IN VIVO*

HARRY SMITH

The Medical School, University of Birmingham,
Birmingham B15 2TT, UK

Bacterial pathogens cause disease in man and animals. They have unique biological properties, which enable them to colonize mucous surfaces, penetrate them, grow in the environment of the host, inhibit or avoid host defences and damage the host. The bacterial products responsible for these five biological requirements are the determinants of pathogenicity (virulence determinants). Current knowledge comes from studies *in vitro*, but now interest is increasing in how bacteria behave and produce virulence determinants within the infected host. There are three aspects to elucidate: bacterial activities, the host factors that affect them and the metabolic interactions between the two. The first is relatively easy to accomplish and, recently, new methods for doing this have been devised. The second is not easy because of the complexity of the environment *in vivo* and its ever-changing face. Nevertheless, some information can be gained from the literature and by new methodology. The third aspect is very difficult to study effectively unless some events *in vivo* can be simulated *in vitro*.

The objectives of the Discussion Meeting were to describe the new methods and to show how they, and conventional studies, are revealing the activities of bacterial pathogens *in vivo*. This paper sets the scene by raising some questions and suggesting, with examples, how they might be answered.

Bacterial growth *in vivo* is the primary requirement for pathogenicity. Without growth, determinants of the other four requirements are not formed. Results from the new methods are underlining this point. The important questions are as follows. What is the pattern of a developing infection and the growth rates and population sizes of the bacteria at different stages? What nutrients are present *in vivo* and how do they change as infection progresses and relate to growth rates and population sizes? How are these nutrients metabolized and by what bacterial mechanisms? Which bacterial processes handle nutrient deficiencies and antagonistic conditions that may arise? Conventional and new methods can answer the first question and part of the second; examples are described. The difficulties of trying to answer the last two are discussed.

Turning to production *in vivo* of determinants of mucosal colonization, penetration, interference with host defence and damage to the host, here are the crucial questions. Are putative determinants, which have been recognized by studies *in vitro*, produced *in vivo* and are they relevant to virulence? Can hitherto unknown virulence determinants be recognized by examining bacteria grown *in vivo*? Does the complement of virulence determinants change as infection proceeds? Are regulatory processes recognized *in vitro*, such as ToxR/ToxS, PhoP/PhoQ, quorum sensing and type III secretion, operative *in vivo*? What environmental factors affect virulence determinant production *in vivo* and by what metabolic processes? Examples indicate that the answers to the first four questions are 'yes' in most but not all cases. Attempts to answer the last, and most difficult, question are also described.

Finally, sialylation of the lipopolysaccharide of gonococci *in vivo* by host-derived cytidine 5′-monophospho-N-acetyl neuraminic acid, and the effect of host lactate are described. This investigation revealed a new bacterial component important in pathogenicity, the host factors responsible for its production and the metabolism involved.

Keywords: bacteria; pathogens; *in vivo*; gonococci; sialylation; lipopolysaccharide

1. INTRODUCTION

Pathogenicity (virulence) is the capacity to cause disease. Bacterial pathogens form only a small part of the bacterial world but they attract the most attention. They have unique biological properties, which enable them to enter the tissues of man or animals and cause sickness and sometimes death. These biological properties are indicated by the progression of the disease process. The skin is a formidable barrier against bacterial attack. It can be breached by gross trauma or vector bite and a few pathogens, e.g. staphylococci are able to exploit small abrasions to cause skin infections. However, the usual route of entry for most pathogens is not through the skin but over the internal surfaces of the respiratory, alimentary or urogenital tracts. Initially, only a small number of bacteria are deposited on these mucous surfaces, so the first requirement for pathogenicity is survival and growth on them. Survival entails competition with commensals that normally inhabit these surfaces, penetration of mucus that covers them and adherence to epithelial cells. Next, most pathogens need to penetrate into the tissues to be effective, although some, like *Vibrio cholerae*, can cause disease while remaining on the mucous surface. Penetration may be achieved by entry into and egestion from epithelial cells, by passage between them or destruction of them. The third requirement is the ability to grow and multiply in the environment of host tissues, otherwise the pathogen cannot cause harm. On the mucous surfaces and within the tissues, pathogens have to contend with antibacterial substances in body fluids and within the cells they infect. Also, there are polymorphonuclear (PMN) phagocytes and macrophages that can ingest and kill them. These defence mechanisms, which act against any invading pathogen, are present at the site of infection and, within a few hours, are reinforced by the inflammatory response. Clearly, ability to withstand these host defences is the fourth requirement for pathogenicity. This quality is needed again as bacteria spread through the lymphatic system into the bloodstream, where many macrophages line the vessels of the lymph nodes, spleen and liver. A few days after initial infection, the pathogen faces an even greater obstacle, the specific immune response. To survive, it must either suppress or circumvent antibody- and cell-mediated immunity. Finally, entry and growth in the host and inhibition of host defence are not enough. To be pathogenic, the bacteria must damage the host, thus causing disease. This can be achieved directly either by production of toxins or lysis of host cells by intracellular bacteria. It can also be accomplished indirectly by stimulation of host cytokines or immunopathology.

To summarize, for the many pathogens that do not enter the host by direct penetration of skin, there are five essential biological requirements for pathogenicity: colonize mucous surfaces; penetrate them; grow in the tissues; inhibit host defence; and damage the host. The cardinal fact about pathogenicity is that it is multifactorial. Many genes are involved. Their products, the determinants of pathogenicity (virulence determinants), are the molecular bases for the five essential biological properties (Smith 1995). The goal of studies on bacterial pathogens is to recognize these determinants, to identify them and to relate their structure to function.

In the past, most of our knowledge about these determinants has come from experiments with bacteria grown in culture. Almost all pathogens have been investigated and many determinants have been identified. Examples are the pili of gonococci, which aid mucosal colonization; the Ipa proteins of shigellae, which are involved in invasion of colonic epithelium; the enterochelin of *Esherichia coli*, which aids acquisition of iron for growth; the capsular polysaccharides of pneumococci that interfere with host defence; and the lethal toxin of diphtheria bacilli (Smith 1995; Finlay & Falkow 1997).

Now, the situation has changed. There is a burgeoning interest in the activities of bacteria within the infected host. Environmental conditions *in vivo* (osmolarity, pH, E_h, and nutrient and substrate availability) differ from those in laboratory cultures. They are more complex. Pathogens may be in films on mucosal surfaces or free in body fluids, cell cytoplasm or cell vacuoles; in all cases, the environment is neither simple nor defined. Also, the conditions change during the course of infection due to inflammation, tissue breakdown and spread from one anatomical site to another. Within cells, the environment can alter when bacteria invade due to changes in host gene expression. Since environmental conditions affect bacterial growth, metabolism and regulation of gene expression (Busby *et al.* 1998; Marshall *et al.*, this issue) we should expect bacteria taken from infected animals to be different, in some respects, from those grown *in vitro*, a fact now well established for many pathogenic species (Smith 1990, 1996).

Clearly, the activities of bacteria *in vivo* must be explored for a fuller appreciation of pathogenicity. There are three aspects to the full picture. First, there are observations on the bacteria themselves and identification of virulence determinants formed by them at different stages of infection. Second, there is recognition of host factors that affect bacterial behaviour and production of virulence determinants. Third, there is investigation of the underlying metabolic interactions between bacterial and host factors. The first aspect is relatively easy to accomplish and recently new methods for doing so have been devised. The second is not easy to achieve because of the complexity of the environment *in vivo* and the fact that it changes as infection proceeds. Nevertheless, there is relevant information in the literature and some new methods have been evolved. The final aspect is very difficult to study effectively *in vivo*; some progress might be made if events *in vivo* can be simulated *in vitro*. It is not surprising that the first aspect receives most attention. Indeed, one can understand an attitude to concentrate solely on bacterial properties because the environment *in vivo* and its influence are too complex to analyse properly. However, this leaves part of the story untold.

The new methods for studying bacterial pathogens *in vivo* are listed in Table 1. Some require the pathogen to have robust genetics that are easily manipulated, which is not always the case, e.g. for *Campylobacter jejuni*. The objectives of this Discussion Meeting were to describe these methods and to show how they, and a recent surge in conventional studies, are advancing knowledge. This paper sets the scene by posing some questions and suggesting, with examples, methods whereby answers may be forthcoming.

2. QUESTIONS ABOUT THE DETERMINANTS OF BACTERIAL GROWTH *IN VIVO*

The multifactorial nature of pathogenicity means that the determinants of all five requirements are essential for its manifestation. However, growth

Table 1. New methods for studying bacterial behaviour *in vivo*.

Function	Method	Principle	Reference
Following infection in animals	Confocal laser scanning microscopy (CLSM)	Possible to examine a few bacteria in thick slides	Richter-Dahlfors *et al.* 1997
	Fluorescence-activated cell sorting (FACS)	Rapid identification of host cells containing fluorescent bacteria	Valdivia & Falkow 1997a
	Laser microprobe mass spectrometry	Measuring viability of bacteria in biopsies by Na^+/K^+ ratios	Haas *et al.* 1993; Seydel *et al.* 1992
	Photonic and radio detection of pathogens *in vivo*	Pathogens made bioluminescent by a luciferase or radiolabelled by technetium-99 m	Contag *et al.* 1995; Perin *et al.* 1997
Measuring enviromental parameters *in vivo*	Quantitative fluorescence microscopy use of reporter genes	Measurement of fluorescing dyes that react to environmental factors *LacZ* fusions to genes that respond to certain levels of compounds in the environment	Akins *et al.* 1995; Aranda et al 1992 Garcia-del Portillo *et al.* 1992; Pollack 1986
	X-ray microanalytical electron microscopy		Morgan 1985; Spencer *et al.* 1990
Detection of genes expressed *in vivo*	*in vivo* expression technology (IVET)	Genes expressed *in vivo* provide promoters for various reporting systems	Camilli *et al.* 1994; Camilli & Mekalanos 1995; Heithoff *et al.* 1997; Lowe *et al.* 1998; Mahan *et al.* 1994, 1995; Wang *et al.* 1996a, b; Young & Miller 1997
	Differential fluorescence induction (DFI)	Promoters of genes expressed *in vivo* drive expressions of green fluorescent protein	Valdivia & Falkow 1996, 1997a

Table 1. (Continued)

Function	Method	Principle	Reference
	Differential display of cDNAs	cDNAs prepared from mRNAs of genes expressed *in vivo* and compared with cDNAs from organisms *in vitro*	Abu-Kwaik & Pedersen 1996; Plum & Clark-Curtiss 1994
	Reaction with antibodies produced by infection	Reaction of products of gene libraries with antibodies evoked by infection compared to those of antibodies formed against killed bacteria	Akins *et al.* 1995; Suk *et al.* 1995; Wallich *et al.* 1995
	Labelling proteins with diaminopimelate	Diaminopimelate used by bacteria and not by host cells	Burns-Keliher *et al.* 1997
Direct identification of virulence genes	Signature-tagged mutagenesis (STM)	Non-recovery from animals after inoculating individually tagged insertion mutants indicates genes required for infection	Hensel *et al.* 1995; Chiang & Mekalanos 1998; Coulter *et al.* 1998; Mei *et al.* 1997; Shea *et al.* 1996
	Virulence complementation	Complementation of avirulent strains by gene libraries from virulent strains	Collins 1996; Pascopella *et al.* 1994
Global analysis of potential gene expression	Complete genome sequencing	Aids the identification of new genes expressed *in vivo*	Strauss & Falkow 1997; Tang & Holden 1999
	Chip technologies	Microarrays of probes allow monitoring of expression of many genes in parallel	De Saizieu *et al.* 1998; Lockhardt *et al.* 1996; Ramsay 1998; Schena *et al.* 1996

holds the primary position because without it other determinants would not be formed. This is the first reason for dealing with growth separately. The second is that the new methods for recognizing genes expressed *in vivo*

(Table 1) are underlining its importance. Many of the genes detected are involved in the acquisition of nutrients and their metabolism, e.g. members of the 100 or more genes demonstrated by *in vivo* expression technology (IVET) for infections of *Salmonella typhimuriun* in mice and macrophages (Heithoff *et al.* 1997). Finally, compared with other aspects of virulence, growth and metabolism have been neglected because it is difficult to do meaningful experiments. A discussion of these difficulties and possible ways of solving them could encourage more work in the area.

Perhaps the first point to emphasize is that it is not just growth, but rate of growth, that is important in pathogenicity. On epithelial surfaces, receptors for some bacterial adhesins are present in mucus and could delay contact between the pathogen and the surface. Their influence can be overwhelmed by rapid and substantial bacterial growth in mucus (McCormick *et al.* 1988; Mantle & Rombough 1993). At primary lodgement, the few bacterial invaders must multiply rapidly to replace losses inflicted by the powerful host defences of the inflammatory response. In acute disease, rapid growth of the pathogen in tissues is needed to produce harmful effects before a protective immune response is mounted. In chronic disease, slow growth at all stages may lead to less stimulation of immune responses. To form carrier states, a resistant stationary phase of the pathogen (Kolter *et al.* 1993; Kolter 1999) may be an advantage. These different growth rates will be determined by prevailing environmental conditions, which will also influence the size of pathogen populations that can be sustained by different tissues.

The important questions regarding bacterial growth *in vivo* are as follows. What is the pattern of a developing infection and growth rates and population sizes at different stages? What nutrients become available or depleted as infection proceeds and how do they relate to growth rates and population size? How are the nutrients metabolized and by what bacterial determinants? How do bacteria handle nutrient deficiences and antagonistic biochemical conditions? They are discussed in two sections, bacterial activities and host factors.

(a) *Bacterial activities*

The classical method of following pathogenesis is to take samples of body fluids and tissues during the course of infection, either from live animals

or those killed at intervals, and then to examine them outside the host by
in vitro methods. The latter include total and viable counts, and light and
electron microscopy. Recently, confocal laser scanning microscopy (CLSM),
fluorescence-activated cell sorting (FACS), and laser microprobe mass spectrometry (Table 1) have added new dimensions to these classical methods.
For example, CLSM of immunostained sections of livers of mice infected
with realistically small doses of *S. typhimurium* showed that the pathogen
resides intracellularly in macrophages and is cytotoxic to them (Richter-
Dahlfors *et al.* 1997), as occurs with cultured macrophages (Chen *et al.*
1996). Hence, efforts to identify the molecular determinants of cell culture cytotoxicity are now relevant to behaviour *in vivo*. The major advance
has been to use non-invasive methods such as photonic imaging and radiolabelling (Table 1) for following the pattern of infection. For example,
photonic imaging provided a surprise about salmonellosis of mice; after oral
infection, the organisms were concentrated in the caecum rather than the
ileum (Contag *et al.* 1995).

Increases or decreases in bacterial populations have been measured by
counting bacteria in blood, lymph glands, spleen, liver and other relevant
tissues, e.g. Peyer's patches (Curtiss *et al.* 1988; O'Callaghan *et al.* 1988).
Now, the new non-invasive methods can be used for evaluating population
changes in different tissues.

In considering growth rates *in vivo*, it should be remembered that population size is the result of bacterial growth and destruction or removal by
the host. If a population increases rapidly, there is no doubt that bacteria
are multiplying rapidly. But, the precise growth rate is unknown because,
although dominated by the growing pathogen, host defence will have some
effect. When the population increases slowly, or even decreases, as happens
early and late in the disease process, multiplication rates are not clear. A
rapid growth rate may be masked by an equally quick destruction by the
host. Certainly, a stationary population does not necessarily mean that the
pathogen has stopped growing.

Methods for measuring doubling times in tissues are available. The first
methods relied on a genetic marker distributing to only one of two daughter cells in each succeeding generation. The proportion of the bacterial
population carrying non-replicating markers, examined at intervals during
infection, revealed the number of preceding generations (Maw & Meynell
1968; Hormaeche 1980). The method was used for infections of *E. coli* and

S. typhimurium in mice but its scope was limited by the need for a non-replicating marker. The next method, used for mice infected with *E. coli* and *Pseudomonas aeruginosa* (Hooke *et al.* 1985; Sordelli *et al.* 1988), could have wide application. Growth rates were calculated from increases in ratios of wild-type organisms (which multiply *in vivo*) to temperature-sensitive mutants (which should not multiply *in vivo*) during the course of infection. Unfortunately, this method has not been exploited. Recently, a combination of the two methods has been used to compare growth rates in mice of virulent and attenuated strains of *S. typhimurium* (Gulig & Doyle 1993). The marker, inherited by only one of the progeny on division *in vivo*, was the temperature-sensitive Cm^r plasmid pHSG 422, which is maintained on replication at 30°C but not at 37°C. Overall, these methods indicated that growth *in vivo* was slower than *in vitro* in some cases (Maw & Meynell 1968; Hormaeche 1980) and similar in others (Hooke *et al.* 1985; Sordelli *et al.* 1985). But, their use has been limited. Since the 1960s, when the subject was first raised, only three pathogens, *E. coli*, *S. typhimurium* and *P. aeruginosa*, have been examined.

In view of the importance of growth rates in pathogenicity, it would be a great advance if an easily used, non-invasive method for measuring them *in vivo* could be devised. If a method became available, particular attention should be given to growth rates in the early, crucial stage of infection, which are obscured by the bactericidal effects of host defences. Also, the possible occurrence of non-growing bacterial populations in persistent infection and carrier states should be investigated, in view of the knowledge accumulating about stationary bacterial populations *in vitro* (Kolter *et al.* 1993; Kolter 1999).

(b) *Host factors*

The first method to recognize nutrients that might determine growth *in vivo* relies on the fact that most key nutrients, e.g. iron, will also be necessary for growth *in vitro*. Hence, the first step is to grow the pathogen in a defined medium and observe the effects of deleting specific nutrients. Then, the presence *in vivo* of the identified nutrients can be ascertained. Much information on sugars, aliphatic-, hydroxy- and long-chain fatty acids, amino acids, purines, pyrimidines, vitamins and metal ions in blood, body fluids, neutrophils, macrophages and other tissues is known from physiological, pathological and pharmacological studies (Lentner 1981, 1984). Also, tissue samples can be analysed by established biochemical methods.

Some key nutrients may not be revealed by these studies *in vitro*. Erythritol is used preferentially by *Brucella abortus* in a medium containing glucose (Anderson & Smith 1965). It is concentrated in the placenta, foetal fluids and chorions of pregnant cattle, and during brucellosis promotes infection of these tissues leading to abortion (Keppie *et al.* 1965; Smith *et al.* 1962; Williams *et al.* 1964). Its importance in the metabolism of *B. abortus* was discovered by noting growth stimulation when foetal fluids or placental extracts were added to cultures *in vitro* and then purifying the stimulant. This procedure could be applied to other pathogens, particularly those that show tissue tropism in disease.

Auxotrophic mutants of pathogens can be used to check the availability of specific nutrients *in vivo*. For example, auxotrophs unable to synthesize p-aminobenzoic acid, purines, thymine and histidine have been prepared from *Salmonella typhi*, *S. typhimurium* and *Shigella flexneri* (Ahmed *et al.* 1990; Curtiss *et al.* 1988; Fields *et al.* 1986; Karnell *et al.* 1993; Leuing & Finlay 1991; Levine *et al.* 1987; O'Callaghan *et al.* 1988). They have low virulence for mice, rabbits, monkeys or man due to impaired growth *in vivo*, indicating that p-aminobenzoic acid, purines, thymine or histidine are absent or scarce in these animals. Similar auxotrophs for other nutrients could be prepared from different pathogens. The absence or presence of the specific nutrients would be indicated by comparing their multiplication rates *in vivo* with those of wild-types. This method reveals deficiencies in nutrients that may be required by other pathogens. It does not provide information on nutrients that the wild-type uses *in vivo* to synthesize the particular metobolite required by the auxotroph.

Another method for recognizing key nutrients of growth *in vivo* covers the possibility that they may not be the same as those needed for growth *in vitro*. Bacteria grown *in vivo* can be investigated by the new methods (Table 1) for genes that code for enzymes involved in acquiring and metabolizing nutrients. The nature of these enzymes will indicate nutrients used *in vivo*. In a signature-tagged mutagenesis (STM) study of staphylococci in infected mice, a prominent identified virulence gene (i.e. one which when mutated results in reduced virulence) coded for a proline permease, indicating that scavenging for proline is essential for virulence (Schwan *et al.* 1998).

After key nutrients have been identified their concentrations in the tissues can be obtained from the literature (Lentner 1981, 1984) or measured by appropriate methods. The latter is relatively easy for normal uninfected tissues but monitoring nutrient concentration in infected tissues and their changes as disease progresses is extremely difficult, even if good animal models are available.

Measuring environmental parameters within infected cells is not easy but some progress has been made using new methods. Quantitative fluorescence microscopy (Table 1) has been used to measure intraphagosomal pH in macrophages (Aranda *et al.* 1992) and LacZ reporter genes (Table 1) have been used to indicate Ca^{2+}, Fe^{2+} and Mg^{2+} levels in tissue culture cells (Pollack *et al.* 1986; Garcia-del Portillo *et al.* 1992).

The relevance of identified nutrients to infection *in vivo* can be tested by two methods. The virulence of the pathogen may be enhanced by injecting additional nutrient. Also, reduced virulence of mutants unable to use it could be investigated. If it is irreplaceable, e.g. iron, these mutations will usually be lethal. If, however, the nutrient is preferred but replaceable by another, e.g. erythritol by glucose for *B. abortus* (Anderson & Smith 1965), then the required mutant might be obtained.

Some biochemical conditions existing *in vivo*, which might have adverse effects on growth, e.g. high osmolarity, low pH and anaerobic conditions, can be identified and quantified by reference to the literature or by analysis. Antagonistic influences may also be indicated by the functions of genes whose expression *in vivo* is detected by the new methods (Table 1), e.g. those that deal with variations in osmolarity.

The final questions posed at the beginning of this section — How are the nutrients metabolized and by what bacterial determinants? How do bacteria handle nutrient deficiencies and antagonistic biochemical conditions? — are extremely difficult to answer. Trying to investigate the metabolism of pathogens growing in infected tissues is well nigh impossible. If some aspect of growth *in vivo*, such as doubling time or population size, could be simulated *in vitro* by culturing the pathogen in a medium to which nutrients known to be important for growth *in vivo* are added at the appropriate concentrations, some meaningful observations could be made by established methods of bacterial physiology. Similarly, the effects of adverse conditions could be investigated provided the phenomenon *in vivo* could be simulated *in vitro*. In both cases, the enzymic products of genes shown to be expressed *in vivo* by the new methods (Table 1) should be kept in mind. Also, the complete genomes of pathogens will reveal their overall metabolic potential, e.g. the presence, absence or incompleteness of a citric acid cycle (Huynen *et al.* 1999).

In view of the complexity of the experimental systems, it is hardly surprising that there has been little progress in answering questions on the nutrients and metabolism that underpin bacterial growth *in vivo*. However, there are a few examples showing that answers can be obtained. The best is the acquisition of iron by pathogens *in vivo*. Iron is an essential nutrient for all bacteria. First, it was shown that availability of iron *in vivo* is restricted by chelation to host transferrin and lactoferrin, and that injection of iron salts enhanced virulence of many pathogens in various animal models (Bullen 1981). Then, molecular studies were conducted *in vitro* under iron-limiting conditions. These showed that different pathogens adopt numerous strategies to overcome iron restriction (Weinberg 1995). In some cases, siderophores are excreted, which chelate iron and return to bacteria via specially induced cell-surface protein receptors (Brown & Williams 1985; Weinberg 1995). After internalization, the siderophores give up their iron under the influence of reductases (Halle & Meyer 1992). In other cases, transferrin-bearing iron interacts with cell-wall protein receptors and iron is delivered into the bacteria (Cornelissen *et al.* 1992; Anderson *et al.* 1994). In both cases, the cell-wall receptors were shown to be present on bacteria in patients or infected animals (Brown & Williams 1985; Cornelissen *et al.* 1992; Smith 1990, 1996). Finally, mutants deficient in the determinants of iron acquisition were shown to be attenuated in virulence tests, e.g. a gonococcal mutant deficient in the transferrin receptor was less infective for human volunteers than the wild-type (Cornelissen *et al.* 1997).

Two other examples relate to tissue localization by pathogens. Urea is a growth stimulant for *Proteus mirabilis*, which causes severe kidney infections (Braude & Siemienski 1960). The fact that this growth stimulation contributes to localization in the kidney was supported by the impaired growth in the kidneys of mice of a urease-deficient mutant (MacLaren 1968). Also, the following evidence supports the role of erythritol in stimulating massive growth by *B. abortus* in foetal tissues of cattle, resulting in abortion (Keppie *et al.* 1965; Smith *et al.* 1965; Williams *et al.* 1964). Injections of erythritol enhanced infection of *B. abortus* in newborn calves. Erythritol analogues inhibited growth of *B. abortus in vitro* and *in vivo*. A strain (S19) of *B. abortus* unable to use erythritol did not cause abortion.

Finally, there are results emerging from use of the new methods of detecting gene expression *in vivo*. Two previously unrecognized genes of *E. coli*, *guaA* and *argC*, induced in urine appear important in uropathogenesis (Russo *et al.* 1996). Urine contains no guanine and only low levels of arginine and the induced genes allow *E. coli* to synthesize them. Deletion mutants do not grow in urine and in mice are less virulent than the wild-type. In addition to these genes, the osmoregulatory transporter ProP, coupled with osmoprotective betaine, allow *E. coli* to grow in human urine and to colonize the urinary tract of mice (Culham *et al.* 1998). Turning to *V. cholerae*, STM identified an attenuated biotin auxotroph from the intestine of infected mice (Chiang & Mekalanos *et al.* 1998). This suggested that biotin synthesis is a virulence attribute and that there was little available biotin in the infant mouse intestine, a fact supported by enhanced colonization when biotin was added to the inoculum.

3. QUESTIONS ABOUT PRODUCTION *IN VIVO* OF DETERMINANTS OF MUCOSAL COLONIZATION, PENETRATION, INTERFERENCE WITH HOST DEFENCE AND DAMAGE TO THE HOST

Far more is known about these determinants than those responsible for growth.

(a) *Bacterial activities*

The fact that environmental conditions *in vivo* differ from those *in vitro* and change as infection proceeds has the following implications. First, some putative virulence determinants indicated by experiments *in vitro* may not be formed *in vivo*, and even if they are, they may not be necessary for virulence. Second, some determinants formed *in vivo* may not be produced *in vitro*. Third, the complement of determinants may change as infection proceeds and different anatomical sites are affected. The questions relate to the validity of these implications.

(i) Confirming production *in vivo* and relevance to virulence of putative determinants recognized *in vitro*

It is now standard practice in most studies of pathogenicity to confirm the production *in vivo* of putative determinants. Bacteria harvested directly from patients or infected animals can be examined by conventional methods (Smith 1990, 1996). The profiles of homogenates run on sodium dodecyl sulphate polyacrylamide gel electrophoresis (SDSPAGE) can be examined for bands corresponding to those of purified putative determinants. If the latter are antigenic, specific antisera can be used to immunoblot the profiles and for fluorescence microscopy of bacteria in tissue sections. Also, convalescent sera from patients or animals can be examined for appropriate antibodies by using them to immunoblot SDSPAGE profiles of *in vivo*- and *in vitro*-grown organisms. In addition, the new methods for detecting gene expression *in vivo* (Table 1) can confirm production *in vivo* of putative determinants, e.g. *tcp* (the toxin-coregulated pilus) of *V. cholerae* by STM (Chiang & Mekalanos 1998). Usually the putative determinant is produced *in vivo* but not always. For example, capsular polysaccharide type 5 is formed by *Staphylococcus aureus in vitro* in aerated cultures but minimally in the lungs and nasal polyps of cystic fibrosis patients (Herbert *et al.* 1997). It is also usual to prove relevance to virulence of the putative determinant by showing that deficient mutants are less virulent than wild-type strains in animal models of infection, followed by complementing the deficient gene (Falkow 1988). In most cases, relevance to virulence is proven, but not always, even when the putative determinant is present *in vivo*. The 17 kDa product of the *ail* gene of *Yersinia enterocolitica* is formed in Peyer's patches of mice (Wachtel & Miller 1995) but experiments with an *ail*-deficient mutant indicated that the gene is not required, either for primary invasion or to establish systemic infection. In cell cultures, Opa proteins are determinants of cell invasion by gonococci grown *in vitro* (Dehio *et al.* 1998). Gonococcal strain FA1090 produces at least eight antigenically distinct Opa proteins. When an Opa-negative variant was inoculated into volunteers, Opa proteins were expressed in a large proportion of the re-isolates. The predominant Opa variants differed between subjects. Hence, Opa proteins are formed *in vivo*. However, one variant expressing Opa protein F, which was highly represented in the re-isolates, was no more infective for volunteers than an Opa-negative variant (Jerse *et al.* 1994).

Another encouraging trend is that results of cell culture tests for virulence determinants are being viewed in relation to the pathology of disease. In the forefront of such studies is penetration of the mucosa by intestinal pathogens for which the determinants of invasion of epithelial cell lines are known. However, *in vivo* these pathogens do not usually invade epithelial cells directly. They penetrate the mucosa via M cells present in Peyer's patches and elsewhere (Jepson & Clark 1998). Now, the determinants of invasion of cell lines are being investigated for a role in this process. Pioneering studies were done on *S. flexneri* (Sansonetti *et al.* 1999). Experiments with Hela cells showed that entry, intracellular movement and transfer between cells were determined by plasmid gene products IpaB, IpaC, IpaD and T_{CSD}. Then, experiments with confluent polarized colonic epithelial cell lines, e.g. CaCo-2 cells showed that shigellae could not penetrate the intact brush border that exists *in vivo* and is absent from Hela cells. *In vivo*, shigellae are ingested by M cells and delivered to macrophages in the *lamina propria*. IpaB causes apoptic death of the macrophages and IL-1β is released. Inflammation follows with disruption of the epithelial cells so that shigellae can invade the sides and bases of these cells, using the determinants recognized in the Hela cell studies. *In vivo*, *S. typhimurium* causes membrane ruffling of M cells, then is internalized leading to cell destruction as for tissue culture cells (Jones *et al.* 1994; Jepson & Clark 1998). Mutants deficient in salmonella pathogenicity island SPI1 genes, and thus unable to enter tissue culture cells, caused membrane ruffling and entered M cells but to a lesser degree than the wild-type (Clark *et al.* 1996; Jepson & Clark 1998). Also, they killed mice after oral inoculation but the LD_{50}s were higher than for the wild-type. Hence, the determinants of invasion of cell lines by *S. typhimurium* have some role in invasion via the M-cell system *in vivo* but unknown determinants are also involved.

(ii) Examining bacteria grown *in vivo* for hitherto unknown virulence
    determinants

The fact that some virulence determinants may not be formed *in vitro* is generally accepted. Increasingly, the genes and their products in bacteria harvested from patients and infected animals are being compared with those of bacteria grown *in vitro* to reveal differences that may be biologically important. Also, for intracellular pathogens, organisms grown in macrophages are compared with those from cultures. Such studies are aimed at recognizing potential diagnostic aids and immunizing antigens as well as virulence determinants. Over the past ten years, conventional studies using SDS–PAGE have revealed many hitherto unknown bacterial components of numerous different pathogens (Smith 1990, 1996). Now, the new methods for detecting genes expressed *in vivo* (Table 1) are recognizing new genes of potential importance.

Having recognized a previously unknown bacterial component, the next step is to prove that it is a virulence determinant by biological tests related to pathogenicity and virulence tests on appropriate mutants. Unfortunately, in conventional studies, this follow-up has not been as popular as the original demonstration of a new component. For example, in the intestinal lumen of mice, *Y. enterocolitica* produced a plasmid-encoded outer membrane protein (23 kDa), which had not been seen in culture. On invasion of Peyer's patches, this protein and two further proteins (210 and 240 kDA) were formed (Nauman *et al.* 1991). There the matter remains. The functions of these novel proteins in intestinal invasion have not been investigated. Follow-up on hitherto unknown genes revealed by the new methods has been much better. In many cases, virulence tests have been done on appropriate mutants and, for STM, the method itself detects only virulence genes. As a result, many new virulence genes have been recognized, e.g. by IVET and STM for *V. cholerae* (Camilli & Mekalanos 1995; Chiang & Mekalanos 1998), *S. typhimurium* (Hensel *et al.* 1995; Mahan *et al.* 1993), *Y. enterocolitica* (Young & Miller 1997) and *S. aureus* (Lowe *et al.* 1998; Mei *et al.* 1997).

(iii) Identifying virulence determinants as infection proceeds, together with the regulatory processes involved

There are some indications that the requirements for virulence determinants change as infection proceeds. For example, in experiments on gonococcal infection in volunteers (see §4), there was an indication that gonococci lacking a sialylated lipopolysaccharide (LPS) are optimal for initial invasion of epithelial cells but that later, to cope with host-defence mechanisms, production of sialylated LPS is an advantage. However, proof of specific changes in virulence-determinant complements at different stages of infection in animals has not been obtained. In studies with macrophages, switch-on of genes at different times during an infection has been detected. Differential fluorescence induction (DFI) (Table 1) distinguished two classes of macrophage-induced genes, one induced within an hour of *S. typhimurium* entering and the other after four hours (Valdivia & Falkow 1997*b*).

Some observations have been made on the operation *in vivo* of regulatory systems detected *in vitro*. Two examples are discussed. The ToxR/ToxS virulence regulon of *V. cholerae* comprises over 20 genes involved in colonization (e.g. *tcp* genes which code for toxin-coregulated pili) and production of cholera toxin (Skorupski & Taylor 1997; DiRita *et al.*, this volume). It depends on a transcriptional activator ToxR, ToxS (a stabilizer of ToxR) and ToxT, another transcriptional activator that is positively regulated by ToxR. The regulon is modulated *in vitro* by temperature, osmolarity, pH, oxygen status and availability of amino acids. The PhoP/PhoQ regulator of *S. typhimurium* is a two-component regulatory system. PhoQ is the sensor and PhoP is the transcriptional activator that is phosphorylated by PhoQ. It controls expression of more than 40 genes, including those needed to resist killing within phagocytes and those involved in lipid A synthesis (Garcia Vescovi *et al.* 1994; Guo *et al.* 1997). *In vitro*, it is affected by pH, oxygen tension, carbon and nitrogen starvation and phosphate and magnesium concentrations (Garcia Vescovi *et al.* 1996).

In considering whether these and other regulons operate *in vivo*, it should be remembered that production *in vivo* of a virulence determinant whose formation *in vitro* is controlled by a certain regulator does not necessarily mean that control *in vivo* is effected by the same regulator. To confirm that the regulator is involved, expression of its genes should be detected *in vivo*; deletion of the genes should reduce virulence; and relevant environmental parameters should be present at the level at which they are effective *in vivo*.

The IVET method did not detect expression of the cholera toxin gene nor other ToxR/ToxS regulated genes in the intestines of infected mice (Camilli & Mekalanos 1995); but STM in the same model detected mutants with insertions in *tcp* and *toxT* (Chiang & Mekalanos 1995). IVET showed that the genes of the PhoP/PhoQ system were expressed in mice infected with *S. typhimurium* (Heithoff *et al.* 1997; Conner *et al.* 1998). However, in infected macrophages DFI and IVET identified two and one gene, respectively, which were not controlled by PhoP/PhoQ (Valdivia &

Falkow 1997a; Heithoff *et al.* 1999), thus indicating that this system is not the only one operating. With regard to virulence tests on mutants, a ToxR/ToxS-deficient mutant of *V. cholerae* produced less colonization and diarrhoea in volunteers than did the wild-type (Herrington *et al.* 1988), and PhoP/PhoQ-deficient mutants were avirulent for mice (Miller *et al.* 1989; Garcia Vescovi *et al.* 1996). The influence of environmental parameters is discussed later.

Recently, two new regulatory systems have been described, quorum sensing and type III secretion (Williams *et al.*, this issue; Cornelis, this issue). They are summarized here, in order to ask some questions about their operation *in vivo*. In quorum sensing, transcriptional activators of virulence-determinant production only go into operation when a significant cell population has been attained (Guangyong *et al.* 1995; Winson *et al.* 1995). The cell-density dependency reflects a cell-to-cell communication system based on accumulation of signal molecules to a threshold concentration. The signalling molecules for *P. aeruginosa* are N-(3-oxododecanoyl)-L-homoserine lactone and N-butanoyl-L-homoserine lactone, which switch on two transcriptional activators, LasR (regulates expression of the elastase LasB) and RhiR (regulates expression of rhamnolipid), respectively (Lafiti *et al.* 1996; Pesci & Iglewski 1997). Together they regulate virulence determinants, secondary metabolites and survival in the stationary phase. The signalling molecule for *S. aureus* is an octapeptide which activates the accessary gene regulator (Agr), which positively regulates production of extracellular toxins (e.g. α-toxin, β-toxin and toxic shock syndrome toxin 1) and fibronectin-binding proteins (Guangyong *et al.* 1995).

In vivo, quorum sensing cannot operate early in infection because the numbers of pathogens are too small. To be certain that it operates later, evidence is needed for expression of relevant genes, reduced virulence if these genes are mutated, attainment of requisite population densities and detection of signalling molecules. Some of this evidence is emerging. Examination of RNAs from *P. aeruginosa* in the sputum of cystic fibrosis patients indicated that *lasR* transcription occurs and may coordinately regulate virulence genes *lasA*, *lasB* and *toxA* (Storey *et al.* 1988). Also, *lasR*-deficient mutants of *P. aeruginosa* were less able than the wild-type to produce pneumonia in neonatal mice (Tang *et al.* 1996). They were, however, equally able to infect the corneas of mice (Preston *et al.* 1997). IVET demonstrated the expression of the *agrA* gene of *S. aureus* in mice (Lowe *et al.* 1998) and staphylococcal mutants defective in *agrA* were of reduced virulence (Gillaspy *et al.* 1995).

Type III secretion systems respond when bacteria contact eukaryotic cells. They induce secretion of virulence determinants from the bacterial cell and deliver them into host cells (Hueck 1998). A good example is the Yop protein system of *Yersinia* spp. (Cornelis 1998; Cornelis *et al.* 1998). The Yop virulon is encoded by a 70 kb plasmid, pYV. It consists of (i) a secretion apparatus Ysc comprising over 20 proteins, (ii) a delivery (to host cells) system consisting of YopB, YopD, LcrV and YopQ/YopK, (iii) a control element, YopN, TycA and LcrG, and (iv) a set of effector proteins that harm phagocytes, YopE, YopH, YpkA/YopO, YopM and YopT. Transcription of genes is influenced by temperature changes and cell contact.

S. typhimurium has two type III secretion systems comprising many genes dealing with secretion, delivery, regulation and production of effects on host cells. One deals with export and translocation to host cells of invasion proteins that are responsible for membrane ruffling and entry of epithelial cells in culture (Galan 1996; Hueck 1998). The other is involved with virulence in mice and proliferation in macrophages (Shea *et al.* 1996; Hensel *et al.* 1998).

In proving operation *in vivo* of type III secretion systems, it will not be easy to show that they are switched on by contact with relevant host cells in tissue sections or biopsies as has been demonstrated for contact with tissue culture cells (Petterson *et al.* 1996). However, the new methods such as CLSM may help in this respect. It will be easier to demonstrate the expression of the regulatory genes of these complex systems *in vivo* and reduction of virulence when these genes are mutated. Indeed, this has been done for the second type III secretion system of *S. typhimurium*. The system was discovered by STM showing expression of relevant genes *in vivo*. Mutation of one regulatory gene, P_3F_4, resulted in loss of virulence (Shea *et al.* 1996).

(b) *Host factors*

First, as for growth, factors in the environment that could affect production of virulence determinants (osmolarity, pH, E_h and metabolites) should be recognized and their levels or concentrations *in vivo* determined. Then, attempts should be made to mimic *in vitro* the conditions *in vivo* and observe the effect on virulence-determinant production and its regulation. Although receiving far less attention than bacterial activities, such experiments are beginning. Six invasion genes of *S. typhimurium* were maximally expressed *in vitro* at an oxygen tension, osmolarity and pH likely to exist in the ileum (Bajaj *et al.* 1996). Conditions designed to mimic those of the intestine showed that two type III secretion genes of *S. typhi*, *invG* and *prgH* were induced by high osmolarity, anaerobic conditions and pH 6.5, and strongly repressed at pH 5.0 (Leclerc *et al.* 1998). When *S. typhi* was grown at an osmolarity (300 mM NaCl) similar to that of the human intestine, production of flagellin and salmonella invasion proteins increased, and that of the surface Vi antigen, which prevents their secretion, was depressed (Arricau *et al.* 1998). This would facilitate contact with and entry into epithelial cells. The position was reversed at an osmolarity (150 mM NaCl) similar to that within the tissues (Arricau *et al.* 1998). The *Yst* regulated toxin of *Y. enterocolitica* is produced at 37°C *in vivo* but not in culture media unless the temperature is below 30°C. However, if the osmolarity and pH of the medium are adjusted to values normally present in the ileum, toxin production at 37°C occurs (Mikulskis *et al.* 1994). Similarly, *in vitro* expression of the invasin gene *inv* by *Y. enterocolitica* is depressed at 37°C compared with that at lower temperatures, but it increases significantly if the pH is less than 7 (which occurs in the stomach) and when concentrations of Na^+ increase (which occurs near the enterocyte brush border) (Pepe *et al.* 1994). The level of *inv* expression in the mouse intestine is comparable to that at 23°C *in vitro*.

In some cases the environmental conditions that affect regulons *in vitro* have been shown to exist *in vivo*. The PhoP/PhoQ system of *S. typhimurium* appears to be controlled *in vivo* by Mg^{2+} as it is *in vitro*. During infection, *S. typhimurium* resides in phagosomes where the Mg^{2+} concentration (estimated 50–100 μM) is permissive for *phoP/phoQ* expression, unlike the high concentrations (0.5–1.0 mM) found in cytosols and body fluids (Garcia Vescovi *et al.* 1996). Also, a *phoQ* mutant that was less responsive to Mg^{2+} was of attenuated virulence for mice (Garcia Vescovi *et al.* 1996). In contrast, the position on the ToxR/ToxS system is not as clear. Classic strains of *V. cholerae* form toxin maximally *in vitro* at low temperatures and under aerobic conditions at low pH, whereas in the human intestine, toxin production occurs at 37°C and under anaerobic conditions at high pH. The actual environmental control of the ToxR/ToxS system *in vivo* is still under investigation (Skorupski *et al.* 1997; DiRita *et al.*, this volume).

4. SIALYLATION OF GONOCOCCAL LPS BY HOST CMP-NANA AND EFFECT OF LACTATE: A PARADIGM FOR INVESTIGATION OF BEHAVIOUR *IN VIVO*

Sialylation of gonococcal LPS by host-derived cytidine 5'-monophospho-N-acetyl neuraminic acid (CMP-NANA) and lactate has a major influence on many aspects of gonococcal pathogenicity. This was revealed by investigating the cause of a biological property of gonococci in urethral exudates which was lost on subculture *in vitro*. This work shows how bacterial activities *in vivo*, relevant host factors and the metabolism concerned can be identified. References to papers up to 1995 are given in Smith *et al.* (1995).

(a) *Sialylation of LPS by host CMP-NANA affects pathogenicity*

Gonococci in urethral exudates are resistant to complement-mediated killing by fresh human serum. In most cases, resistance is lost on one subculture *in vitro* but it can be restored by incubation with blood cell extracts. Fractionation showed that the resistance inducing activity is due to CMP-NANA. After growing gonococci with CMP-^{14}CNANA, autoradiography of LPS bands separated by SDS–PAGE showed that some, but not all, LPS components are sialylated. One sialylated component of 4.5 kDa is conserved in many strains and its side chain is Galβ1-4GlcNacβ1-Galβ1-4Glc. The sialylated LPS forms an irregular surface coat, which is seen on gonococci in urethral exudates whose LPS was shown to be sialylated. A previously unknown gonococcal sialyltransferase was demonstrated in gonococcal extracts. LPS sialylation is responsible for serum resistance since conversion to resistance accompanies sialylation by CMP-NANA and reversion to sensitivity occurs when sialyl groups are removed by neuraminidase. The crucial importance of LPS sialylation in serum resistance of gonococci is now generally accepted (Vogel & Frosch 1999). Sialylation of LPS also interferes with the following host-defence mechanisms: absorption of complement component C3; ingestion and killing by PMN phagocytes; killing by antisera against gonococcal proteins; and stimulation of the immune response. On the other hand, sialylation prevents invasion of epithelial cells. Observations on volunteers were consistent with these results. When they were inoculated with a strain whose LPS could not be sialylated, a variant was selected *in vivo* whose LPS could be sialylated (Schneider *et al.* 1991). This variant, recovered from the volunteers, was more virulent than the parent strain provided it was inoculated after being grown in a medium without CMP-NANA (i.e. its LPS was not sialylated so that it could invade epithelial cells) (Schneider *et al.* 1995, 1996). If the inoculum was grown with CMP-NANA, it did not infect volunteers so well, consistent with inhibition by LPS sialylation of the ability to invade epithelial cells (Schneider *et al.* 1996).

A sialyltransferase-deficient mutant, in contrast to its parent strain, did not become serum resistant when incubated with either CMP-NANA or blood cell extracts (Bramley *et al.* 1996). Hence, the latter do not contain a mechanism for sialylating gonococcal LPS, which is independent of CMP-NANA. Also, unlike the wild-type, incubation of the mutant with CMP-NANA did not increase resistance to ingestion and killing by PMN phagocytes, killing by antiserum to porin I and human complement, binding of C3 of complement, and invasion of epithelial cells (Gill *et al.* 1996). The mutant was unable to sialylate any of its LPS components, which were shown by mass spectrometry to be similar to those of the parent strain. They included components sialylatable by the sialyltransferase from the parent strain, such as the 4.5 kDa conserved component mentioned above (Crooke *et al.* 1998). Hence, loss of ability of the mutant to be converted by CMP-NANA to resistance to serum killing, and all the other facets of pathogenicity mentioned above, is attributed to loss of sialyltransferase activity rather than inability to synthesize the LPS substrate for sialylation. Final confirmation would have come from complementation of the mutant by the gene for the sialyltransferase which has been cloned and sequenced (Gilbert *et al.* 1996). This could not be achieved (Crooke *et al.* 1998) so it seems that multiple genetic loci may be essential for LPS sialylation. The mutant has not been examined in human volunteers. A sialyltransferase-negative mutant of gonococcal strain FA 1090 was as infective as the wild-type for volunteers (Cannon *et al.* 1998), but this is not unexpected because this strain is fully resistant to serum killing (Cohen *et al.* 1994) and does not require LPS sialylation to make it so.

(b) *Host lactate enhances LPS sialylation, and LPS and protein synthesis*

Another blood cell factor, which enhances the ability of CMP-NANA to sialylate gonococcal LPS and to induce serum resistance, has been identified as lactate (Parsons *et al.* 1996*a*). The enhancement occurs with minute quantities of lactate in a defined medium containing high concentrations of glucose. The action of lactate is separate from that of CMP-NANA because enhanced sialylation occurred when gonococci were pretreated with lactate (Parsons *et al.* 1996*b*). Lactate did not increase the gonococcal content of sialyltransferase (Gao *et al.* 1998). On the other hand, there was a marked increase in LPS synthesis (10–20%), which could explain the enhancement of sialylation because additional receptors for sialyl groups are provided. The increase in LPS synthesis was paralleled by increases in protein synthesis and ribose content, presumably reflecting additional ribosome production (Gao *et al.* 1998).

(c) *Metabolic effects of lactate on gonococci growing in a medium containing glucose, as occurs* in vivo

The increases in LPS, protein and ribose synthesis, first noticed under the conditions for detecting enhancement of sialylation by lactate, also occurred when both glucose and lactate concentrations in the defined medium were adjusted to levels akin to those occurring *in vivo* (Gao *et al.* 1998). Hence, there appears to be a general stimulation of gonococcal metabolism when lactate is added to media containing glucose, and there is other evidence. Lactate increased oxygen consumption by gonococci in a solution containing glucose (Britigan *et al.* 1988). Also, in the above-defined medium, growth rate was faster and lactate was metabolized side-by-side with glucose and more rapidly (Regan *et al.* 1999). When gonococci were grown with ^{14}C-labelled lactate in this medium (Yates *et al.* 1999), tricene SDS–PAGE on homogenates showed that lactate is not a general carbon source. Label was concentrated in a low M_r component, LPS and a few proteins. N-terminal sequencing of the three most heavily labelled proteins showed one (M_r *ca.* 58 kDa) to be the chaperone, GroEl and another (M_r *ca.* 35 kDa) porin 1B. Nuclear magnetic resonance after ^{13}C labelling, and thin layer chromatography following ^{14}C labelling (Yates *et al.* 1999), showed the low M_r

component to be lipid. Gonococcal membrane lipids consist mainly of phosphatidyl ethanolamine and glycerol esterified to palmitic, myristic, a 16:1 and a 18:1 acid (Sud & Feingold 1975). In the glucose-containing medium, the carbon atoms from the ^{13}C lactate were incorporated specifically into the fatty acid portions, in contrast to both glycerol and fatty acid moeities when ^{13}C glucose was used without lactate present. The location in the LPS of the ^{14}C label from the lactate is not yet known but the fatty acid residues seem likely.

The incorporation of lactate carbon into GroEl is interesting in two respects. GroEl would ensure correct folding of the products of the large increase in protein synthesis (Gao et al. 1998). Also, it could contribute to the inflammation seen in gonorrhoea, since it is a potent stimulator of relevant cytokines (Coates & Henderson 1998); and it is significant that patients have high levels of antibody to GroEl (Demarco de Hormaeche et al. 1991). Porin 1B plays a major metabolic role, membrane transport, in gonococci (Gotschlich et al. 1987) and can contribute to pathogenicity by inserting into host-cell membranes (Bjerknes et al. 1995). Stimulation of lipid formation would aid membrane synthesis and therefore metabolism and growth. It is fascinating that gonococci have adapted to use lactate and glucose, which are ubiquitous in vivo, to produce a vibrant metabolism and a large content of virulence determinants such as LPS and GroEl. The marked effects of minute amounts of lactate on LPS, protein and ribose synthesis in a medium containing large quantities of glucose suggests that lactate may have a signalling as well as a metabolic role.

(d) Extension of the work to meningococci

The work on gonococci stimulated similar investigations on meningococci. Some strains contain LPS components that are endogenously sialylated (serogroups B, C, W and Y) and others have components that can be sialylated exogenously by host CMP-NANA (groups A and 29E) (Smith et al. 1995). An LPS sialyltransferase is present (Smith et al. 1995). LPS sialylation affects facets of pathogenicity but not as markedly as for gonococci because capsular polysaccharide is the more powerful virulence determinant. It is sometimes difficult to distinguish between their respective roles. LPS sialylation inhibits meningococcal invasion of epithelial cell lines, endothelial cells and mono- and PMN phagocytes (McNeil & Virgi 1997; Virgi et al.

1993). Also, it interferes with opsonophagocytosis of some strains (Smith *et al.* 1995). The position regarding serum resistance is equivocal; some papers indicate that LPS sialylation is important (Esterbrook *et al.* 1997; Kahler *et al.* 1998) and others that it is less so (Vogel *et al.* 1997; Vogel & Frosch 1999). In an outbreak of group B meningitis, an immunotype capable of LPS sialylation was associated with invasive disease and an immunotype incapable of LPS sialylation with the carrier state (Smith *et al.* 1995). The effect of lactate on meningococci has not yet been investigated.

5.   CONCLUDING REMARKS

I hope this paper has made clear the pertinent questions about the behaviour of bacterial pathogens *in vivo*; and has indicated how they might be answered, despite difficulties in some areas, by a combination of conventional and newly devised methods.

I am indebted to J. A. Cole, M. J. Gill, N. J. Parsons C. W. Penn and E. Yates for critical reading of the manuscript.

REFERENCES

Abu-Kwaik, Y. & Pedersen, L. L. 1996 The use of differential display-PCR to isolate and characterize a *Legionella pneumophila* locus induced during intracellular infection of macrophages. *Mol. Microbiol.* **21**, 543–556.

Ahmed, Z. U., Sarker, M. R. & Sack, D. A. 1990 Protection of adult rabbits and monkeys from lethal shigellosis by oral immunization with thymine requiring and sensitive mutants of *Shigella flexneri. Vaccine* **8**, 153–158.

Akins, D. R., Porcella, S. F., Popova, T. G., Shevchenko, D., Buker, S. I, Li, M., Norgard, M. V. & Radolf, J. D. 1995 Evidence for *in vivo* but not *in vitro* expression of *Borrelia burgdorferi* outer surface protein F(OspF) homologue. *Mol. Microbiol.* **18**, 507–520.

Anderson, J. D. & Smith, H. 1965 The metabolism of erythritol by *Brucella abortus. J. Gen. Microbiol.* **38**, 100–124.

Anderson, J. E., Sparling, P. F. & Cornelissen, C. N. 1994 Gonococcal transferrin-binding protein 2 facilitates but is not essential for transferrin utilization. *J. Bacteriol.* **176**, 3162–3170.

Aranda, C. M. A., Swanson, J. A., Loomis, W. P. & Miller, S. I. 1992 *Salmonella typhimurium* activates virulence gene transcription within acidified macrophage phagolysomes. *Proc. Natl Acad. Sci. USA* **89**, 10 079–10 083.

Arricau, N., Hermant, D., Waxin, H., Ecobichon, C., Duffey, P. S. & Popoff, M. Y. 1998 The RcsB–RcsC regulatory system of *Salmonella typhi* differentially modulates the expression of invasion proteins, flagellin and Vi antigen in response to osmolarity. *Mol. Microbiol.* **29**, 835–850.

Bajaj, V., Lucas, R. L., Hwang, C. & Lee, C. A. 1996 Coordinate regulation of *Salmonella typhimurium* invasion genes by environmental and regulatory factors is mediated by control of *hilA* expression. *Mol. Microbiol.* **22**, 703–714.

Bjerknes, R., Guttormsen, H., Selberg, C. O. & Wetzler, L. M. 1995 Neisserial porins inhibit human neutrophil actin polymerization, degranulation, opsonin receptor expression and phagocytosis but prime the neutrophils to increase their oxidative burst. *Infect. Immun.* **63**, 160–167.

Bramley, J., Demarco de Hormaeche, R., Constantinidou, C., Nassif, X., Parsons, N. J., Jones, P., Cole, J. A. & Smith, H. 1996 A serum-sensitive sialyltransferase-deficient mutant of *Neisseria gonorrhoeae* defective in conversion to serum resistance by CMP-NANA or blood cell extracts. *Microb. Pathogen.* **18**, 187–195.

Braude, A. I. & Siemienski, J. 1960 Role of bacterial urease in experimental pylonephritis. *J. Bacteriol.* **80**, 171–179.

Britigan, B. E., Klapper, D., Svendsen, T. & Cohen, M. S. 1988 Phagocyte-derived lactate stimulates oxygen consumption by *Neisseria gonorrhoeae*. *J. Clin. Invest.* **81**, 318–324.

Brown, M. R. W. & Williams, P. 1985 The influence of environment on envelope properties affecting survival of bacteria in infections. *A. Rev. Microbiol.* **39**, 527–556.

Bullen, J. J. 1981 The significance of iron in infection. *Rev. Infect. Dis.* **3**, 1127–1138.

Burns-Keliher, L. L., Portteus, A. & Curtiss III, R. 1997 Specific detection of *Salmonella typhimurium* proteins synthesized intracellularly. *J. Bacteriol.* **179**, 3604–3612.

Busby, S. J. W., Thomas, M. C. & Brown, N. L. 1998 *Molecular biology. Cell biology*, series H, Vol. 103. NATO ASI series. Heidelberg, Germany: Springer.

Camilli, A. & Mekalanos, J. J. 1995 Use of recombinase gene fusions to identify *Vibrio cholerae* genes induced during infection. *Mol. Microbiol.* **18**, 671–683.

Camilli, A., Beattie, D. T. & Mekalanos, J. J. 1994 Use of genetic recombination as a reporter of gene expression. *Proc. Natl Acad. Sci. USA* **91**, 2634–2638.

Cannon, J. G., Johannsen, D., Hobbs, J. D., Hoffmann, H., Dempsey, J. A. F., Johnstone, D., Kayman, H. & Cohen M. S. 1998 Studies with the human challenge model of gonococcal infections; population dynamics of gonococci during experimental infection and infectivity of isogenic mutants. In *Eleventh Pathogenic Neisseria Conference* (ed. X. Nassif, M. J. Quentin-Millet & M. K. Taha), p. 48. Paris: EDK.

Chen, L. M., Kaniga, K. & Galan, J. E. 1996 *Salmonella* spp. are cytotoxic for cultured macrophages. *Mol. Microbiol.* **21**, 1101–1115.

Chiang, S. L. & Mekalanos, J. J. 1998 Use of signature-tagged transposon mutagenesis to identify *Vibrio cholerae* genes critical for colonization. *Mol. Microbiol.* **27**, 797–805.

Clarke, M. A., Reece, K. A., Lodge, J., Stephen, J., Hirst, B. H. & Jepson, M. A. 1996 Invasion of murine intestinal M cells by *Salmonella typhimurium* inv mutants severely deficient for invasion of cultured cells. *Infect. Immun.* **64**, 4363–4368.

Coates, A. R. M. & Henderson, B. 1998 Chaperonins in health and disease. *Annls NY Acad. Sci.* **851**, 48–53.

Cohen, M. S., Cannon, J. G., Jerse, A. E., Charniga, L. M., Isbey, S. F. & Whicker, L. G. 1994 Human experimentation with *Neisseria gonorrhoeae*: rationale, method and implication for biology of infection and vaccine development. *J. Infect. Dis.* **169**, 532–537.

Collins, D. M. 1996 In search of tuberculosis virulence genes. *Trends Microbiol.* **4**, 426–430.

Conner, C. P., Heithoff, D. M. & Mahan, M. J. 1998 *In vivo* gene expression: contributions to infection, virulence, and pathogenesis. *Curr. Top. Microbiol. Immunol.* **225**, 1–12.

Contag, C. H., Contag, P. R., Mullins, J. I., Spilman, S. D., Stevenson, D. K. & Benaron, D. A. 1995 Photonic detection of bacterial pathogens in living hosts. *Mol. Microbiol.* **18**, 593–603.

Cornelis, G. R. 1998 The *Yersinia* deadly kiss. *J. Bacteriol.* **180**, 5495–5504.

Cornelis, G. R., Boland, A., Boyd, A. P., Geuijen, C., Iriate, M., Neyt, C., Sory, M. P. & Stainier, I. 1998 The virulence plasmid of *Yersinia*; an antihost genome. *Microbiol. Mol. Biol. Rev.* **62**, 1315–1352.

Cornelissen, C. N., Biswas, G. D., Tsai, J., Parachuri, D. K., Thompson, S. A. & Sparling, P. F. 1992 Gonococcal transferrin-binding protein 1 is required for transferrin utilization and is homologous to Ton-B-dependent outer membrane receptors. *J. Bacteriol.* **174**, 5788–5797.

Cornelissen, C. N., Kelly, M., Hobbs, M. M., Anderson, J. F., Cannon, J. G., Cohen, M. S. & Sparling, P. F. 1997 The transferrin receptor expressed by gonococcal strain FA1090 is required for experimental infection of human male volunteers. *Mol. Microbiol.* **27**, 611–616.

Coulter, S. N., Schwan, W. R., Ng, E. Y. W., Langhorne, M. H., Ritchie, H. D., Westbrook-Wadman, S., Hufnagle, W. O., Folger, K. R., Bayer, A. S. & Stover, C. K. 1998 Staphylococcus aureus genetic loci impacting growth and survival in multiple infection environments. *Mol. Microbiol.* **30**, 393–404.

Crooke, H., Griffiss, J. M., John, C. M., Lissenden, S., Bramley, J., Regan, T., Smith, H. & Cole, J. A. 1998 Characterization of a sialyltransferase-deficient mutant of *Neisseria gonorrhoeae* strain F62; instability of transposon Tn1595Δ3 in gonococci and evidence that multiple genetic loci are essential for lipooligosaccharide sialylation. *Microb. Pathogen.* **25**, 237–252.

Culham, D. E., Dalgado, C., Gyles, C. L., Mamelah, D., MacLellan, S. & Wood, J. M. 1998 Osmoregulatory transporter ProP influences colonization of the urinary tract by *Escherichia coli*. *Microbiology* **144**, 91–102.

Curtiss III, R., Kelly, S. M., Gulig, P. A., Gentry-Weeks, C. R. & Galan, J. E. 1988 Avirulent salmonellae expressing virulence antigens from other pathogens for use as orally administered vaccines. In *Virulence mechanisms of bacterial pathogens* (ed. J. A. Roth), pp. 311–328. Washington, DC: American Society for Microbiology.

De Saizieu, A., Certa, U., Warrington, J., Gray, C., Keck, W. & Mous, J. 1998 Bacterial transcript imagery by hybridization of total RNA to oligonucleotide arrays. *Nat. Biotech.* **16**, 45–48.

Dehio, C., Gray-Owen, S. D. & Meyer, T. F. 1998 The role of Opa proteins in interactions with host cells. *Trends Microbiol.* **6**, 489–495.

Demarco de Hormaeche, R., Mehlert, A., Young, D. B. & Hormaeche, C. E. 1991 Antigenic homology between the 65 kDa heat shock proteins of *Mycobacterium tuberculosis*, GroEl of *E. coli* and proteins of *Neisseria gonorrhoeae* expressed during infection. In *Neisseria 1990* (ed. M. Achtman, P. Kohl, C. Marchal, G. Morelli, A. Seiler & B. Thiesen), pp. 199–203. New York: Walter de Gruyter.

Esterbrook, M. E., Griffiss, J. M. & Jarvis, G. A. 1997 Sialylation of *Neisseria meningitidis* lipooligosaccharide inhibits serum bactericidal activity by masking lacto-N-tetraose. *Infect. Immun.* **65**, 4436–4444.

Falkow, S. 1988 Molecular Koch's postulates applied to microbial pathogenicity. *Rev. Infect. Dis.* **10**, S274–S276.

Fields, P. I., Swanson, R. V., Haidaris, C. G. & Heffron, F. 1986 Mutants of *Salmonella typhimurium* that cannot survive within macrophages are avirulent. *Proc. Natl Acad. Sci. USA* **83**, 5189–5193.

Finlay, B. B. & Falkow, S. 1997 Common themes in microbial pathogenicity revisited. *Microbiol. Mol. Biol. Rev.* **61**, 136–169.

Galan, J. E. 1996 Molecular genetic bases of salmonella entry into host cells. *Mol. Microbiol.* **20**, 263–271.

Gao, L., Parsons, N. J., Curry, A., Cole, J. A. & Smith, H. 1998 Lactate causes changes in gonococci including increased lipopolysaccharide synthesis during short-term incubation in media containing glucose. *FEMS Microbiol. Lett.* **169**, 309–316.

Garcia-del Portillo, F., Foster, J. W., Maguire, M. E. & Finlay, B. B. 1992 Characterization of the microenvironment of *Salmonella typhimurium* containing vacuoles within MDCK epithelial cells. *Mol. Microbiol.* **6**, 3289–3297.

Garcia Vescovi, E., Soncini, F. C. & Groisman, E. A. 1994 The role of the PhoP/PhoQ regulon in *Salmonella* virulence. *Res. Microbiol.* **145**, 473–480.

Garcia Vescovi, E., Soncini, F. C. & Groisman, E. A. 1996 Mg^{2+} as an extracellular signal: environmental regulation of *Salmonellae* virulence. *Cell* **84**, 165–174.

Gilbert, M., Watson, D. C., Cunningham, A.-M., Jennings, M. P., Young, N. M. & Wakarchuk, W. W. 1996 Cloning of the lipooligosaccharide α-2,3-sialyltransferase from the bacterial pathogens *Neisseria meningitidis* and *Neisseria gonorrhoeae*. *J. Biol. Chem.* **271**, 28 271–28 276.

Gill, M. J., McQuillen, D. P., Van Putten, J. P. M., Wetzler, L. M., Bramley, J., Crooke, H., Parsons, N. J., Cole, J. A. & Smith, H. 1996 Functional characterization of a sialyltransferase deficient mutant of *Neisseria gonorrhoeae*. *Infect. Immun.* **64**, 3374–3378.

Gillaspy, A. F., Hickmon, S. G., Skinner, R. A., Thomas, T. R., Nelson, C. L. & Smeltzer, M. S. 1995 Role of the accessory gene regulator (*agr*) in pathogenesis of staphylococcal osteomyelitis. *Infect. Immun.* **63**, 3373–3380.

Gotschlich, E. M., Seiff, M. E., Blake, M. S. & Koomey, M. 1987 Porin protein of *Neisseria gonorrhoeae*: cloning and gene structure. *Proc. Natl Acad. Sci. USA* **84**, 8135–8139.

Guangyong, J., Beavis, R. C. & Novick, R. P. 1995 Cell density control of staphylococcal virulence mediated by an octapeptide pheromone. *Proc. Natl Acad. Sci. USA* **92**, 12 055–12 059.

Gulig, P. A. & Doyle, T. J. 1993 The *Salmonella typhimurium* virulence plasmid increases the growth rate of salmonellae in mice. *Infect. Immun.* **61**, 504–511.

Guo, L., Lim, K. B., Gunn, J. S., Bainbridge, B., Darveau, R. P., Hackett, M. & Miller, S. I. 1997 Regulation of lipid A modification by *Salmonella typhimurium* virulence genes *phoP-phoQ*. *Science* **276**, 250–253.

Haas, M., Lindner, B., Seydel, U. & Levy, L. 1993 Comparison of the intrabacterial Na^+K^+ ratio and multiplication in the mouse foot pad as measures of the proportion of viable *Mycobacterium leprae*. *Int. J. Antimicrobiol. Agents* **2**, 117–128.

Halle, F. & Meyer, J. M. 1992 Iron release from ferrisiderophores. A multistep mechanism involving a NADH/FMN oxidoreductase and chemical reduction by FMNH. *Eur. J. Biochem.* **209**, 621–627.

Heithoff, D. M., Conner, C. P., Hanna, P. C., Julio, S. M., Henschel, U. & Mahan, M. J. 1997 Bacterial infection assessed by *in vivo* gene expression. *Proc. Natl Acad. Sci. USA* **94**, 934–939.

Heithoff, D. M., Conner, C. P., Hentschel, U., Govantes, F., Hanna, P. C. & Mahan, M. J. 1999 Coordinate intracellular expression of Salmonella genes induced during infection *in vivo*. *J. Bacteriol.* **181**, 799–807.

Hensel, M., Shea, J. E., Gleeson, C., Jones, M. D., Dalton, E. & Holden, D. W. 1995 Simultaneous identification of bacterial virulence genes by negative selection. *Science* **269**, 400–403.

Hensel, M., Shea, J. E., Waterman, S. R., Mundy, R., Nikolaus, T., Banks, G., Vazquez-Torres, A., Gleeson, C., Fang, F. C. & Holden, D. W. 1998 Genes encoding putative effector proteins of the type III secretion system of *Salmonella* pathogenicity island 2 are required for bacterial virulence and proliferation in macrophages. *Mol. Microbiol.* **30**, 163–174.

Herbert, S., Werlitzsch, D., Dussy, B., Nontonnier, A., Fourner, J. M., Belton, G., Dolhoff, A. & Daring, G. 1997 Regulation of *Staphylococcus aureus* capsular polysaccharide type 5: CO_2 inhibition *in vitro* and *in vivo*. *J. Infect. Dis.* **176**, 431–438.

Herrington, D. A., Hall, R. H., Lasonsky, G., Mekalanos, J. J., Taylor, R. K. & Levine, M. A. 1988 Toxin, toxin-coregulated pili and the *toxR* regulon are essential for *Vibrio cholerae* pathogenesis in humans. *J. Exp. Med.* **168**, 1487–1492.

Hooke, A. M., Sordelli, D. O., Cerquetti, H. C. & Vogt, A. J. 1985 Quantitative determination of bacterial replication *in vivo*. *Infect. Immun.* **49**, 424–427.

Hormaeche, C. E. 1980 The *in vivo* division and death rate of *Salmonell typhimurium* in the spleens of naturally resistant and susceptible mice measured by the superinfecting phage technique of Meynell. *Immunology* **41**, 973–979.

Hueck, C. J. 1998 Type III protein secretion systems in bacterial pathogens of animals and plants. Microbiol. *Mol. Biol. Rev.* **62**, 379–433.

Huynen, M. A., Dandekar, T. & Bork, P. 1999 Variation and evolution of the citric-acid cycle: a genomic perspective. *Trends Microbiol.* **7**, 281–291.

Jepson, M. A. & Clark, M. A. 1998 Studying M cells and their role in infection. *Trends Microbiol.* **6**, 359–365.

Jerse, A. E., Cohen, M. S., Drown, P. M., Whicker, L. G., Isbey, S. F., Seifert, H. S. & Cannon, J. G. 1994 Multiple gonococcal opacity proteins are expressed during experimental urethral infection in the male. *J. Exp. Med.* **179**, 911–920.

Jones, B. D., Ghori, N. & Falkow, S. 1994 *Salmonella typhimurium* initiates murine infection by penetrating and destroying the specialized epithelial M cells of the Peyer's patches. *J. Exp. Med.* **180**, 15–23.

Kahler, C. M., Martin, L. E., Shih, G. C., Rahman, M. M., Carlson, R. W. & Stephens, D. S. 1998 The (α2-8)linked polysialic capsule and lipopolysaccharide structure both contribute to ability of serogroup N *Neisseria meningitidis* to resist the bactericidal activity of normal human serum. *Infect. Immun.* **66**, 5939–5945.

Karnell, A., Cam, P. D., Verna, N. & Lindberg, A. A. 1993 *AroD* deletion attenuates *Shigella flexneri* strain 2457T and makes it a safe and efficaceous oral vaccine in monkeys. *Vaccine* **8**, 830–836.

Keppie, J., Williams, A. E., Witt, K. & Smith, H. 1965 The role of erythritol in the tissue localization of the brucellae. *Br. J. Exp. Pathol.* **46**, 104–108.

Kolter, R. 1999 Evolution of microbial diversity during prolonged starvation. *Proc. Natl Acad. Sci. USA* **96**, 4023–4027.

Kolter, R., Siegele, D. A. & Tormo, A. 1993 The stationary phase of the bacterial life cycle. *A. Rev. Microbiol.* **47**, 855–874.

Latifi, A., Foglino, M., Tanaka, K., Williams, P. & Lazdunski, A. 1996 A heirachical quorum-sensing cascade in *Pseudomonas aeruginosa* links the transcriptional activators LasR and RhiR (VsmR) to the stationary phase sigma factor RpoS. *Mol. Microbiol.* **21**, 1137–1146.

Leclerc, G. J., Tartera, C. & Metcalf, E. S. 1998 Environmental regulation of *Salmonella typhi* invasion-defective mutants. *Infect. Immun.* **66**, 682–691.

Lentner, C. 1981 Units of measurement, body fluids, composition of the body, nutrition. In *Geigy scientific tables*, Vol. 1. Basle, Switzerland: Ciba Geigy.

Lentner, C. 1984 Physical chemistry, composition of the blood, haematology, sonatometric data. In *Geigy scientific tables*, Vol. 3. Basle, Switzerland: Ciba Geigy.

Leuing, K. Y. & Finlay, B. B. 1991 Intracellular replication is essential for virulence of *Salmonella typhimurium*. *Proc. Natl Acad. Sci. USA* **88**, 11 470–11 474.

Levine, M. M., Herrington, D., Murphy, J. R., Morris, J. G., Losonsky, G., Tall, B., Lindberg, A. A., Svenson, S., Bagar, S., Edwards, M. F. & Stocker, B. 1987 Safety, infectivity, immunogenicity and *in vivo* stability of two attenuated auxotrophic mutant strains of *Salmonella typhi* 541Ty and 543Ty used as oral vaccines for man. *J. Clin. Invest.* **79**, 888–902.

Lockhardt, D. J., Dong, H., Byrn, M. C., Follettie, M. T., Gallo, M. W., Chie, M. S., Mittmann, M., Wang, C., Kobayashi, M., Horton, H. & Brown, E. L. 1996 Expression monitoring by hybridisation to high density oligonucleotide arrays. *Nat. Biotech.* **14**, 1675–1680.

Lowe, A. M., Beattie, D. T. & Deresiewicz, R. C. 1998 Identification of novel staphylococcal virulence genes by *in vivo* expression technology. *Mol. Microbiol.* **27**, 967–976.

McCormick, B. A., Stocker, B. A. D., Laux, D. C. & Cohen, P. S. 1988 Roles of motility, chemotaxis and penetration through and growth in intestinal mucus in the ability of an avirulent strain of *Salmonella typhimurium* to colonize the large intestine of streptomycin treated mice. *Infect. Immun.* **56**, 2209–2217.

MacLaren, D. M. 1968 The significance of urease in proteus pylonephritis. Bacteriological study. *J. Pathol. Bact.* **96**, 45–51.

McNeil, G. & Virgi, M. 1997 Phenotypic variants of meningococci and their potential in phagocytic interactions: the influence of opacity proteins, pili, PilC and surface sialic acids. *Microb. Pathogen.* **22**, 295–304.

Mahan, M. J., Slauch, J. M. & Mekalanos, J. J. 1993 Selection of bacterial virulence genes that are specifically induced in the host tissues. *Science* **259**, 686–688.

Mahan, M. J., Slauch, J. M., Hanna, P. C., Camilli, A., Tobias, J. W., Waldor, M. K. & Mekalanos, J. J. 1994 Selection for bacterial genes that are specifically induced in host tissues: the hunt for virulence factors. *Infect. Agents Dis.* **2**, 263–268.

Mahan, M. J., Tobias, J. W., Slauch, J. M., Hanna, P. C., Collier, R. J. & Mekalanos, J. J. 1995 Antibiotic-based selection for bacterial genes that are specifically induced during infection of a host. *Proc. Natl Acad. Sci. USA* **92**, 669–673.

Mantle, M. & Rombough, C. 1993 Growth in and breakdown of purified rabbit small intestinal mucus by *Yersinia enterocolitica. Infect. Immun.* **61**, 4131–4138.

Maw, J. & Meynell, G. G. 1968 The true division and death rates of *Salmonella typhimurium* in the mouse spleen determined with superinfecting phage P22. *Br. J. Exp. Path.* **49**, 597–613.

Mei, J. M., Nourbakhsh, F., Ford, C. W. & Holden, D. W. 1997 Identification of *Staphylococcus aureus* genes in a murine model of bacteraemia using signature-tagged mutagenesis. *Mol. Microbiol.* **26**, 399–407.

Mikulskis, A. V., Delor, I., Thi., V. H. & Cornelis, G. R. 1994 Regulation of the *Yersinia enterocolitica* enterotoxin Yst gene. Influence of growth phase, temperature, osmolarity, pH and bacterial host factors. *Mol. Microbiol.* **14**, 905–915.

Miller, J. F., Kukral, A. M. & Mekalanos, J. J. 1989 A two-component regulatory system (*phoP/phoQ*) controls *Salmonella typhimurium* virulence. *Proc. Natl Acad. Sci. USA* **86**, 5054–5058.

Morgan, A. J. 1985 *X-ray microanalysis: electron microscopy for biologists.* Oxford University Press.

Nauman, M., Hanski, C. & Reichen, E. O. 1991 Expression *in vivo* of additional plasmid-mediated proteins during intestinal infection with *Yersinia enterocolitica* serotype O8. *J. Med. Microbiol.* **35**, 257–261.

O'Callaghan, D., Maskell, D., Liew, F. Y., Easmon, C. S. F. & Dougan, D. 1988 Characterization of aromatic- and purine-dependent *Salmonella typhimurium*; attenuated persistence, and ability to induce protective immunity in BALB/c mice. *Infect. Immun.* **56**, 419–423.

Parsons, N. J., Boons, G. J., Ashton, P. R., Redfern, P. D., Quirk, P., Gao, Y., Constantinidou, C., Patel, J., Bramley, J., Cole, J. A. & Smith, H. 1996*a* Lactic acid is the factor in blood cell extracts which enhances the ability of CMP-NANA to sialylate gonococcal lipopolysaccharide and induce serum resistance. *Microb. Pathogen.* **20**, 87–100.

Parsons, N. J., Emond, J. P., Goldner, M., Bramley, J., Crooke, H., Cole, J. A. & Smith, H. 1996*b* Lactate enhancement of sialylation of gonococcal lipopolysaccharide and induction of serum resistance by CMP-NANA is not due to direct activation of the sialyltransferase: metabolic events are involved. *Microb. Pathogen.* **21**, 193–204.

Pascopella, L., Collins, F. M., Martin, J. M., Lee, M. H., Hatfull, G. F., Stover, C. K., Bloom, B. R. & Jacobs, W. R. 1994 Use of *in vivo* complementation in *Mycobacterium* tuberculosis to identify a genome fragment associated with virulence. *Infect. Immun.* **62**, 1313–1319.

Pepe, J. C., Badger, J. L. & Miller, V. L. 1994 Growth phase and low pH affect the thermal regulation of the *Yersinia enterocolitica inv.* gene. *Mol. Microbiol.* **11**, 123–135.

Perin, F., Laurence, D., Savary, I., Bernard, S. & Le Pape, A. 1997 Radioactive technetium-99m labelling of *Salmonella abortusovis* for assessment of bacterial dissemination in sheep by *in vivo* imaging. *Vet. Microbiol.* **51**, 171–180.

Pesci, E. C. & Iglewski, B. H. 1997 The chain of command in *Pseudomonas* quorum sensing. *Trends Microbiol.* **5**, 133–135.

Petterson, J., Nordfelth, R., Dubinina, E., Bergman, T., Gustafsson, M., Magnusson, K. E. & Wolf-Watz, H. 1996 Modulation of virulence factor expression by pathogen target cell contact. *Science* **273**, 1231–1233.

Plum, G. & Clark-Curtiss, J. E. 1994 Induction of *Mycobacterium avium* gene expression following phagocytosis by human macrophages. *Infect. Immun.* **62**, 476–483.

Pollack, C., Straley, S. C. & Klempner, M. S. 1986 Probing the phagolysosome environment of human phagocytes with a Ca^{2+} responsive operon fusion in *Yersinia pestis*. *Nature* **332**, 834–836.

Preston, M. J., Seed, P. C., Toder, D. S., Iglewski, B. H., Ohman, D. E., Gustin, J. K., Goldberg, J. B. & Pier, G. P. 1997 Contributions of proteases and LasR to virulence of *Pseudomonas aeruginosa* during corneal infection. *Infect. Immun.* **65**, 3086–3090.

Ramsay, G. 1998 DNA chips: state of the art. *Nat. Biotech.* **16**, 40–44.

Regan, T., Watts, A., Smith, H. & Cole, J. A. 1999 Regulation of the lipopolysaccharide-specific sialyltransferase activity of gonococci by growth state of the bacteria, but not by carbon source, catabolite repression or oxygen supply. *Anton von Leeuwenhoek* **75**, 369–379.

Richter-Dahlfors, A., Buchan, A. M. J. & Finlay B. B. 1997 Murine salmonellosis studied by confocal microscopy: *Salmonella typhimurium* resides intracellularly inside macrophages and exerts a cytotoxic effect on phagocytes *in vivo*. *J. Exp. Med.* **186**, 569–580.

Russo, T. A., Jadush, S. T., Brown, J. J. & Johnson, J. R. 1996 Identification of two previously unrecognised genes (*guaA* and *argC*) important for uropathogenesis. *Mol. Microbiol.* **22**, 217–228.

Sansonetti, P. J., Nhieu, G. T. V. & Egile, C. 1999 Rupture of the intestinal epithelial barrier and mucosal invasion by *Shigella flexneri*. *Clin. Infect. Dis.* **28**, 466–475.

Schena, M., Shalon, D., Heller, R., Chai, A., Brown, P. O. & Davis, R. W. 1996 Parallel human genome analysis: microarray-based expression monitoring of 1000 genes. *Proc. Natl Acad. Sci. USA* **93**, 10 614–10 619.

Schneider, H., Griffiss, J. M., Boslego, J. W., Hitchcock, P. J., Zahos, K. M. & Apicella, M. A. 1991 Expression of paragloboside-like lipooligosaccharides may be a necessary component of gonococcal pathogenesis in men. *J. Exp. Med.* **174**, 1601–1605.

Schneider, H., Cross, A. S., Kuschner, R. R., Taylor, D. N., Sadoff, J. C., Boslego, J. W. & Deal, C. D. 1995 Experimental human gonococcal urethritis: 250 *Neisseria gonorrhoeae* MSIImkC are infective. *J. Infect. Dis.* **172**, 180–185.

Schneider, H., Schmidt, K. A., Skillman, D. R., Van De Verg, L., Warren, R. L., Wylie, H. J., Sadoff, J. C., Deal, C. D. & Cross, A. S. 1996 Sialylation lessens the infectivity of *Neisseria gonorrhoeae* MSIImkC. *J. Infect. Dis.* **173**, 1422–1427.

Schwan, W. R., Coulter, S. N., Ng, E. Y. W., Langhorne, M. H., Ritchie, H. D., Brody, L. L., Westbrock-Wadman, S., Bayer, A. S., Folger, K. R. & Stover, C. K. 1998 Identification and characterization of the PutP proline permease that contributes to *in vivo* survival of *Staphylococcus aureus* in animal models. *Infect. Immun.* **66**, 567–572.

Seydel, U., Haas, M., Rietschel, E. T. & Lindner, B. 1992 Laser probe mass spectrometry of individual bacteria organisms and of isolated bacterial compounds; a tool in microbiology. *J. Microbiol. Meth.* **15**, 167–181.

Shea, J. E., Hensel, M., Gleeson, C. & Holden, D. W. 1996 Identification of a virulence locus encoding a second type III secretion system in *Salmonella typhimurium. Proc. Natl Acad. Sci. USA* **93**, 2593–2597.

Skorupski, K. & Taylor, R. K. 1997 Control of the ToxR virulence regulon in *Vibrio cholerae* by environmental stimuli. *Mol. Microbiol.* **25**, 1003–1009.

Smith, H. 1990 Pathogenicity and the microbe *in vivo. J. Gen. Microbiol.* **136**, 377–393.

Smith, H. 1995 The revival of interest in mechanisms of bacterial pathogenicity. *Biol. Rev.* **70**, 277–316.

Smith, H. 1996 What happens *in vivo* to bacterial pathogens? *Annls NY Acad. Sci.* **797**, 77–92.

Smith, H., Anderson, J. D., Keppie, J., Kent, P. W. & Timmis, C. M. 1965 The inhibition of the growth of *Brucellae in vitro* and *in vivo* by analogues of erythritol. *J. Gen. Microbiol.* **38**, 101–108.

Smith, H., Williams, A. E., Pearce, J. H., Keppie, J., Harris-Smith, P. W., Fitzgeorge, R. B. & Witt, K. 1962 Foetal erythritol: a cause of the tissue localization of *Brucella abortus* in bovine contagious abortion. *Nature* **193**, 47–49.

Smith, H., Cole, J. A. & Parsons, N. J. 1995 Sialylation of neisserial lipopolysaccharide: a major influence on pathogenicity. *Microb. Pathogen.* **19**, 365–377.

Sordelli, D. O., Cerquetti, M. C. & Hooke, A. M. 1985 Replication rate of *Pseudomonas aeruginosa* in the murine lung. *Infect. Immun.* **50**, 388–391.

Spencer, A. J., Osbourne, M. P., Haddon, S. J., Collins, J., Starkey, W. G., Candy, D. C. A. & Stephen, J. 1990 Xray microanalysis of rotavirus-infected mouse intestine: a new concept of diarrhoeal secretion. *J. Pediatr. Gastroenterol. Nutr.* **10**, 516–529.

Storey, D. G., Ujack, E. E., Rabin, H. R. & Mitchell, I. 1998 *Pseudomonas aeruginosa lasR* transcription correlates with the transcription of *lasA, lasB,* and *toxA* in chronic lung infections associated with cystic fibrosis. *Infect. Immun.* **66**, 2521–2528.

Strauss, E. J. & Falkow, S. 1997 Microbial pathogenesis, genomes and beyond. *Science* **276**, 701–712.

Sud, I. J. & Feingold, D. S. 1975 Phospholipids and fatty acids of *Neisseria gonorrhoeae*. *J. Bacteriol.* **124**, 713–717.

Suk, K., Das, S., Sun, W., Jwang, B., Barthold, S. W., Flavell, R. A. & Fikrig, E. 1995 *Borrelia burgdorferi* genes selectively expressing in the infected host. *Proc. Natl Acad. Sci. USA* **92**, 4269–4273.

Tang, C. & Holden, D. W. 1999 Pathogen virulence genes-implications for vaccines and drug therapy. *Br. Med. Bull.* **55**, 387–400.

Tang, H. B., DiMango, E., Bryan, R., Gambello, M., Iglewski, B. H., Goldberg, J. P. & Prince, A. 1996 Contribution of specific *Pseudomonas aeruginosa* virulence factors to pathogenesis of pneumonia in a neonatal mouse model of infection. *Infect. Immun.* **64**, 37–43.

Valdivia, R. H. & Falkow, S. 1996 Bacterial genetics by flow cytometry: rapid isolation of *Salmonella typhimurium* acid inducible promoters by differential fluorescence induction. *Mol. Microbiol.* **23**, 367–378.

Valdivia, R. H. & Falkow, S. 1997*a* Fluorescence-based isolation of bacterial genes expressed within host cells. *Science* **277**, 2007–2011.

Valdivia, R. H. & Falkow, S. 1997*b* Probing bacterial gene expression within host cells. *Trends Microbiol.* **5**, 360–363.

Virgi, M., Makepeace, K., Ferguson, D. J. P., Acktman, M. & Moxon, E. R. 1993 Meningococcal Opa and Opc proteins; their role in colonization and invasion of epithelial and endothelial cells. *Mol. Microbiol.* **10**, 499–510.

Vogel, U. & Frosch, M. 1999 Mechanisms of neisserial serum resistance. *Mol. Microbiol.* **32**, 1133–1139.

Vogel, U., Claus, H., Heinze, G. & Frosch, M. 1997 Functional characterization of an isogenic meningococcal α2-3sialyltransferase mutant: the role of lipopolysaccharide sialylation for serum resistance in serogroup B meingococci. *Med. Microbiol. Immunol.* **186**, 159–166.

Wachtel, M. R. & Miller, V. L. 1995 *In vitro* and *in vivo* characterization of an ail mutant of *Yersinia enterocolitica*. *Infect. Immun.* **63**, 2541–2548.

Wallich, R., Brenner, C., Kramer, M. D. & Simon, M. M. 1995 Molecular cloning and immunological characterization of a novel linear-plasmid-encoded gene *pG* of *Borrelia burgdorferi* expressed only *in vivo*. *Infect. Immun.* **63**, 3327–3335.

Wang, J., Lory, S., Ramphal, R. & Jin, S. 1996*a* Isolation and characterization of *Pseudomonas aeruginosa* genes inducible by respiratory mucus derived from cystic fibrosis patients. *Mol. Microbiol.* **22**, 1005–1012.

Wang, J., Mushegian, A., Lory, S. & Jin, S. 1996*b* Large scale isolation of candidate virulence genes of *Pseudomonas aeruginosa* by *in vivo* selection. *Proc. Natl Acad. Sci. USA* **93**, 10 434–10 439.

Weinberg, E. D. 1995 Acquisition of iron and other nutrients *in vivo*. In *Virulence mechanisms of bacterial pathogens*, 2nd edn (ed. J. A. Roth, C. A. Bolin, R. A. Brogden, F. C. Minion and M. J. Wannemuehler), pp. 79–93. Washington, DC: American Society for Microbiology.

Williams, A. E., Keppie, J. & Smith, H. 1964 The relation of erythritol usage to virulence in *Brucellas*. *J. Gen. Microbiol.* **37**, 265–292.

Winson, M. K. (and 12 others) 1995 Multiple N-acyl-L-homoserinelactone signal molecules regulate production of virulence determinants and secondary metabolites in *Pseudomonas aeruginosa*. *Proc. Natl Acad. Sci. USA* **92**, 9427–9431.

Yates, E., Gao, L., Parsons, N. J., Cole, J. A. & Smith, H. 1999 When lactate is used by gonococci growing in a medium containing glucose as occurs *in vivo*, its carbon is incorporated into lipid, LPS, GroEl and porin 1B. In *Abstracts of a Royal Society Meeting on the 'Activities of bacterial pathogens in vivo'*. Reading, UK: Society for General Microbiology.

Young, G. M. & Miller, V. L. 1997 Identification of novel chromosome loci affecting *Yersinia enterocolitica* pathogenesis. *Mol. Microbiol.* **25**, 319–328.

DNA TOPOLOGY AND ADAPTATION OF SALMONELLA TYPHIMURIUM TO AN INTRACELLULAR ENVIRONMENT

DAVID G. MARSHALL and CHARLES J. DORMAN*

*Department of Microbiology, Moyne Institute of Preventive Medicine,
University of Dublin, Trinity College, Dublin 2, Republic of Ireland*

FRANCES BOWE, CHRISTINE HALE and GORDON DOUGAN

*Department of Biochemistry, Imperial College of Science,
Technology and Medicine, Imperial College Road,
South Kensington, London SW7 2AY, UK*

The expression of genes coding for determinants of DNA topology in the facultative intracellular pathogen *Salmonella typhimurium* was studied during adaptation by the bacteria to the intracellular environment of J774A.1 macrophage-like cells. A reporter plasmid was used to monitor changes in DNA supercoiling during intracellular growth. Induction of the *dps* and *spv* genes, previously shown to be induced in the macrophage, was detected, as was expression of genes coding for DNA gyrase, integration host factor and the nucleoid-associated protein H-NS. The *topA* gene, coding for the DNA relaxing enzyme topoisomerase I, was not induced. Reporter plasmid data showed that bacterial DNA became relaxed following uptake of *S. typhimurium* cells by the macrophage. These data indicate that DNA topology in *S. typhimurium* undergoes significant changes during adaptation to the intracellular environment. A model describing how this process may operate is discussed.

Keywords: intracellular growth; DNA topology; gene expression; adaptation; *Salmonella typhimurium*

1. INTRODUCTION

Bacterial adaptation to stressful environments is the subject of intense investigation at present and studies performed *in vitro* have helped to identify many of the components used by prokaryotes to survive when under stress. Much of this work has been carried out with the facultative intracellular pathogen, *Salmonella typhimurium* (see §5). The central importance of gene regulation to stress responses is obvious and a great deal of detailed information is available about the processes that control gene expression at the transcriptional and post-transcriptional levels. Almost all of this

---

*Author for correspondence. (cjdorman@tcd.ie).

39

information has been acquired through work with the bacteria under *in vitro* conditions, although attempts have been made to extrapolate from the *in vitro* work to the *in vivo* situation. The recent development of techniques that permit stress responses to be studied directly in the *in vivo* situation has accelerated the pace of this field.

An overview of gene regulatory processes reveals controls that operate at a 'local' level (to regulate individual promoters, etc.) and others that play a more general or 'global' role. This report deals with regulatory mechanisms of the latter class. It is concerned with determinants of the topology of bacterial DNA, each of which has been shown *in vitro* to be capable of modulating the transcriptional profile of the cell. These are DNA gyrase, a type II topoisomerase that introduces negative supercoils into DNA, an activity that is unique to prokaryotes; DNA topoisomerase I, an enzyme that relaxes DNA and which acts antagonistically to gyrase; integration host factor (IHF), a sequence-specific DNA-binding protein that places 1808 bends into DNA and modulates the function of many bacterial promoters, as well as influencing transposition, site-specific recombination, DNA replication and other DNA transactions; H-NS, a nucleoid-associated protein that binds DNA in a sequence-independent manner (it is thought to recognize and bind at, or close to, regions of intrinsic curvature in DNA) and can regulate transcription from a large number of promoters (almost always negatively) as well as influencing other DNA reactions such as site-specific recombination. In addition to the genes coding for these global regulators, this study also encompasses two well-characterized stress response promoters, those of the plasmid-located *spv* virulence genes and of the chromosomally linked *dps* gene that expresses a DNA protection system in starved cells.

2. DNA TOPOISOMERASES

Gyrase is composed of two subunit proteins, GyrA and GyrB, and has an A_2B_2 tetrameric structure. It requires ATP to supercoil DNA negatively (Gellert *et al.* 1976) and this requirement links gyrase activity to the physiological state of the cell. Thus, when bacteria experience certain stresses, such as an upshift in osmolarity or a transition from aerobic to anaerobic growth, gyrase activity is altered and the result is a change in the level of supercoiling in the DNA (Dorman *et al.* 1988; Higgins *et al.* 1988; Hsieh *et al.* 1991*a*,*b*; Jensen *et al.* 1995; Van Workum *et al.* 1996). Supercoiling imparts free energy to DNA and this drives structural transitions in the DNA helix, including open complex formation at certain promoters (Drlica 1992). In this way, many promoters located in different parts of the genome can respond simultaneously to an alteration in the external environment. The promoters of the gyrase genes are supercoiling responsive. They are inhibited by increases in negative superhelicity but become induced when DNA is relaxed (Menzel & Gellert 1983, 1987a, b). Thus, gyrase expression can respond to changes in supercoiling at the level of transcription and gyrase activity can respond to fluctuations in the [ATP]: [ADP] ratio at the level of topoisomerase activity.

Another gene with a supercoiling-sensitive promoter is *topA*, which codes for topoisomerase I (Tse-Dinh 1985). This monomeric type I topoisomerase relaxes DNA that has become supercoiled past a critical point (Wang 1971). It is thought that the countervailing activities of topoisomerase I and DNA gyrase establish a homeostatic supercoiling balance in the cell (DiNardo *et al.* 1982; Menzel & Gellert 1983). Topoisomerase I does not have an ATP requirement and uses energy stored in the negatively supercoiled DNA to drive the DNA relaxation reaction (Drlica 1992). The *topA* promoter is activated when DNA supercoiling increases and is inhibited by relaxation (the opposite of the *gyr* gene promoters) (Tse-Dinh 1985).

3. INTEGRATION HOST FACTOR

The IHF is encoded by two unlinked genes, *ihfA* and *ihfB*. The protein has a heterodimeric AB structure and binds to the consensus sequence WATCAANNNNTTR (where W is a pyrimidine, R is a purine and N is any base). IHF binding introduces a bend of up to 1808 at the binding site and this is thought to be central to its biological role (Ellenberger & Landy 1997; Rice *et al.* 1996; Travers 1997). It promotes long-range interactions in DNA and between proteins bound at distant sites on the same DNA molecule (Goosen & Van de Putte 1995; Nash 1996). IHF is required for the formation of nucleoprotein complexes such as the lambda intasome (Goodman & Nash 1989; Snyder *et al.* 1989) and the type 1 fimbrial invertasome (Blomfield *et al.* 1997; Dorman & Higgins 1987; Eisenstein *et al.* 1987), both in *Escherichia coli*. Expression of the *ihf* genes is subject to complex control; they respond to growth phase, RpoS, guanosine tetraphosphate and are subject to autoregulation (Aviv *et al.* 1994). (RpoS is an alternative sigma factor used by RNA polymerase when bacteria undergo stress (Hengge-Aronis 1996); guanosine tetraphosphate is an alarmone that is synthesized by bacteria during starvation (Cashel *et al.* 1996).) Furthermore, the *ihfA* gene is part of an operon with the phenylalanine tRNA synthetase genes, *pheST*. Although *ihfA* has its own promoter, it is also subject to coregulation with *pheST* (Mechulam *et al.* 1987; Miller 1984).

4.   PROTEIN H-NS

Protein H-NS is a component of the bacterial nucleoid. It binds to DNA and can constrain supercoils. Its oligomeric structure is a matter of disagreement in the literature, although it is probably at least tetrameric (Spurio *et al.* 1997). H-NS can form heteromeric complexes with a closely related paralogue, StpA (Cusick & Belfort 1998), and this may be important for the biological roles of both proteins (Dorman *et al.* 1999; Zhang *et al.* 1996). H-NS influences the transcription of many genes. Usually it acts as a repressor and the genes under its control usually have other, specific regulators (Atlung & Ingmer 1997; Bertin *et al.* 1999; Ussery *et al.* 1994; Williams & Rimsky 1997). Some of these are transcription activators, which directly oppose the negative influence of H-NS (Jordi *et al.* 1992). H-NS-responsive genes have little in common, apart from a general contribution to the ability of the bacterium to adapt to environmental stress. The mechanism of action of H-NS is a matter of controversy, and it is likely that more than one mechanism of gene regulation is employed. In several cases an association of H-NS binding with a region of DNA curvature has been reported (Yamada *et al.* 1991), but data from *in vitro* experiments have caused the importance of curvature in H-NS binding to be questioned (Jordi *et al.* 1997).

5.   *SALMONELLA TYPHIMURIUM*

S. typhimurium is a facultative intracellular pathogen and is a useful model for studying gene expression during *in vivo* growth (Finlay & Falkow 1997; Gulig 1996; Jones 1997). A great deal of research has been conducted *in vitro* into the response of *S. typhimurium* to stress at the level of gene expression (Alpuche-Aranda *et al.* 1992; Foster & Spector 1995; Groisman & Saier 1990). Recently, this work has been extended by a number of *in vivo* studies using reporter gene fusions to stress-regulated promoters. These investigations have identified several genes as being activated during *in vivo* growth or as coding for products that are essential for survival while the bacterium is within the host (Heitoff *et al.* 1997; Hensel *et al.* 1995; Valdivia & Falkow 1997).

Investigations performed with bacteria grown *in vitro* have illustrated the contributions made to stress responses by genes whose products modulate DNA topology (Dorman 1995). We wished to study the responses of these genes to intracellular growth as a first step in elucidating their involvement in bacterial adaptation to *in vivo* growth. The work was carried out in the murine macrophage-like J774A.1 cell line (American Type Culture Collection, Manassas, VA, USA), which has been used extensively for the study of intracellular growth (Buchmeier *et al.* 1993; Francis & Gallagher 1993; Rhen *et al.* 1993; Uchiya *et al.* 1999; Wilson *et al.* 1997). Survival in macrophage requires the bacteria to survive several environmental assaults, including oxidative stress, acid stress and cationic peptides (Foster & Spector 1995; Francis & Gallagher 1993; Groisman 1994; Para-Lopez *et al.* 1994). The virulent SL1344 strain of *S. typhimurium* was used because it was fully virulent and was therefore capable of expressing all of the factors required for macrophage survival (Hoiseth & Stocker 1981). The response of *S. typhimurium* to signals encountered in macrophage includes a role for the pleiotropic regulatory proteins PhoP/PhoQ. This two-component signal transduction system controls members of a large regulon of genes negatively or positively in response to intracellular signals (García Véscovi *et al.* 1996, 1997; Groisman 1994; Groisman *et al.* 1997; Gunn & Miller 1996; Soncini & Groisman 1996; Waldburger & Sauer 1996). Under *in vitro* growth conditions, specific regulators of this type cooperate with the more global influences of DNA topology to modulate the transcriptional profile of the cell (Dorman 1995). Therefore, we wished to ascertain if determinants of DNA topology formed part of the bacterium's response to intracellular growth.

6.　*spv* AND *dps* GENES

In addition to the genes involved in the regulation of DNA topology, this study included two *S. typhimurium* promoters shown previously to be activated in J774A.1 cells. These were from the *spv* and the *dps* genes, each of which has been studied in detail previously *in vitro* and *in vivo* (Marshall *et al.* 2000). The *spv* genes are located on a 90 kb virulence plasmid in the non-typhoid serovars of *Salmonella* (Guiney *et al.* 1995; Gulig 1996; Libby *et al.* 1997). They are required for the establishment of a systemic infection in the host and have been shown by signature-tagged mutagenesis and *in vivo* expression technology to be required for full virulence and to be expressed during infection (Heitoff *et al.* 1997; Hensel *et al.* 1995). The *spv* locus is composed of a regulatory gene, *spvR*, coding for a LysR-like transcription activator that positively regulates both its own gene and the *spvABCD* operon of structural genes (Fang *et al.* 1991, 1992; Krause *et al.* 1992; Guiney *et al.* 1995; Pullinger *et al.* 1989; Sheehan & Dorman 1998). Activation of *spv* transcription occurs during stationary phase *in vitro* and during *in vivo* growth (Chen *et al.* 1995; Kowartz *et al.* 1994). It requires the RpoS stress-response sigma factor, and is modulated by IHF, the leucine-responsive regulatory protein (Lrp), L-leucine, H-NS and cAMP-Crp (the latter probably acting indirectly through its effect on RpoS expression) (Marshall *et al.* 1999; O'Byrne & Dorman 1994*a, b*; Robbe-Saule *et al.* 1997). In addition, *spv* requires a negatively supercoiled DNA template for transcription *in vitro* (O'Byrne & Dorman 1994*b*; Marshall *et al.* 1999).

The *dps* gene is located on the bacterial chromosome, and like *spv*, it requires the RpoS sigma factor and IHF for full expression; unlike *spv*, *dps* is under the control of the OxyR regulator, a redox-sensitive DNA-binding protein (Altuvia *et al.* 1994). Dps protein is produced by starving bacteria, it co-crystallizes with the bacterial DNA and this is thought to protect the nucleic acid from damage while the bacteria are in stationary phase (Wolf *et al.* 1999). The *dps* promoter has a similar induction profile to that of *spv* when *S. typhimurium* is growing intracellularly and both promoters have been used successfully to express heterologous antigens in live attenuated vaccine strains (Marshall *et al.* 2000).

7.   EXPERIMENTAL RESULTS AND DISCUSSION

Each promoter was cloned from the *S. typhimurium* genome by the polymerase chain reaction using standard methods, it was sequenced to ensure its structural integrity, and then fused to promoterless copies of the *gfp*-M2 gene from *Aequorea victoria* (Cormack *et al.* 1996) and the *lacZ*

plasmid	promoter insert	insert size (bp)		maximum reporter *in vitro* TCM	gene induction J774A.1
pGfp2			*gfp* *lacZ*		
pGyrB	*gyrB*	1600		× 0.8 ± 0.05	× 5.1 ± 0.8
pTopA	*topA*	3300		× 1.6 ± 0.05	× 1.3 ± 0.1
pPheST	*pheST-ihfA*	4700		× 1.2 ± 0.08	× 5.7 ± 1.0
pIhfA	*ihfA*	860		× 1.1 ± 0.01	× 6.7 ± 0.8
pIhfB	*ihfB*	4000		× 0.6 ± 0.04	× 2.8 ± 0.3
pHns	*hns*	1600		× 1.5 ± 0.09	× 7.1 ± 1.0
pDps	*dps*	714		× 0.6 ± 0.05	× 3.0 ± 0.5
pSpv	*spv*	1574		× 0.5 ± 0.06	× 5.1 ± 1.0

☐ open boxes represent coding sequences of genes

Fig. 1. A schematic representation of the different promoters assessed for macrophage gene expression in plasmid pGfp2 is shown (not to scale). Plasmid pGfp2 contains contiguous *lacZ* and *gfp* reporter genes downstream of a multiple cloning site (Marshall *et al.* 2000). Promoter fragments were amplified with the primer combinations detailed in Table 1 (sites engineered into primers are underlined and named in parentheses). Following cleavage with restriction endonucleases the fragments were cloned into similarly cleaved pGfp2 in a direction driving reporter gene expression. Open reading frames of genes assessed are indicated by open boxes. Maximal induction levels are shown as fold increase in LacZ expression, similar induction profiles were recovered when assaying for GFP intensity (data not shown). The LacZ levels were assessed using a chemiluminesence assay as previously described (Marshall *et al.* 2000). GFP intensity levels were determined using a FACScan (Becton Dickenson, Oxford, UK) with argon lasers emitting at 488 nm and bacteria were detected by side scatter as previously described (Valdiva *et al.* 1996). The mean fold induction and standard deviations were calculated from a minimum of three independent experiments.

reporter gene from *E. coli*, carried in the low copy number plasmid pQF50 (Farinha & Kropinski 1990) derivative, pGfp2 (Fig. 1). This plasmid offered two reporters, green fluoresence encoded by *gfp* and β-galactosidase encoded by *lacZ*. In the experiments described here, data obtained from monitoring *lacZ* expression are presented (Fig. 2), although all of the results were validated by *gfp* expression (data not shown). *S. typhimurium* is naturally deficient in the *lacZ* gene so it was possible to monitor the levels of its product, β-galactosidase, from pGfp2 derivatives harboured in SL1344 strains while these were growing *in vitro* or *in vivo*. β-galactosidase

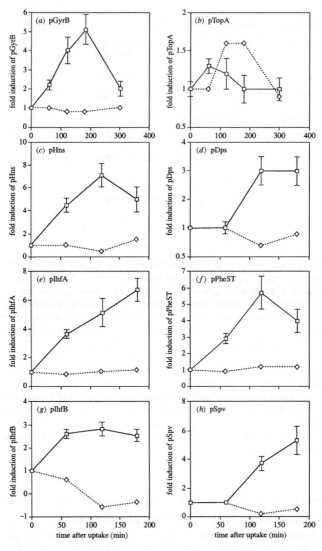

Fig. 2. Induction kinetics for *S. typhimurium* promoters studied *in vivo* following uptake by J774A.1 macrophages. Macrophage infection and LacZ assays were performed as described previously (Marshall *et al.* 2000). β-galactosidse activity was expressed as light units released per viable bacterium (light units per colony forming unit) following infection relative to the activity in the inoculum sample, and is the mean of at least three independent experiments for studies with J774A.1 cells. Squares, data points for intracellular bacteria; diamonds, data points for bacteria in TCM alone.

Table 1. Primers (5′–3′) used for the PCR amplification of plasmids.

Plasmid	Primer name	Primer sequence	Restriction enzyme recognition site[a]
pDps	dps up	CGCGGATCCTATATATTCTTACCGG	(*Bam*HI)
	dps start	CGCGGATCCAATCTCATATCCTCTTGATG	(*Bam*HI)
pTopA	topA cds	CGGGGTACCGGATCAATCCCCATACG	(*Kpn*I)
	topA up2	CGCGGATCCTTTGCCGGTATGTACGACGCC	(*Bam*HI)
pGyrB	gyrB up	ACATGCATGCTGCTTTCACAACGAAGCC	(*Sph*I)
	gyrB start	CGGGGATCCCTTGTCGAAGCGCGCTTTCTCG	(*Bam*HI)
pHns	hns up	CGCGGATCCACTGTCTGAAGATGCCTTCG	(*Bam*HI)
	hns cds	CGCGGATCCCTTCAACGCTTTCCAGAGTAC	(*Bam*HI)
pPheST	ihfA cds	CGCGGATCCAACAGATATTCTGACATTTCAGC	(*Bam*HI)
	ihfC	ACATGCATGCAGGCCGTCAGATGATCATGG	(*Sph*I)
pIhfB	ihfB cds	CGCGGATCCCGTCTTGGCGGGAATGTGC	(*Bam*HI)
	ihfB up	CGCGGATCCTGGCACGTACAACGACCACC	(*Bam*HI)
pIhfA	ihfA2 val up	CGCGGATCCGTTTACGTTCCAGTTCAGGG	(*Bam*HI)
	ihfA2 val cds	CGCGGATCCAACAGATATTCTGACATTTCAGC	(*Bam*HI)
pSpv	spv up	CGCGGATCCAACAGGTCAATTAAATCC	(*Bam*HI)
	spv cds	CCCGGATCCCCTGAAAATAAACAGAATGAAATCC	(*Bam*HI)

[a]The location of the site in the primer sequence is shown by underlining.

expression was measured using the chemiluminescent substrate Galacton Star (Clontech, Basingstoke, UK). The cloning strategy and the standard β-galactosidase assay employed have been described elsewhere (Marshall *et al.* 2000). Details of individual plasmid constructions are summarized in Fig. 1.

J774A.1 macrophage-like cells grown in tissue culture medium (TCM) were infected with SL1344 bacteria at a ratio of ten bacteria per macrophage. Following infection, treatment with gentamycin (20 μgml^{-1}) was used to kill bacteria that had not been engulfed by the J774A.1 cells. Procedures for the growth of J774A.1 and SL1344 cells and the monitoring of *in vivo* gene expression have been described elsewhere (Marshall *et al.* 2000). The bacteria were also inoculated into macrophage-free TCM to determine background levels of gene expression.

Topoisomer analysis of reporter plasmid supercoiling was used to determine the effects of intracellular growth on DNA topology. The multicopy plasmid pUC18 was introduced into strain SL1344 and this strain was used to infect J774A.1 macrophage-like cells. The pUC18 DNA was recovered from the infected cells and examined electrophoretically for changes in supercoiling. Due to the low yield of plasmid DNA from the intracellular bacteria, the chloroquineagarose gels were hybridized with a digoxygenin-labelled pUC18 probe using the labelling and detection procedures of Free & Dorman (1997). For electrophoresis in one dimension, gels contained 2.5 μgml^{-1} chloroquine; under these conditions those topoisomers that had been more supercoiled in the bacteria migrated fastest in the gel (Fig. 3). For electrophoresis in a second dimension, the topoisomers were passed through a gel containing 20 μgml^{-1} chloroquine. Here, topoisomers that were more relaxed in the first dimension formed an arc of positively supercoiled topoisomers migrating above the arc of negatively supercoiled topoisomers (Higgins *et al.* 1988; Wu *et al.* 1988) (Fig. 3).

Plasmid DNA isolated from *S. typhimurium* SL1344 cells that had been incubated in TCM alone showed a broad distribution of topoisomers similar to that seen in extracts from bacteria grown in standard laboratory media (Dorman *et al.* 1988; Higgins *et al.* 1988) and this distribution varied by just one or two topoisomers over time (Fig. 3*b*). In contrast, plasmid DNA removed from SL1344 that had infected J774A.1 macrophage-like cells showed a distribution that became progressively relaxed with time and included a novel fast-migrating species that increased in intensity with time spent in the macrophage (Fig. 3*a*). Electrophoresis with different concentrations of the chloroquine intercalator and two-dimensional electrophoresis (Fig. 3*c*) showed that this species was composed of positively supercoiled DNA, i.e. plasmid DNA that had been highly relaxed in the bacteria.

The discovery that *in vivo* growth led to the production of a population of highly relaxed plasmids raised the possibility that bacterial DNA lost negative supercoils as a result of life within the macrophage. This finding is analogous to the effect of starvation on bacteria grown *in vitro* (Balke & Gralla 1987) and normally elicits a response at the level of gene transcription. In particular, genes required to compensate for the change in DNA topology might have been expected to have been induced, whereas genes whose products are unhelpful would not. We studied the response to

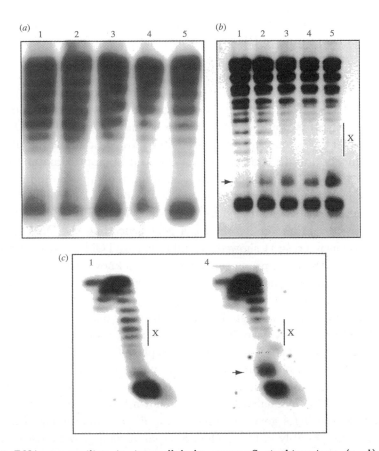

Fig. 3. DNA supercoiling in intracellularly grown *S. typhimurium*. (*a*, *b*) One-dimensional agarose gel electrophoresis through 1% (w/v) agarose gel in 1×tris borate (TBE) containing 2.5 mg chloroquine ml^{-1}. Topoisomer distribution in inoculum sample (lanes 1); after 40 min (lanes 2); after 80 min (lanes 3); after 120 min (lanes 4); after 180 min (lanes 5). (*a*) Topoisomer distribution of pUC18 following incubation of SL1344(pUC18) in TCM in 5% CO_2 at 37°C. (*b*) Topoisomer distribution of plasmid pUC18 before and following infection of macrophages with SL1344(pUC18). (*c*) Two-dimensional agarose gel electrophoresis through 1% (w/v) agarose gel in 1×TBE, containing 2.5 μg chloroquine ml^{-1} in the first dimension and 20 μg chloroquine ml^{-1} in the second dimension. Sample 1, pUC18 topoisomer distribution in *S. typhimurium* inoculum sample (corresponds to lane 1 of *b*); Sample 4, pUC18 topoisomer distribution in *S. typhimurium* following infection of J774A.1 macrophages for 120 min (corresponds to lane 4 of *b*). The distribution of more negatively supercoiled topoisomers present in the inoculum but absent in the intracellular bacteria is highlighted (X) in *b* and *c*. The novel fast migrating species unique to the intracellularly recovered bacteria is arrowed in *b* and *c*.

intracellular growth of a panel of *S. typhimurium* promoters likely to be involved in adaptation to DNA relaxation.

DNA gyrase would be expected to play a prominent role in the restoration of lost negative supercoiling. To do this it would require an upward shift in the [ATP]:[ADP] ratio. Even in the absence of sufficient ATP (as in starving cells) to allow gyrase to function in negative supercoiling, the *gyr* gene promoters would be expected to be induced directly by DNA relaxation. In TCM alone, the *gyrB* promoter showed no induction during a 180 min incubation. In contrast, it was induced by approximately fivefold when the bacteria were inside macrophage for the same period (Fig. 1). This was in agreement with the known behaviour of the equivalent promoter from *E. coli* when studied *in vitro* (Menzel & Gellert 1983, 1987a, b). In contrast, the promoter for the *topA* gene (encoding the DNA relaxing enzyme topoisomerase I) showed no induction in TCM or in macrophage (Fig. 1). Since the *topA* promoter from *E. coli* is known to be inhibited by DNA relaxation, this finding was in keeping with the plasmid topoisomer data described above.

Other studies have suggested that the IHF may play a role in maintaining DNA topology in a form favourable to certain promoters under conditions where a general DNA relaxation takes place (Porter & Dorman 1997). *In vitro*, this relaxation occurs in stationary phase, and IHF levels have been shown to increase during this period of the growth curve, at least in *E. coli* (Ditto *et al.* 1994). An earlier study had shown that one of the genes coding for IHF, the *ihfA* gene (designated *mig-23* for 'macrophage-inducible gene 23'), was possibly induced when *S. typhimurium* grows intracellularly (Valdivia & Falkow 1987). In that study, the recombinant plasmid tested contained just a segment of the *pheT* gene where the *ihfA* promoter was located; it had not included the complete *pheST* operon and its promoter as well as that of *ihfA*. We constructed a plasmid that included the complete *pheST-ihfA* operon and obtained an approximately fivefold induction in the J774A.1 cells with no induction in TCM alone (Fig. 1). To resolve the issue of the possible contribution of the *pheST* promoter, a derivative plasmid was made in which the expression of the reporter genes was under the control of the *ihfA* promoter alone. The data obtained with this plasmid showed that *ihfA* was induced approximately sixfold during intracellular growth but not in TCM alone (Fig. 1). This shows that the *ihfA* promoter responds to growth in macrophage and this response is not modulated by a contribution from the *pheST* promoter. The promoter from the *ihfB* gene was also tested and found to be inducible intracellularly but not in growth medium alone (Fig. 1). (The response of the *ihfB* promoter to intracellular growth had not been assessed previously.) These data for *ihf* gene expression are in keeping with results obtained from *in vitro* studies showing that these genes are induced when bacteria are starved or undergo environmental stress (Aviv *et al.* 1994; Ditto *et al.* 1994).

Successful establishment of an infection requires the bacterium to multiply in the intracellular state (Finlay & Falkow 1997). Previous work performed *in vitro* with *E. coli* has shown that expression of the nucleoid-associated protein H-NS is coupled to DNA synthesis such that in rapidly growing bacteria the *hns* gene is induced while in quiescent cells it is repressed (Free & Dorman 1995). The *hns* promoter was induced strongly in *S. typhimurium* cells growing in macrophage and not induced in TCM alone (Fig. 1). This suggests that while the bacteria may have been stressed when associated with the J774A.1 cells, they were still able to operate an efficient cell cycle, thus creating a demand for *hns* gene expression. A previous study has indicated that intracellular *S. typhimurium* cells grow rapidly and require protein synthesis for survival. As the bacteria adapt to the intracellular environment of the macrophage, they switch from this rapid growth mode to a survival state that does not require protein synthesis (Abshire & Neidhardt 1993). The kinetics of *hns* induction revealed here, with expression declining 2 h post-infection (Fig. 2), concurs with the findings of this earlier study.

As positive controls for the experiments described here, the *dps* and *spv* promoters were also studied. Each had been shown previously to be induced during intracellular growth (Marshall *et al.* 2000) and induction was also detected in this study (Figs. 1 and 2).

This study addresses a specific aspect of bacterial physiology during infection, namely what are the consequences of the infection process for the topology of bacterial DNA and for the expression of genes coding for proteins that modulate DNA topology? It appears that infection of the J774A.1 cell line causes bacterial DNA to become relaxed. As one might predict, the relaxation-inhibited promoter of the *topA* gene remains repressed under these conditions. In contrast, the *gyrB* gene, coding for one of the subunits of DNA gyrase, is induced strongly. This is consistent with the production of extra copies of gyrase to restore supercoiling levels to values favourable to most DNA-based molecular transactions. The level of *gyrB* expression declines after 3 h, suggesting that by this stage post-infection supercoiling is restored at the *gyrB* promoter (Fig. 2).

The *spv* promoter used in this study illustrates some of the complexities arising from our data. It is supercoiling responsive and has been shown previously to be inhibited when DNA becomes relaxed (Marshall *et al.* 1999). However, *in vitro* data show that this inhibition requires the IHF protein; in the absence of this accessory factor, *spv* expression remains high even when DNA is relaxed (Marshall *et al.* 1999). This points to a subtle interplay between IHF and gyrase in the regulation of *spv* transcription. The *in vivo* data presented here show that *spv* is induced when the global level of DNA supercoiling is low. This may point to a shortage of IHF *in vivo* (consistent with the strong induction of the genes that code for this protein) or simply point to additional complexities inherent in the *in vivo* situation. For example, the leucine-responsive regulatory protein (Lrp) also regulates *spv* expression *in vitro* (Marshall *et al.* 1999) and this may have a positive modulatory influence when the bacterium is growing in association with macrophage. Alternatively, the negative influence of H-NS on *spv* expression described previously *in vitro* (O'Byrne & Dorman 1994*b*; Robbe-Saule *et al.* 1997) may not operate under the *in vivo* conditions studied here, and this is consistent with our data on *hns* gene induction in macrophage (Figs. 1 and 2) suggesting that not all H-NS-binding sites in the bacterial genome are occupied.

The data presented here allow a tentative model to be advanced to describe early events in intracellular adaptation from the perspective of DNA topology. Three lines of evidence suggest that DNA becomes relaxed early in the process. These are a direct observation of reporter plasmid topology by electrophoresis in the presence of a DNA intercalator, the positive response of the DNA relaxation-activated *gyrB* promoter, and the lack of a response from the DNA relaxation-inhibited *topA* promoter. Why does the DNA become relaxed? The most likely explanation is that the negative supercoiling activity of bacterial gyrase is inhibited as a result of interactions with the mammalian cells. This could arise from an unfavourable shift in the [ATP]:[ADP] ratio. As ATP levels fall and ADP levels rise, the negative supercoiling activity of gyrase is inhibited, although it can continue to contribute to ATP-independent relaxation of DNA (Drlica 1992). DNA topoisomerase I, and to some extent the other two topoisomerases in the cell, might also be expected to contribute to the general relaxation of genomic DNA. The induction of the genes encoding IHF may be regarded as a mechanism for dealing selectively with unfavourable alterations in DNA topology. Those promoters with appropriately located IHF-binding sites (such as *spv*) may be protected from inhibition as the DNA relaxes through the creation of a nucleoprotein complex that preserves promoter function. Other IHF-dependent processes, such as DNA replication, transposition and site-specific recombination, may also continue to function. Other evidence suggesting that DNA replication continues under the conditions of our experiments comes from the observation of *hns* gene expression. A demand for *hns* transcription has been shown *in vitro* to be created by DNA synthesis (Free & Dorman 1995) and the *in vivo* data presented here suggest that this process operates during the macrophage infection. This is consistent with the data of Abshire & Neidhardt (1993) showing that vigorous protein synthesis occurs for approximately 2 h following macrophage engulfment of *S. typhimurium.* The overall picture that emerges is one of stressed bacteria with abnormally relaxed levels of DNA supercoiling that continue to grow and express genes that are critical for adaptation to this environment. This situation is in marked contrast to that described for invasion of epithelia by *S. typhimurium.* Their increased negative supercoiling is an essential prerequisite for expression of invasion functions (Galán & Curtiss 1990). Thus, DNA topology contributes to different types of hostpathogen interaction, and it does so in a dynamic way. Understanding how transitions between the different phases of the infection affect DNA topology and gene expression will form the next phase of the work.

This work was supported by a grant (BIO4-CT96–0144) from the European Union.

REFERENCES

Abshire, K. Z. & Neidhardt, F. C. 1993 Growth rate paradox of *Salmonella typhimurium* within host macrophages. *J. Bacteriol.* **175**, 3744–3748.

Alpuche-Aranda, C. M. A., Swanson, J. A., Loomis, W. P. & Miller, S. I. 1992 *Salmonella typhimurium* activates virulence gene transcription within acidified macrophage phagosomes. *Proc. Natl Acad. Sci. USA* **89**, 10 079–10 083.

Altuvia, S., Almiron, M., Huisman, G., Kolter, R. & Storz, G. 1994 The *dps* promoter is activated by OxyR during growth and by IHF and σ^s in stationary phase. *Mol. Microbiol.* **13**, 265–272.

Atlung, T. & Ingmer, H. 1997 H-NS: a modulator of environmentally-regulated gene expression. *Mol. Microbiol.* **24**, 7–17.

Aviv, M., Giladi, H., Schreiber, G., Oppenheim, A. B. & Glaser, G. 1994 Expression of the genes coding for the *Escherichia coli* integration host factor are controlled by growth phase, *rpoS*, ppGpp and by autoregulation. *Mol. Microbiol.* **14**, 1021–1031.

Balke, V. L. & Gralla, J. D. 1987 Changes in linking number of supercoiled DNA accompany growth transitions in *Escherichia coli*. *J. Bacteriol.* **169**, 4499–4506.

Bertin, P., Benhabiles, N., Krin, E., Laurent-Winter, C., Tendeng, C., Turlin, E., Thomas, A., Danchin, A. & Brasseur, R. 1999 The structural and functional organization of the H-NS-like proteins is evolutionarily conserved in Gramnegative bacteria. *Mol. Microbiol.* **31**, 319–329.

Blomfield, I. C., Kulasekara, D. H. & Eisenstein, B. I. 1997 Integration host factor stimulates both FimB- and FimE-mediated site-specific DNA inversion that controls phase variation of type 1 fimbriae in *Escherichia coli*. *Mol. Microbiol.* **23**, 705–717.

Buchmeier, N. A., Lipps, C. J., So, M. Y. & Heffron, F. 1993 Recombination-deficient mutants of *Salmonella typhimurium* are avirulent and sensitive to the oxidative burst of macrophages. *Mol. Microbiol.* **7**, 933–936.

Cashel, M. Gentry, D. R., Hernandez, V. J. & Vinella, D. 1996 The stringent response. In Escherichia coli *and* Salmonella: *cellular and molecular biology* (ed. F. C. Neidhardt, R. Curtiss III, J. L. Ingraham, E. C. C. Lin, K. B. Low, B. Magasanik, W. S. Reznikoff, M. Riley, M. Schaechter & H. E. Umbarger), pp. 1458–1496. Washington, DC: American Society for Microbiology.

Chen, C.-Y., Eckmann, L., Libby, S. J., Fang, F. C., Okamoto, S., Kagnoff, M. F., Fierer, J. & Guiney, D. G. 1996 Expression of *Salmonella typhimurium rpoS* and *rpoS*-dependent genes in the intracellular environment of eukaryotic cells. *Infect. Immun.* **64**, 4739–4743.

Cormack, B. P., Valdivia, R. H. & Falkow, S. 1996 FACS-optimized mutants of the green fluorescent protein (GFP). *Gene* **173**, 33–38.

Cusick, M. E. & Belfort, M. 1998 Domain structure and RNA annealing activity of the *Escherichia coli* regulatory protein StpA. *Mol. Microbiol.* **28**, 847–857.

DiNardo, S., Voekel, K. A., Sternglanz, R., Reynolds, A. E. & Wright, A. 1982 *Escherichia coli* DNA topoisomerase I mutants have compensatory mutations in DNA gyrase genes. *Cell* **31**, 43–51.

Ditto, M. D., Roberts, D. & Weisberg, R. A. 1994 Growth phase variation of integration host factor level in *Escherichia coli*. *J. Bacteriol.* **176**, 3738–3748.

Dorman, C. J. 1995 DNA topology and the global control of bacterial gene expression: implications for the regulation of virulence gene expression. *Microbiology* **141**, 1271–1280.

Dorman, C. J. & Higgins, C. F. 1987 Fimbrial phase variation in *Escherichia coli*: dependence on integration host factor and homologies with other site-specific recombinases. *J. Bacteriol.* **169**, 3840–3843.

Dorman, C. J., Barr, G. C., Ni Bhriain, N. & Higgins, C. F. 1988 DNA supercoiling and the anaerobic and growth phase regulation of *tonB* gene expression. *J. Bacteriol.* **170**, 2816–2826.

Dorman, C. J., Hinton, J. C. D. & Free, A. 1999 Homo- and hetero-oligomerization among H-NS-like nucleoid-associated proteins in bacteria. *Trends Microbiol.* **7**, 124–128.

Drlica, K. 1992 Control of bacterial DNA supercoiling. *Mol. Microbiol.* **6**, 425–433.

Eisenstein, B. I., Sweet, D. S., Vaughn, V. & Friedman, D. I. 1987 Integration host factor is required for the DNA inversion that controls phase variation in *Escherichia coli*. *Proc. Natl Acad. Sci. USA* **84**, 6506–6510.

Ellenberger, T. & Landy, A. 1997 A good turn for DNA: the structure of integration host factor bound to DNA. *Structure* **5**, 153–157.

Fang, F. C., Krause, M., Roudier, C., Fierer, J. & Guiney, D. G. 1991 Growth regulation of a *Salmonella* plasmid gene essential for virulence. *J. Bacteriol.* **173**, 6783–6789.

Fang, F. C., Libby, S. J., Buchmeier, N. A., Loewen, P. C., Switala, J., Harwood, J. & Guiney, D. G. 1992 The alternative sigma factor KatF (RpoS) regulates *Salmonella* virulence. *Proc. Natl Acad. Sci. USA* **89**, 11 978–11 982.

Farinha, M. A. & Kropinski, A. M. 1990 Construction of broad-host-range plasmid vectors for easy visible selection and analysis of promoters. *J. Bacteriol.* **172**, 3496–3499.

Finlay, B. B. & Falkow, S. 1997 Common themes in microbial pathogenicity revisited. *Microbiol. Mol. Biol. Rev.* **61**, 136–169.

Foster, J. W. & Spector, M. P. 1995 How *Salmonella* survive against the odds. *A. Rev. Microbiol.* **49**, 145–174.

Francis, K. P. & Gallagher, M. P. 1993 Light emission from a mud-*lux* transcriptional fusion in *Salmonella typhimurium* is stimulated by interaction with the mouse macrophage cell line J774.2. *Infect. Immun.* **61**, 640–649.

Free, A. & Dorman, C. J. 1995 Coupling of *Escherichia coli hns* mRNA levels to DNA synthesis by autoregulation: implications for growth phase control. *Mol. Microbiol.* **18**, 101–113.

Free, A. & Dorman, C. J. 1997 The *Escherichia coli stpA* gene is transiently expressed during growth in rich medium and is induced in mimimal medium and by stress conditions. *J. Bacteriol.* **179**, 909–918.

Galán, J. E. & Curtiss, R. 1990 Expression of *Salmonella typhimurium* genes required for invasion is regulated by changes in DNA supercoiling. *Infect. Immun.* **58**, 1879–1885.

García Véscovi, E., Soncini, F. C. & Groisman, E. A. 1996 Mg^{2+} as an extracellular signal: environmental regulation of *Salmonella typhimurium*. *Cell* **84**, 165–174.

García Véscovi, E., Ayala, Y. M., Di Cera, E. & Groisman, E. A. 1997 Characterization of the bacterial sensor protein PhoQ. Evidence for distinct binding sites for Mg^{2+} and Ca^{2+}. *J. Biol. Chem.* **272**, 1440–1443.

Gellert, M., Mizuuchi, K., O'Dea, M. H. & Nash, H. A. 1976 DNA gyrase: an enzyme that introduces superhelical turns into DNA. *Proc. Natl Acad. Sci. USA* **73**, 3872–3876.

Goodman, S. D. & Nash, H. A. 1989 Functional replacement of a protein-induced bend in a DNA recombination site. *Nature* **341**, 251–254.

Goosen, N. & Van de Putte, P. 1995 The regulation of transcription initiation by integration host factor. *Mol. Microbiol.* **16**, 1–7.

Groisman, E. A. 1994 How bacteria resist killing by host-defense peptides. *Trends Microbiol.* **2**, 444–449.

Groisman, E. A. 1998 The ins and outs of virulence gene expression: Mg^{2+} as a regulatory signal. *BioEssays* **20**, 96–101.

Groisman, E. A. & Saier, M. H. 1990 *Salmonella* virulence: new clues to intracellular survival. *Trends Biochem. Sci.* **15**, 30–33.

Groisman, E. A., Kayser, J. & Soncini, F. C. 1997 Regulation of polymyxin resistance and adaptation to low-Mg^{2+} environments. *J. Bacteriol.* **179**, 7040–7045.

Guiney, D. G., Libby, S., Fang, F. C., Krause, M. & Fierer, J. 1995 Growth-phase regulation of plasmid virulence genes in *Salmonella*. *Trends Microbiol.* **3**, 275–279.

Gulig, P. A. 1996 Pathogenesis of systemic disease. In Escherichia coli *and* Salmonella: *cellular and molecular biology* (ed. F. C. Neidhardt, R. Curtiss III, J. L. Ingraham, E. C. C. Lin, K. B. Low, B. Magasanik, W. S. Reznikoff, M. Riley, M. Schaechter & H. E. Umbarger), pp. 2774–2787. Washington, DC: American Society for Microbiology.

Gunn, J. S. & Miller, S. I. 1996 PhoP-PhoQ activates transcription of *pmrAB*, encoding a two-component regulatory system involved in *Salmonella typhimurium* antimicrobial peptide resistance. *J. Bacteriol.* **178**, 6857–6864.

Heitoff, D. M., Conner, C. P., Hanna, P. C., Julio, S. M., Hentschel, U. & Mahan, M. J. 1997 Bacterial infection as assessed by *in vitro* gene expression. *Proc. Natl Acad. Sci. USA* **94**, 934–939.

Hengge-Aronis, R. 1996 Regulation of gene expression during entry into stationary phase. In Escherichia coli *and* Salmonella: *cellular and molecular biology* (ed. F. C. Neidhardt, R. Curtiss III, J. L. Ingraham, E. C. C. Lin, K. B. Low, B. Magasanik, W. S. Reznikoff, M. Riley, M. Schaechter & H. E. Umbarger), pp. 1497–1512. Washington, DC: American Society for Microbiology.

Hensel, M., Shea, J. E., Gleeson, C., Jones, M. D., Dalton, E. & Holden, D. W. 1995 Simultaneous identification of bacterial virulence genes by negative selection. *Science* **269**, 400–403.

Higgins, C. F., Dorman, C. J., Stirling, D. A., Waddell, L. Booth, I. R., May, G. & Bremer, E. 1988 A physiological role for DNA supercoiling in the osmotic regulation of gene expression in *S. typhimurium* and *E. coli. Cell* **52**, 569–584.

Hoiseth, S. K. & Stocker, B. A. D. 1981 Aromatic-dependent *S. typhimurium* are non-virulent and are effective as live vaccines. *Nature* **291**, 238–239.

Hsieh, L.-S., Burger, R. M. & Drlica, K. 1991a Bacterial DNA supercoiling and [ATP]/[ADP] changes associated with a transition to anaerobic growth. *J. Mol. Biol.* **219**, 443–450.

Hsieh, L.-S., Rouvierere-Yaniv, J. & Drlica, K. 1991b Bacterial DNA supercoiling and [ATP]/[ADP] ratio: changes associated with salt shock. *J. Bacteriol.* **173**, 3914–3917.

Jensen, P. R., Loman, L., Petra, B., Van der Weijden, C. & Westerhoff, H. V. 1995 Energy buffering of DNA structure fails when *Escherichia coli* runs out of substrate. *J. Bacteriol.* **177**, 3420–3426.

Jones, B. D. 1997 Host responses to pathogenic *Salmonella* infection. *Genes Dev.* **11**, 679–687.

Jordi, B. J. A. M., Dagberg, B., de Haan, A. A. M., Hamers, A. M., Van der Zeijst, B. J. A. M., Gaastra, W. & Uhlin, B. E. 1992 The positive regulator CfaD overcomes the repression mediated by histone-like protein H-NS (H1) in the CFA/I fimbrial operon of *Escherichia coli. EMBO J.* **11**, 2627–2632.

Jordi, B. J. A. M., Fielder, A., Burns, C. M., Hinton, J. C. D., Dover, N., Ussery, D. W. & Higgins, C. F. 1997 DNA binding is not sufficient for H-NS-mediated repression of *proU* expression. *J. Biol. Chem.* **272**, 12 083–12 090.

Kowartz, L., Coynault, C., Robbe-Saule, V. & Norel, F. 1994 The *Salmonella typhimurium katF* (*rpoS*) gene: cloning, nucleotide sequence, and regulation of *spvR* and *spvABCD* virulence plasmid genes. *J. Bacteriol.* **176**, 6852–6860.

Krause, M., Fang, F. C. & Guiney, D. G. 1992 Regulation of plasmid virulence gene expression in *Salmonella dublin* involves an unusual operon structure. *J. Bacteriol.* **174**, 4482–4489.

Libby, S. J., Adams, L. G., Ficht, T. A., Allen, C., Whitford, H. A., Buchmeier, N. A., Bossie, S. & Guiney, D. G. 1997 The *spv* genes on the *Salmonella dublin* virulence plasmid are required for severe enteritis and systemic infection in the natural host. *Infect. Immun.* **65**, 1786–1792.

Marshall, D. G., Sheehan, B. J. & Dorman, C. J. 1999 A role for the leucine-responsive regulatory protein and integration host factor in the regulation of the *Salmonella* plasmid virulence (*spv*) locus in *Salmonella typhimurium. Mol. Microbiol.* **34**, 134–145.

Marshall, D. G., Hague, A., Fowler, R., del Guidice, G., Dorman, C. J., Dougan, G. & Bowe, F. 2000 Use of the stationary phase inducible promoters, *spv* and *dps*, to drive heterologous antigen expression in *Salmonella* vaccine strains. *Vaccine* **18**, 1298–1306.

Mechulam, Y., Blanquet, S. & Fayat, G. 1987 Dual level control of the *Escherichia coli pheST − himA* operon expression: tRNA^{phe}-dependent attenuation and transcriptional operatorrepressor control by *himA* and the SOS network. *J. Mol. Biol.* **197**, 453–470.

Menzel, R. & Gellert, M. 1983 Regulation of the genes for *E. coli* DNA gyrase: homeostatic control of DNA supercoiling. *Cell* **34**, 105–113.

Menzel, R. & Gellert, M. 1987a Fusions of the *Escherichia coli gyrA* and *gyrB* control regions to the galactokinase gene are inducible by coumermycin treatment. *J. Bacteriol.* **169**, 1272–1278.

Menzel, R. & Gellert, M. 1987b Modulation of transcription by DNA supercoiling: a deletion analysis of the *Escherichia coli gyrA* and *gyrB* promoters. *Proc. Natl Acad. Sci. USA* **84**, 4185–4189.

Miller, H. I. 1984 Primary structure of the *himA* gene of *Escherichia coli*: homology with DNA-binding protein HU and association with the phenylalanyl-tRNA synthetase operon. *Cold Spring Harbor Symp. Quant. Biol.* **49**, 691–698.

Nash, H. A. 1996 The HU and IHF proteins: accessory factors for complex protein–DNA assemblies. In *Regulation of gene expression in* Escherichia coli (ed. E. C. C. Lyn & A. S. Lynch), pp. 149–179. Austin, TX: R. G. Landes Co.

O'Byrne, C. P. & Dorman, C. J. 1994a The *spv* virulence operon of *Salmonella typhimurium* LT-2 is regulated negatively by the cyclic AMP (cAMP)–cAMP receptor protein system. *J. Bacteriol.* **176**, 905–912.

O'Byrne, C. P. & Dorman, C. J. 1994b Transcription of the *Salmonella typhimurium spv* virulence locus is regulated negatively by the nucleoid-associated protein H-NS. *FEMS Microbiol. Lett.* **121**, 99–106.

Para-Lopez, C., Lin, R., Aspedon, A. & Groisman, E. A. 1994 A *Salmonella* protein that is required for resistance to antimicrobial peptides and transport of potassium. *EMBO J.* **13**, 3964–3972.

Porter, M. E. & Dorman, C. J. 1997 Positive regulation of *Shigella flexneri* virulence genes by integration host factor. *J. Bacteriol.* **179**, 6537–6550.

Pullinger, G. D., Baird, G. D., Williamson, C. M. & Lax, A. J. 1989 Nucleotide sequence of a plasmid gene involved in the virulence of salmonellas. *Nucl. Acids Res.* **17**, 7983.

Rhen, M., Riikonen, P. & Taira, S. 1993 Transcriptional regulation of *Salmonella enterica* virulence plasmid genes in cultured macrophages. *Mol. Microbiol.* **10**, 45–56.

Rice, P. A., Yang, S.-W., Mizuuchi, K. & Nash, H. A. 1996 Crystal structure of an IHF-DNA complex: a protein-induced DNA U-turn. *Cell* **87**, 1295–1306.

Robbe-Saule, V., Schaeffer, F., Kowartz, L. & Norel, F. 1997 Relationships between H-NS, sigma S, SpvR and growth phase in the control of *spvR*, the regulatory gene of the *Salmonella dublin* virulence plasmid. *Mol. Gen. Genet.* **256**, 333–347.

Sheehan, B. J. & Dorman, C. J. 1998 *In vivo* analysis of the interactions of the LysR-like regulator SpvR with the operator sequences of the *spvA* and *spvR* virulence genes of *Salmonella typhimurium*. *Mol. Microbiol.* **30**, 91–105.

Snyder, U. K., Thompson, J. F. & Landy, A. 1989 Phasing of protein-induced DNA bends in a recombination complex. *Nature* **341**, 255–257 (erratum 342, 206).

Soncini, F. C. & Groisman, E. A. 1996 Two-component regulatory systems can interact to process multiple environmental signals. *J. Bacteriol.* **178**, 6796–6801.

Spurio, R., Falconi, M., Brandi, A., Pon, C. L. & Gualerzi, C. O. 1997 The oligomeric structure of nucleoid protein H-NS is necessary for recognition of intrinsically curved DNA and for DNA bending. *EMBO J.* **16**, 1795–1805.

Travers, A. 1997 DNA-protein interactions: IHF—the master bender. *Curr. Biol.* **7**, R252–R254.

Tse-Dinh, Y.-C. 1985 Regulation of the *Escherichia coli* DNA topoisomerase I gene by DNA supercoiling. *Nucl. Acids Res.* **13**, 4751–4763.

Uchiya, K., Barbieri, M. A., Funato, K., Shah, A. H., Stahl, P. D. & Groisman, E. A. 1999 A *Salmonella* virulence protein that inhibits cellular trafficking. *EMBO J.* **18**, 3924–3933.

Ussery, D. W., Hinton, J. C. D., Jordi, B. J. A. M., Granum, P. E., Seirafi, A., Stephen, R. J., Tupper, A. E., Berridge, G., Sidebotham, J. M. & Higgins, C. F. 1994 The chromatin-associated protein H-NS. *Biochimie* **76**, 968–980.

Valdivia, R. H. & Falkow, S. 1997 Fluorescence-based isolation of bacterial genes expressed within host cells. *Science* **277**, 2007–2011.

Van Workum, M., Van Dooren, S. J. M., Oldenburg, N., Molenaar, D., Jensen, P. R., Snoep, J. L. & Westerhoff, H. V. 1996 DNA supercoiling depends on the phosphorylation potential in *Escherichia coli*. *Mol. Microbiol.* **20**, 351–360.

Waldburger, C. D. & Sauer, R. T. 1996 Signal detection by the PhoQ sensor-transmitter: characterization of the sensor domain and a response-impaired mutant that identifies ligand-binding determinants. *J. Biol. Chem.* **271**, 26 630–26 636.

Wang, J. C. 1971 Interaction between DNA and an *Escherichia coli* protein ω. *J. Mol. Biol.* **55**, 523–533.

Williams, R. M. & Rimsky, S. 1997 Molecular aspects of the *E. coli* nucleoid protein, H-NS: a central controller of gene regulatory networks. *FEMS Microbiol. Lett.* **156**, 175–185.

Wilson, J. A., Doyle, T. J. & Gulig, P. A. 1997 Exponential-phase expression of *spvA* of the *Salmonella typhimurium* virulence plasmid: induction in intracellular salts medium and intracellularly in mice and cultured mammalian cells. *Microbiology* **143**, 3827–3839.

Wolf, S. G., Frenkiel, D., Arad, T., Finkel, S. E., Kolter, R. & Minsky, A. 1999 DNA protection by stress-induced biocrystallization. *Nature* **400**, 83–85.

Wu, H.-Y., Shyy, S., Wang, J. C. & Liu, L. F. 1988 Transcription generates positively and negatively supercoiled domains in the template. *Cell* **53**, 433–440.

Yamada, H., Yoshida, T., Tanaka, K. I., Sasakawa, C. & Mizuno, T. 1991 Molecular analysis of the *Escherichia coli hns* gene encoding a DNA-binding protein, which preferentially recognises curved DNA sequences. *Mol. Gen. Genet.* **230**, 332–336.

Zhang, A., Rimsky, S., Reaban, M. E., Buc, H. & Belfort, M. 1996 *Escherichia coli* protein analogs StpA and H-NS: regulatory loops, similar and disparate effects on nucleic acid dynamics. *EMBO J.* **15**, 1340–1349.

New Methods for Studying Bacterial Behaviour *in Vivo*

THE PATHOGENESIS OF *SHIGELLA FLEXNERI* INFECTION: LESSONS FROM *IN VITRO* AND *IN VIVO* STUDIES

DANA J. PHILPOTT, JONATHAN D. EDGEWORTH and
PHILIPPE J. SANSONETTI*

*Unité de Pathogénie Microbienne Moléculaire,
Institut Pasteur, 28 rue du Docteur Roux,
75724 Paris, France*

Shigella flexneri is a Gram-negative facultatively intracellular pathogen responsible for bacillary dysentery in humans. More than one million deaths occur yearly due to infections with *Shigella* spp. and the victims are mostly children of the developing world. The pathogenesis of *Shigella* centres on the ability of this organism to invade the colonic epithelium where it induces severe mucosal inflammation. Much information that we have gained concerning the pathogenesis of *Shigella* has been derived from the study of *in vitro* models of infection. Using these techniques, a number of the molecular mechanisms by which *Shigella* invades epithelial cells and macrophages have been identified. *In vivo* models of shigellosis have been hampered since humans are the only natural hosts of *Shigella*. However, experimental infection of macaques as well as the murine lung and rabbit ligated ileal loop models have been important in defining some of the immune and inflammatory components of the disease. In particular, the murine lung model has shed light on the development of systemic and local immune protection against *Shigella* infection. It would be naive to believe that any one model of *Shigella* infection could adequately represent the complexity of the disease in humans, and more sophisticated *in vivo* models are now necessary. These models require the use of human cells and tissue, but at present such models remain in the developmental stage. Ultimately, however, it is with such studies that novel treatments and vaccine candidates for the treatment and prevention of shigellosis will be designed.

Keywords: *Shigella*; pathogenicity; mucosal immunity; immune response; animal models; *in vitro* models

1. INTRODUCTION

Infection with *Shigella* spp. is a serious cause of morbidity and mortality especially in children of the developing world. Recently, the World Health Organization estimated that 1.1 million deaths per year are attributed to shigellosis (Kotloff *et al.* 1999). There are four species of *Shigella* that cause these infections, with *S. flexneri* and, to a lesser extent, *S. sonnei*,

---

*Author for correspondence (psanson@pasteur.fr).

accounting for most of the endemic disease. Epidemic disease is usually due to *S. dysenteriae*, which displays the same invasive capacity as the other species but in addition, secretes a potent cytotoxin, Shiga toxin, that can cause haemolytic uraemic syndrome. Existing antimicrobial treatments are becoming increasingly compromised because of the growing occurrence of antibiotic resistance among *Shigella* spp. In addition, the cost of treating shigellosis with antibiotics, particularly in the developing world, is impractical and stresses the need for an efficient vaccine against this disease. Currently, however, there is no vaccine available that can provide adequate protection against the many different serotypes of *Shigella*. Therefore, both the development of new treatments and the design of innovative vaccines for the prevention of shigellosis rely on an improved understanding of the pathogenesis of the disease. Our knowledge of the pathogenesis of *Shigella* infection thus far and what we hope to learn in the future has and continues to depend on our ability to model the infection *in vitro* and to validate these models with *in vivo* studies. This review outlines our current understanding of the pathogenesis of *Shigella* infection. Specifically, findings from *in vitro* systems will be compared to those gained from animal models of shigellosis, while keeping in mind essential features of the disease in humans. This is followed by a discussion of the possibilities for future research and where we believe further studies are required.

2. *SHIGELLA* INFECTION — OVERVIEW

Shigella flexneri is a Gram-negative facultatively intracellular pathogen that invades the colonic and rectal mucosae of humans, causing bacillary dysentery. Shigellosis is highly infectious, with ingestion of as few as 100 organisms resulting in disease (Dupont *et al.* 1989), and is transmitted by person-to-person contact or indirectly through contaminated food or water. Shigellosis produces a spectrum of clinical outcomes ranging from watery diarrhoea to classic dysentery characterized by fever, violent intestinal cramps and discharge of mucopurulent and bloody stools. Inflammation of the infected tissue is a key feature of shigellosis. Histopathological studies of colonic biopsies from infected patients reveal inflammatory cell infiltration into the epithelial layer, tissue oedema and eroded regions of the colonic epithelium (Mathan & Mathan 1991).

Since this organism is unable to invade epithelial cells through the apical route, *Shigella* exploits M cells, the specialized epithelial cells in the follicular associated epithelium (FAE) that overlie lymphoid tissue, to gain entry into the colonic epithelium (Wassef *et al.* 1989). M cells allow intact *Shigella* to traverse into the underlying subepithelial pocket where macrophages reside. Macrophages engulf *Shigella*, but instead of successfully destroying the bacteria in the phagosome, the macrophage succumbs to apoptotic death (Zychlinsky *et al.* 1992). Prior to cell death, infected macrophages release IL-1β through the direct activation of caspase-1 by *Shigella* (Zychlinsky *et al.* 1994). The pro-inflammatory nature of this cytokine results in the recruitment of polymorphonuclear cells (PMNs) that infiltrate the infected site and destabilize the epithelium (Perdomo *et al.* 1994*a, b*). Loss of integrity of the epithelial barrier allows more bacteria to traverse into subepithelial space and gives these organisms access to the basolateral pole of the epithelial cells (Mounier *et al.* 1992). *Shigella* can then invade the epithelial cells lining the colon, spread from cell to cell and disseminate throughout the tissue. Cytokines released by infected epithelial cells attract increased numbers of immune cells to the infected site, thus compounding and exacerbating the inflammation.

3.　INVASION OF EPITHELIAL CELLS

The link between epithelial cell invasion and expression of the virulent phenotype of *Shigella* was first made in 1964 (LaBrec *et al.* 1964). The Serény test, which is the oldest animal model of shigellosis, was used as a model to test *Shigella* invasiveness (Serény 1955). This assay consists of inoculating a suspension of bacteria into the keratoconjunctival sac of guinea-pigs or mice. Pathogenic *Shigella* invade the conjunctival epithelium causing conjunctivitis and keratitis. This model proved useful for identifying avirulent mutants of *Shigella* that are incapable of expressing the invasive phenotype. However, the lack of specificity of the response makes it impossible to discriminate among the various phenotypes of *Shigella* including invasion of epithelial cells, cell-to-cell spread and the initiation of an inflammatory response.

Cultured epithelial cell lines have greatly aided the study of the host-cell events involved in cell invasion by *Shigella*. Examination of *Shigella*-infected cells by microscopic methods has defined the entry event as a macropinocytic-like process that results in massive induction of host cell membrane ruffling — changes which are reminiscent of those elicited by growth factors. In the case of invading *Shigella*, however, the membrane ruffles are confined to the site of bacterium–cell interaction. Cytoskeleton-mediated membrane extensions are observed to rise up from the surface of the cell and these projections eventually fuse to engulf the bacterial body (Fig. 1).

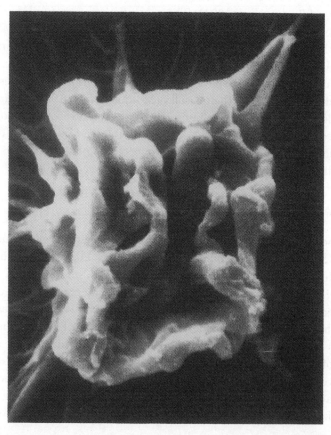

Fig. 1. Scanning electron micrograph of *Shigella flexneri* inducing membrane ruffles on the surface of an epithelial cell prior to its uptake. Photograph is courtesy of Dr Ariel Blocker (Institut Pasteur, France) and Dr Roger Webf (European Molecular Biology Laboratory, Heidelberg, Germany).

Studies of epithelial-cell–*Shigella* interactions often use poorly differentiated and non-polarized epithelial cell lines, such as HeLa or HEp-2 cells, grown in tissue culture flasks. However, more sophisticated systems using human intestinal cell lines grown on permeable filter supports with distinct upper (lumenal) and lower (basal) chambers have been employed. Growing intestinal Caco-2 or T84 cell lines in this way allows the cells to grow as columnar epithelial cells with a more or less organized brush border (depending on the cell line) and to polarize with distinct apical and basolateral membranes separated by intercellular tight junctions. Bacterial infection of the apical surface of cultured intestinal cells grown in this way more closely mimics infection of the human intestinal epithelium. Using this system, a surprising observation was noted as apically infecting *Shigella* cannot invade polarized cells. Only when intercellular junctions are disrupted by treatment of the cells with ethylene glycol-bis (beta-aminoethyl ether)-N,N,N',N'-tetraacetic acid (EGTA) are *Shigella* able to invade the filter-grown Caco-2 cells. These studies indicated that *Shigella* enter polarized Caco-2 cells almost exclusively from the basolateral pole (Mounier *et al.* 1992).

A methodological step forward was made in the study of molecular mechanisms of bacterial invasion when it was realized that the aminoglycoside antibiotic gentamicin is membrane impermeable and thus bacteria that are able to enter host cells survive antibiotic treatment of an infected monolayer. This lack of accessibility of gentamicin to intracellular bacteria forms the basis of the 'gentamicin protection assay' whereby the capacity of an organism to invade eukaryotic cells can be assessed reproducibly and quantitatively. Using this assay, a number of genes necessary for *Shigella* entry have been identified by analysing mutants defective in surviving gentamicin treatment. Genes encoding bacterial factors required for *Shigella* entry reside on a 200 kb virulence plasmid of wild-type *S. flexneri*. Strains lacking the plasmid are non-invasive *in vitro* and also avirulent in animal models of shigellosis. The effectors of *Shigella* entry are the so-called 'invasion plasmid antigens' or Ipa proteins which are encoded in a 30 kb 'entry region'. This region is composed of two adjacent loci transcribed in opposite directions. One locus is essentially composed of the ipa operon, which encodes four secreted proteins, IpaB, IpaC, IpaD and IpaA, which are the effectors of bacterial entry *in vitro*. Mutations in the genes encoding IpaB, IpaC and IpaD proteins render the bacteria non-invasive in cell culture systems and are also avirulent in animal models; these *Shigella* mutants are unable to provoke keratoconjunctivitis in guinea pigs (Ménard *et al.* 1993). An IpaA mutant of *Shigella* maintains a 10% invasion efficiency as assessed *in vitro*; however, it is unable to induce fluid accumulation in rabbit ligated loops, suggesting that the full complement of Ipa proteins are necessary for efficient translocation of *Shigella* across the epithelial barrier and the initiation of an inflammatory response. The other locus in the entry region, the mxi/spa locus, comprises genes that encode for a type-III secretion apparatus, an evolutionary conserved bacterial system that is responsible for the host-cell contact-dependent secretion of the Ipa proteins, presumably into the host cell's cytoplasm (for a review, see Hueck 1998). Mutations in the genes encoding the type-III secretion system are also avirulent based on the Serény test, due to their inability to invade (Sasakawa *et al.* 1988).

The detailed mechanisms by which the Ipa effector proteins bring about *Shigella* invasion have not yet been fully defined. The Ipa proteins are synthesized and stored within the bacterial body and are secreted through the type-III secretion system upon contact with the host cell (Ménard *et al.* 1994). A complex formed by the association of IpaB and IpaD is thought to regulate the flux of Ipa proteins through the secretion system (Ménard *et al.* 1996). Once secreted, IpaB and IpaC form a complex interacting with the epithelial cell membrane. This complex forms a pore through which it is presumed the other Ipa proteins are translocated into the host cytoplasm (Blocker *et al.* 1999). IpaC and IpaA appear to orchestrate the cytoskeletal rearrangements necessary to direct uptake of the organism into the normally non-phagocytic epithelial cell (Tran Van Nhieu *et al.* 1997, 1999; Bourdet-Sicard *et al.* 1999). Once the *Shigella*-containing vacuole is formed within the infected cell, IpaB mediates lysis of the vacuole and the bacterium is then free in the cytosol (High *et al.* 1992).

4.   CELL ADHESION RECEPTORS AND *SHIGELLA* ENTRY

A number of cell adhesion receptors have been implicated in *Shigella* entry into epithelial cells. A secreted complex of IpaBCD has been shown to bind $\alpha 5\beta 1$ integrins *in vitro* and this interaction appears to play a role in *Shigella* entry since overexpression of $\alpha 5\beta 1$ in Chinese hamster ovary cells leads to efficient invasion compared to non-transfected cells (Watari *et al.* 1996). $\alpha5\beta1$ integrins are present on the basolateral surface of epithelial cells where they mediate interaction with the extracellular matrix. Thus, the location of integrins is in agreement with studies indicating that the basolateral membrane is the point of entry of *Shigella* into epithelial cells. Since $\beta1$ integrins interact with the actin cytoskeleton through the carboxy-terminal moiety of the $\beta1$ subunit, it was suggested that the binding of *Shigella* to integrins induces cytoskeletal rearrangements leading to the formation of focal adhesion-like structures. Consistent with this idea, the small GTPase Rho, which is important in stress fibre and focal adhesion formation, was shown to be necessary for invasion of epithelial cells by *Shigella* (Adam *et al.* 1996; Watari *et al.* 1997). Additionally, a number of proteins normally associated with focal adhesions are recruited to the site of *Shigella* entry. The focal adhesion components vinculin and ezrin have been shown to be associated with the *Shigella*-induced entry structure (Tran Van Nhieu *et al.* 1997; Skoudy *et al.* 1999*b*).

More recently, another cell adhesion receptor was shown to play a role in *Shigella* entry into epithelial cells. The IpaBC complex binds to CD44 during *Shigella* entry of HeLa epithelial cells and this interaction also appears to be important for invasion since blocking antibodies to CD44 significantly reduce the uptake of *Shigella* into cells (Skoudy *et al.* 1999*a*). CD44 is the receptor for hyaluronan, a component of the extracellular matrix. Thus, CD44, like $\beta1$ integrins, is likely to be expressed on the basolateral membrane of epithelial cells, putting it in an optimal position for the putative interaction with translocated *Shigella*. Through its cytoplasmic domain, CD44 interacts with ezrin, a protein belonging to the ezrin-radixinmoesin (ERM) family of proteins that act to crosslink the plasma membrane and the actin cytoskeleton. ERM proteins are thought to be important in the dynamic regulation of cell shape as they accumulate underneath the plasma membrane in subcellular structures such as microvilli, cell–cell contact sites as well as membrane ruffles, filopodia, microspikes and lamellipodia. Ezrin is also enriched in the cellular protrusions that engulf invading *Shigella* (Skoudy *et al.* 1999*b*). Moreover, it was shown that the dynamic regulation of the cytoskeleton potentially through ezrin is important for *Shigella* entry. Transfection of cells with a dominant negative form of ezrin significantly reduced the ability of *Shigella* to invade. A role for Rho GTPases is again indicated here since Rho can regulate the association of ERM proteins with the plasma membrane (Takahashi *et al.* 1997).

Unfortunately, *in vivo* validation of the above-mentioned *in vitro* experiments is lacking. Therefore, the role played by either of $\alpha5\beta1$ integrins or CD44 in *Shigella* invasion *in vivo* is unknown and difficult to test directly. However, it is likely that *Shigella* entry into host epithelial cells is the result of a coordinate action of many different signal transduction pathways and the use of any particular receptor may be redundant. In the case of integrins, only the Ipa complex itself and not the bacterium bind to integrins, questioning the role of this interaction in *Shigella* entry. Additionally, cells that are deficient in either integrins or CD44 are only partially defective in their ability to be invaded by *Shigella* (Skoudy *et al.* 1999a). It has been speculated that the IpaBC complex transiently associates with either integrins or CD44 and this increases the efficiency by which these proteins are inserted into the host membrane where they then act as a pore through which the effector Ipa proteins travel into the host cytoplasm. Clearly, however, these proteins can be inserted into host membranes and *Shigella* can invade cells even in the absence of these cell-adhesion receptors. This brings about the question of whether or not adhesion is a necessary prerequisite to epithelial cell invasion by *Shigella*. So far, adherence of *Shigella* to epithelial cells has not been fully described and the recent sequencing of the virulence plasmid has not identified any putative adhesins in *Shigella* (C. Parsot, personal communication). Moreover, the ability of *Shigella* to enter cells of many different species argues against any particular species-specific receptor necessary for invasion. Secretion and insertion of the IpaBC pore into host membranes may be the rate-limiting step in *Shigella* invasion and a receptor as such is potentially unnecessary. This, however, is speculation since an exhaustive search for a putative *Shigella* adhesin awaits further research.

5.   INTRA- AND INTERCELLULAR DISSEMINATION

Once inside the host cell cytoplasm, *Shigella* lyse the membrane-bound vacuole and escape into the cytoplasm. A direct consequence of this contact with the intracellular milieu is intracellular motility. The outer membrane protein, IcsA, is necessary and sufficient to direct actin-based motility of *Shigella* within the host cytoplasm (Bernardini *et al.* 1989). The functional role of IcsA in actin-based motility and the cellular partners involved

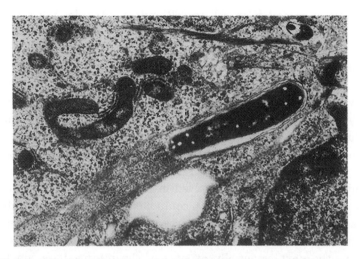

Fig. 2. Transmission electron micrograph of a *Shigella flexneri*-mediated protrusion be-ing taken up by a neighbouring cell. Note the dense accumulation of actin behind the moving bacterium. Photograph is courtesy of Dr Michelle Rathman (Institut Pasteur, France).

been recently reviewed (Sansonetti *et al.* 1999a). Intracellular *Shigella* use cytoskeletal components to propel themselves inside the infected cell and when contact occurs between the moving organism and the host cell mem-brane, cellular protrusions are formed. These protrusions are then engulfed by the neighbouring cell thus permitting cell-to-cell spread of *Shigella* with-out the bacterium ever leaving the confines of the host epithelial layer (Fig. 2).

Assays to study cell-to-cell spread have centred on two techniques, the plaque assay and the infectious foci assay. In the plaque assay, epithelial cells are infected with wild-type *Shigella* and following a period of incuba-tion, medium containing gentamicin and agarose is added to the infected cell monolayer in order to restrict reinfection of cells from bacteria in the culture media. In this way, bacteria must spread through the epithelial layer by passing from one cell to the next. Two to three days later, the agarose plug is removed and plaques can be observed in the epithelial monolayer. These plaques correspond to points of initial cellular infection and the re-sulting destruction and clearing of infected cells (Oaks *et al.* 1985). Using this assay, a number of mutants have been identified that are deficient in their ability to spread from cell to cell and of these, IcsA has been best characterized. The ability of IcsA to induce actin polymerization is equally required for intracellular and intercellular spread. Moreover, the IcsA phenotype is extremely relevant during infection *in vivo*. Monkeys infected with an IcsA mutant of *Shigella* develop only mild dysenteric symptoms and show limited histopathological lesions of the colonic and rectal mucosae (Sansonetti *et al.* 1991). These findings stress the requirement for intercellular spread for full virulence of *Shigella* during infection *in vivo* and perhaps point to the role of the epithelial cell in the development of widespread inflammation (discussed in §9).

E-cadherin, a key protein involved in intercellular adhesion, has been shown to be an important cellular component involved in the intercellular spread of *Shigella*. Cell–cell contacts were thought to be necessary for intercellular spread because of the observation that *Shigella* passed from one cell to the next essentially at sites of the intermediate junctions in Caco-2 cells (Vasselon *et al.* 1992). In addition, transmission electron microscopic observations of various epithelial cell lines showed that passage of *Shigella* protrusions from one cell to the next occurred at sites where the two cells were closely apposed, suggesting that cell–cell contacts were involved. To test this directly, the infectious foci assay was developed. In this assay, cells are infected with *Shigella* for a period of time and subsequently trypsinized and seeded at a very low density with a population of uninfected cells that are either cadherin-negative or stably expressing cadherin. In the cells expressing cadherins, *Shigella* is efficiently transmitted from the originally infected cells to the neighbouring cells such that large areas of the monolayer are observed to be infected. In contrast, cells that are deficient in cadherins do not transmit *Shigella* and the infection remains limited to the index cells only. Therefore cell–cell contacts, dependent on the expression of cadherins, are necessary for intercellular spread of *Shigella* (Sansonetti *et al.* 1994). Further research is required to identify whether or not *Shigella* interacts directly with proteins at this junction to bring about protrusion formation. Additionally, the role played by intermediate junctions in the pathogenesis of *Shigella* during infection *in vivo* needs to be addressed. Again, because of the complexity of the system, such *in vivo* evidence is difficult to obtain and will rely on the development of novel model systems. In the meantime, however, complementary techniques, such as the expression of dominant negative proteins, the use of specific inhibitors as well as cell lines deficient in certain proteins, certainly lends credence to these *in vitro* findings.

6.   M CELLS: PORTS OF ENTRY INTO THE HOST EPITHELIUM

One of the key events in the pathogenesis of enteroinvasive bacterial infections is the penetration of the intestinal epithelium. Since *Shigella* cannot enter epithelial cells via the apical pole, it uses M cells to gain entry into the host epithelium. In fact, many Gram-negative bacteria that cause enteric disease, including *Salmonella* and *Yersinia*, have been shown to preferentially cross the epithelium via specialized antigen sampling cells called M cells (for a review, see Sansonetti & Phalipon 1999). M cells, which stands for membranous or microfold cells, are modified epithelial cells found within the FAE overlying lymphoid follicles. These follicular lymphoid structures are scattered throughout the small intestine in aggregates known as Peyer's patches and in the colon and rectum as isolated solitary nodules. M cells are relatively rare, constituting less than 0.1% of epithelial cells present in the lining of the intestine and can be identified morphologically due to the fact that they display (i) a poorly differentiated brush border compared with neighbouring absorptive epithelial cells, and (ii) an irregular basolateral membrane border containing invaginated lymphocytes. M cells have a high endocytic activity which serves to transport soluble and particulate lumenal antigens across the cytoplasm and deliver them intact to the antigen-processing and -presenting cells in the underlying follicle (Neutra *et al.*). It is perhaps surprising that a cell so rare in the intestinal epithelium can be the target of entry for many different pathogens. How then do these pathogens seek out M cells and use these cells to enter the host epithelium? It has been suggested that the lack of both mucus and a well-developed glycocalyx over the FAE facilitate non-specific interactions of pathogenic organisms with M cells. Increased hydrophobic interactions may be favoured and this could be the primary step that precedes what is likely to be a non-specific transport mechanism. In fact, it has been shown that lectins, positively charged particles and hydrophobic beads, all of which bind to the membrane surface of M cells, are transported with increased efficiency (reviewed in Jepson *et al.* 1996). M cells may also express characteristic surface molecules that could serve as specific receptors for pathogens. For example,

M cells express characteristic glycoconjugates, which vary depending on the species and the location in the intestine (Giannasca *et al.* 1994; Lelouard *et al.* 1999). Although no specific receptor engaged by a bacterial adhesin or invasin has been identified, such a receptor may account for tissue tropism of a particular pathogen as well as its efficient uptake into the FAE.

7.   M CELLS AND ENTRY OF *S. FLEXNERI* INTO THE HOST: *IN VIVO* EVIDENCE

The first indication that *S. flexneri* exploits M cells to enter the host epithelial layer came from studies using a rabbit ligated ileal loop model. In this model, animals are anaesthetized and the intestine is externalized at laparotomy. Sections of ileum are carefully ligated to preserve the existing vasculature and, subsequently, a large inoculum of bacteria, usually 109 colony forming units (CFU) ml^{-1}, is injected into the intestinal loop. Invasive and inflammatory properties of the organisms can be observed following sacrifice of the animals at a given time post-infection (usually 2–8 h). Histological studies and measurements of fluid accumulation within the infected loop can be conducted. By isolating ileal loops with grossly identifiable Peyer's patches, the role of the FAE in the initial steps of epithelial translocation by *Shigella* has been assessed. Wild-type *S. flexneri* was readily detected in the dome epithelium of the FAE, whereas very few organisms were observed within the villus epithelium. In addition, when the infected loops were incubated for longer time periods, ulcerations were observed preferentially over the dome regions of Peyer's patches suggesting that the FAE was the primary site of entry (Wassef *et al.* 1989). These findings were reconfirmed using the rabbit ileal loop model in a study indicating that threefold more bacteria were present within the infected tissue if Peyer's patches were present within the loop compared with those loops that lacked lymphoid follicles (Perdomo *et al.* 1994*b*). Figure 3 shows wild-type *Shigella* crossing the epithelial barrier by an M cell.

These findings were also confirmed in the macaque monkey model of shigellosis. Macaques in particular are one of the few animals that develop a dysentery-like disease following oral or gastric inoculation of *Shigella*, although a dose of 10^{10} organisms is typically required for the development of disease. Using this model, it was observed that when monkeys were infected with an *icsA* mutant of *S. flexneri*, which does not spread intra- or

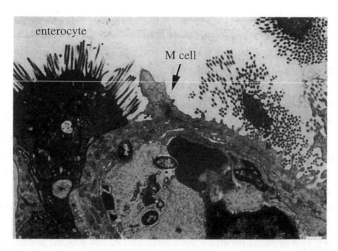

Fig. 3. Transmission electron micrograph of *Shigella flexneri* crossing the intestinal epithelium by an M cell. Photograph adapted from Sansonetti & Phalipon (1999).

intercellularly, animals do not develop clinical symptoms but small ulcers corresponding to the presence of lymphoid follicles are observed on the colonic lining (Sansonetti *et al.* 1991). These findings suggest that the *icsA* mutant is capable of entry into the FAE but owing to its inability to spread from the initial entry site, only a local ulceration at the point of the FAE is observed. These findings confirm that the FAE serves as the site of bacterial entry into the epithelium and also reiterates the role of intra- and intercellular spread in the development of widespread inflammation during wild-type *Shigella* infection.

Another important observation made using the ileal loop model was that a significantly greater number of wild-type *Shigella* were found in the dome epithelium compared with non-pathogenic strains or heat-killed organisms, suggesting that the presence of virulence factors play a role in the increased uptake of the wild-type strain into the FAE (Wassef *et al.* 1989). This observation was further characterized using strains of *S. flexneri* expressing either an invasive or an adhesive but non-invasive phenotype in the rabbit ileal loop model. The adhesiveness of the latter strain is mediated by the expression of an *Escherichia coli* adhesin that mediates attachment of the organism to rabbit M cells (Inman & Cantey 1984). By immunostaining for lipopolysaccharide (LPS), it was shown that the amount of bacterial material associated with the FAE and the dome of the lymphoid follicle was essentially equivalent in loops infected with either the adhesivenon-invasive or the invasive strain, whereas very few control organisms, i.e. non-adhesive and non-invasive, could be isolated from similarly infected loops (Sansonetti *et al.* 1996). These data suggest that either specific adhesion to M cells or an invasive capacity to enter M cells is required for an organism to be transported through the epithelium and into the subepithelial space. What was also clear from this study was that once the bacteria gain access to the subepithelial space they have very different fates depending on their virulence capacities. Whereas infection with wild-type, fully invasive *Shigella* results in rapid inflammation and subsequent destruction of the FAE, the adhesive yet non-invasive strain is sequestered and destroyed within lysosomes of macrophages present within the dome.

Although the animal models discussed above have been useful for studying some aspects of *Shigella*–M-cell interactions, there are significant drawbacks and their direct relevance to human infection can be questioned. Most obvious is the fact that oral or gastric inoculation of rabbits does not lead to dysenteric symptoms. Also, shigellosis in humans is a disease of the distal colon and rectum, whereas in the rabbit model it is the ileum that is studied. Therefore, this model does not take into account the tissue specificity that is seen in human infection thus ignoring the potential of a specific interaction between *Shigella* and M cells of the human colon. In addition to having ethical and financial drawbacks, the macaque model is also not ideal since the infectious dose required for the animals to develop dysentery is ten million to 100 million times higher that the infectious dose in humans. The questionable relevance of such a model is particularly apparent in the context of testing the tolerance of attenuated vaccine candidates or doing challenge experiments in vaccinated animals.

Despite these drawbacks, however, many of the observations made in animal studies do correlate with what we know from human infections with *Shigella*. In fact, clinical observations of patients suffering from shigellosis support the idea that the FAE is the primary route of entry of *Shigella* into the host tissues. In patients examined endoscopically within two days of the start of infection, early inflammatory lesions resembling aphtoid ulcers are present in the rectum and distal colon, and on histopathological observation these lesions are found to correspond to lymphoid follicles (Mathan & Mathan 1991). Additionally, inflammation is observed to be confined to follicular regions of the rectum and distal colon early in the course of infection, but is later detected in the surrounding villi and can be seen to extend proximally (Islam *et al.* 1994).

There are a number of aspects of M cell–*Shigella* interactions, however, that cannot be addressed adequately using these *in vivo* systems and therefore require *in vitro* modelling. Recently, such a model was developed in which Caco-2 cells, a human intestinal cell line, were induced to switch to an M-cell phenotype when co-cultured with lymphocytes isolated from the Peyer's patch (Kernéis *et al.* 1997). Using this model, it was shown that *Vibrio cholerae* O:1, a non-invasive pathogen transported exclusively by M cells (Owen *et al.* 1986), could also be transported across the model epithelium by the *in vitro*-induced M cells. This model will assist studies into *Shigella*–M-cell interactions and may help to identify specific receptors or adhesive factors on M cells that facilitate the uptake of *Shigella* by these cells. Additionally, this model will allow the determination of the virulence factors of *Shigella* that are necessary for entry and transport across M cells.

At later time-points of infection, inflammation disrupts the integrity of epithelium and this may be a secondary means by which pathogenic *Shigella* translocate across the epithelial barrier in order to reach the basolateral pole of epithelial cells, which they can then efficiently invade. The possibility of this mode of *Shigella* translocation was modelled *in vitro* (Perdomo *et al.* 1994*a*). This system was again based on the culture of monolayers of human intestinal cells on filter supports; however, another layer of complexity to this system was added so that the early immune responses to invasive bacteria could be investigated. In this system, isolated human PMNs are added to the basolateral side of polarized T84 cells, which are then infected apically with pathogenic *Shigella*. A rapid paracellular transmigration of the PMNs which disrupts the barrier function of the epithelium is observed, as measured by a drop in transepithelial electrical resistance. Subsequent to these events, *Shigella* is able to pass through the disrupted tight junctions and thus gains access to the basolateral pole of the cells (Perdomo *et al.* 1994*a*). These studies suggest that lumenal *Shigella* can induce epithelial cells to produce potent chemotactic signals that elicit transepithelial transmigration of PMNs. In fact, the epithelial cell has been shown to play a significant role in innate immunity against enteroinvasive bacterial infections (Jung *et al.* 1995) and, in the case of *Shigella* infection, is important in intiating the inflammatory response (Sansonetti *et al.* 1999*b*; see §9).

8.   MACROPHAGE APOPTOSIS IN RESPONSE TO *SHIGELLA* INVASION

Bacteria that have crossed the epithelial layer via M cells are likely to be phagocytosed by resident macrophages within the subepithelial dome overlying the lymphoid follicles (see Fig. 4(a)). The uptake of *Shigella* by macrophages *in vitro* does not require the virulence plasmid and presumably occurs by normal phagocytic mechanisms. The fate of *S. flexneri* following phagocytosis was first studied in the late 1980s using the murine macrophage J774 cell line (Clerc *et al.* 1987). It was noted that uptake of *Shigella* resulted in lysis of the phagocytic vacuole and rapid killing of the infected cell. It was not until five years later, however, that macrophage cell death was shown to occur by apoptosis (Zychlinsky *et al.* 1992; Fig. 4(b)). Apoptosis was not seen with plasmid-cured strains, which led to the identification of the plasmid-encoded protein, IpaB, as the mediator of cell death (Zychlinsky *et al.* 1994). IpaB gains access to the cytosol, where it binds and activates caspase-1, also known as interleukin(IL)-1-converting enzyme (Chen *et al.* 1996). Activation of caspase-1 is absolutely required for *Shigella*-induced apoptosis, since cell death is not seen in caspase-1 knockout mice (Hilbi *et al.* 1998). The downstream events promoting apoptosis following caspase-1 activation by IpaB are unknown.

Apoptosis is generally considered to be an immunologically silent cell death process unaccompanied by inflammation, however, this is not the case with caspase-1-dependent apoptosis. Caspase-1 cleaves and activates the pro-inflammatory cytokines IL-1β and IL-18 (Ghayur *et al.* 1997). Murine macrophages have been shown to release large amounts of mature IL-1β during the *Shigella*-induced apoptotic process (Zychlinsky *et al.* 1994). Given that mucosal inflammation is the hallmark of shigellosis, these observations made largely in murine macrophage cell lines prompted the search for evidence that apoptosis and the consequent cytokine production play a role in *Shigella* infection *in vivo*.

Apoptotic cells have been identified in the subepithelial dome and lymphoid follicles in a rabbit ligated ileal loop model of *Shigella* infection (Zychlinsky *et al.* 1996). Apoptotic cells were not seen in the mucosa when challenged with plasmid-cured *Shigella* or plasmid-cured strains transfected with an *E. coli* adhesin, which allowed the bacteria to penetrate into the

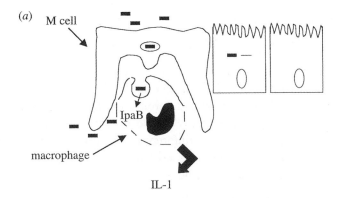

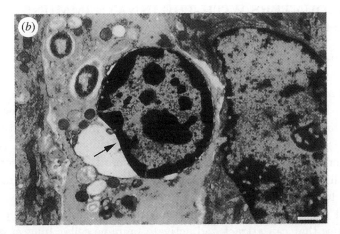

Fig. 4. (a) Model of *Shigella flexneri* penetration of the intestinal epithelium by M cells and subsequent contact with the underlying macrophages at the site of the follicular lymphoid tissue. *Shigella* is phagocytosed by resident macrophages; however, the organism escapes from the phagocytic vacuole and induces macrophage apoptosis via interaction of bacterial IpaB with host cell caspase-1. Activated caspase-1 cleaves and activates pro-IL-1 that is released in large quantities from the dying macrophage. (b) Apoptosis of a *Shigella*-infected macrophage *in vivo*. Arrow points to the condensation of chromatin at the periphery of the nucleus which is a characteristic of apoptotic cell death. Photograph adapted from Sansonetti & Phalipon (1999).

subepithelial space in comparable numbers to the wild-type *Shigella*. Apoptotic cells have also been seen in rectal mucosal biopsies from patients acutely infected with *Shigella* (Islam *et al.* 1997). Together these observations provide evidence for apoptosis *in vivo* during *Shigella* infection, and suggest that this phenomenon is due to the presence of the virulence plasmid.

There have recently been some reports that *Shigella* can kill macrophages by an alternative mechanism termed oncosis, and that this process does not involve caspase-1 (Fernandez-Prada *et al.* 1997; Nonaka *et al.* 1999). In the latter report a differentiated human monocyte-like cell line, U937, was used to show that *Shigella* infection could result in apoptosis or oncosis depending on the differentiation stimulus used. Evidence of oncosis *in vivo* in *Shigella* infection and whether it contributes towards the disease manifestations has yet to be investigated.

9. ENCOUNTERS WITH THE INNATE IMMUNE RESPONSE

The innate immune response provides an early defence against bacterial infection, which serves to limit bacterial proliferation, localize the infection and also both activate and regulate the subsequent adaptive immune response. Many cell types and soluble proteins, including phagocytic cells (neutrophils, macrophages and dendritic cells), lymphocytes (natural killer (NK) cells and $\gamma\delta$ T cells), cytokines (most notably, IL-1, IL-6, IL-12, tumour necrosis factor α (TNFα) and IFNγ) and liver-derived serum proteins such as complement factors contribute towards innate immunity. In addition to these classical immune components, non-immune cells such as epithelial cells recognize and respond to bacterial invasion by producing chemokines that can attract and activate immune cells (Jung *et al.* 1995). The net result of the complex interaction between these many factors is usually manifested as acute inflammation.

The study of the innate immune response in shigellosis has largely focused on the mechanisms involved in regulating the influx of neutrophil into the infected site. The neutrophil response can be separated into two stages: an initial influx focused in the region of the lymphoid follicles; and a later phase of massive neutrophil influx into intestinal villi and crypts, which produces large areas of epithelial destruction and mucosal ulceration that extend far beyond the initial site of bacterial entry (Mathan & Mathan 1991). It is likely that these two stages reflect different processes. IL-1 released from *Shigella*-infected apoptotic macrophages may be responsible for the first stage; in the rabbit ileal loop infection model, treatment with an IL-1 receptor antagonist prior to infection with virulent *S. flexneri* significantly decreased the inflammation and tissue destruction within the lymphoid follicles (Sansonetti *et al.* 1995).

Observations that the second stage of inflammation occurs at some distance from the follicles and that infected epithelial cells secrete pro-inflammatory cytokines prompted an investigation into the role of IL-8 as a mediator of inflammation in this second phase (Sansonetti *et al.* 1999*b*; see Fig. 5(*a*) for model). Again using the *in vivo* rabbit ileal loop model, a neutralizing anti-IL-8 monoclonal antibody was found to considerably reduce the neutrophil influx entering via the lamina propria into the intestinal villi and to attenuate the consequent epithelial destruction. *In vitro* studies have shown that IL-8 production by epithelial cells induces neutrophil migration across polarized epithelial monolayers and this can occur with or without invasion of the epithelial cells by *Shigella* (Beatty & Sansonetti 1997; McCormick *et al.* 1998). Thus, bacterial interaction with epithelial cells appears to be a requirement for this second phase. The rapid extension of inflammation to sites distant from the follicles stresses the importance of cell-to-cell spread by *Shigella*, and is further supported by experimental *Shigella* infection of macaque monkeys with the *icsA* mutant, capable of epithelial cell invasion but unable to spread from one cell to another (Sansonetti *et al.* 1991). The inability of the *icsA* mutant to spread through the epithelial layer restricts the contribution of epithelial chemokine release and consequently limits the inflammation seen in infected animals.

It has been noted that blocking the neutrophil influx using anti-β1-integrin antibodies, IL-1 receptor antagonists or anti-IL-8 antibodies, all result in decreased epithelial destruction implicating the neutrophil rather than *Shigella* as the direct cause of mucosal damage (Perdomo *et al.* 1994*b*; Sansonetti *et al.* 1995, 1999*b*). Neutrophils can kill opsonized *Shigella in vitro* (Mandic-Muleg *et al.* 1997), and the neutrophil inflammatory response localizes the bacteria to the epithelium. When neutrophil influx is blocked, bacteria migrate deep into the lamina propria and mesenteric blood vessels, confirming the importance of neutrophils in localizing bacterial infection (Sansonetti *et al.* 1999*b*). Thus, neutrophil influx appears to be responsible for the majority of tissue destruction associated with shigellosis, and yet is vital for preventing the systemic spread of bacteria (Fig. 5(*b*)).

The murine lung model of shigellosis, although not relevant with regard to the organ specificity of the disease, has been useful for exploring

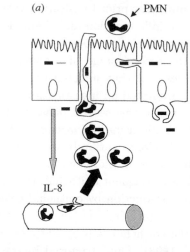

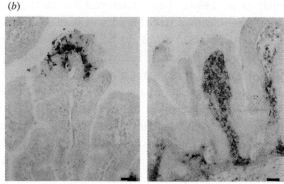

Fig. 5. (a) *Shigella flexneri*-infected epithelial cells are a source of interleukin-8 (IL-8), a potent chemotactic chemokine that is responsible for the recruitment of PMNs into the infected site. PMNs migrate between adjacent epithelial cells, break intercellular junctions and thus compromise the integrity of the epithelial barrier. This causes destruction of the mucosal surface by allowing invasion of further organisms from the colonic lumen. Conversely, the neutrophil influx is necessary in order to control the proliferation of organisms locally and prevent systemic bacterial dissemination. (b) Photographs of intestinal sections stained for LPS from *Shigella flexneri*-infected rabbit ileal loops. The left panel shows a tissue section from an ileal loop infected with *S. flexneri* in a rabbit pre-treated with a control antibody showing abscess formation and local tissue destruction. The right panel shows a tissue section from a rabbit in which IL-8 was neutralized using specific antibodies prior to infection. Although the epithelium is spared, bacterial diffusion into the lamina propria is observed. This stresses the important role for IL-8-mediated neutrophil influx in preventing bacterial translocation. Scale bars, 10 m. Photographs adapted from Sansonetti *et al.* (1999b).

details of the immune and inflammatory components, as well as some aspects of the systemic and local immune response against *Shigella* infection (Mallett *et al.* 1993; Verg *et al.* 1995). In this model, an inoculum of wild-type *Shigella* is administered intranasally resulting in invasion of the tracheo-bronchial tract resulting in an inflammatory broncho-tracheo-alveolitis (Voino-Yasenetsky & Voino-Yasenetskaya 1961). Using this model, the role of cytokines in the innate immune response has been investigated. TNFα and IFNγ are both produced locally during the first 24 hours of infection. Sublethal inoculation into IFNγ knockout mice results in overwhelming local proliferation of bacteria and death, compared with a steady decline in bacterial numbers in wild-type controls (Way *et al.* 1998). Histology of lungs from knockout mice showed an obliterative neutrophilic bronchiolitis suggesting that neutrophils alone, in the absence of IFNγ are unable to clear the infection. The course of infection in αβ and γδT-cell knockout mice was the same as wild-type controls suggesting neither cell type acted as a source for this cytokine. Conversely, NK-cell-deficient mice had an increased susceptibility implicating the NK cell as the source of IFNγ. There are probably multiple roles for IFNγ, including inhibition of bacterial proliferation within epithelial cells (Way *et al.* 1998), enhanced macrophage killing of bacteria and perhaps inhibition of macrophage apoptosis induced by *Shigella* (Hilbi *et al.* 1997).

10. THE ADAPTIVE IMMUNE RESPONSE

During the course of infection, *Shigella* exists in both an extracellular and intracellular location. This implies a requirement for both humoral and cellular immune responses for effective sterilizing immunity. The fact that mice do not display intestinal infection following challenge with *Shigella* has hampered study of the adaptive immune response. Some information, however, has been obtained from study of the murine pulmonary inoculation model and serological studies of infected humans. In the pulmonary model, isotype-specific secretory IgA-anti LPS antibodies targeted to the mucosa from subcutaneous hybridomas, provides protection against challenge with a lethal dose of organisms (Phalipon *et al.* 1995). This underscores the importance of local IgA in providing protection, and supports observations in humans suggesting that protective immunity is isotype specific and therefore directed predominantly against LPS (DuPont *et al.* 1972). Using the pulmonary challenge model, immunized mice were used to define the characteristics of a protective humoral response. Sublethal infection induces local IgG and IgA responses directed against LPS, and some Ipa proteins, but responses are slow to develop (Verg *et al.* 1995). Short-lived, protective, serotype-specific humoral responses have been generated, although this predominantly consists of an IgM response and is T-cell independent (Way *et al.* 1999*a*, *b*). Again, it is not clear whether these results accurately reflect the situation in the intestinal mucosa.

Our understanding of how the mucosal immune system manages any Gram-negative bacteria, including *Shigella*, to obtain LPS in a form that can be presented to B, and perhaps T, cells for induction of a high-affinity IgA response is minimal. Recently, it was shown that *Shigella* LPS can be trafficked through polarized intestinal cells and thus potentially processed and presented in an immunologically active form (Beatty *et al.* 1999). Lipoglycans such as LPS are clearly dealt with differently from proteins, but apart from observations that CD1-restricted CD4CD8 double negative T cells can be generated, reacting with mycobacterial lipoglycans, there is little information (Porcelli & Modlin 1999). There is an urgent need for studies into the immune responses against bacterial LPS. The situation with cellular immunity in shigellosis is equally uncertain. T-cell clones have been produced against *Shigella* (Zwillich *et al.* 1989) and activated T cells have been isolated from the blood of patients with shigellosis, but their function is unknown (Islam *et al.* 1995, 1996).

11.   CONCLUSION AND PERSPECTIVES

Our understanding of the pathophysiology of shigellosis is largely based on studying the invasion of epithelial cell monolayers and macrophages *in vitro*, and the experimental infection of exteriorized rabbit ileal loops. Information has also been obtained from the murine pulmonary model and from rectal biopsies of macaque monkeys and humans after experimental and natural infection, respectively. The fact that *Shigella* does not cause intestinal infection in mice, which denies the use of the many murine-specific reagents and genetic manipulations, has probably inhibited a more detailed investigation of the cytokine and cellular mechanisms involved. Nonetheless, the application of knockout mice in the murine lung model of shigellosis has added, and will continue to add, to our knowledge of the innate and adaptive immune responses to *Shigella* infection. Future studies using transgenic animals expressing human-specific factors will also open up new possibilities for investigating the immune response in shigellosis.

In vitro and *in vivo* studies have allowed the formulation of a detailed model of the disease process, however, many questions still remain to be addressed. An analysis into the timing of the inflammatory response in terms of the cell types and mediators that are recruited and secreted at the site of infection needs to be conducted. For example, it will be important to determine the relative contribution of resident macrophages versus newly recruited monocytes/macrophages to the disease process and which inflammatory mediators are responsible for this induction during infection *in vivo*. Improved techniques that combine multiple immune staining and confocal microscopy of infected tissue sections will help to identify the early players in the development of inflammation following infection with *Shigella*. These techniques could also be used to observe the fate of bacterial virulence factors, such as LPS, in the infected tissue during the course of infection. Techniques to measure cytokines *in situ* with placement of microdialysis probes in the infected site (Bruce *et al.* 1999) have the potential to identify new inflammatory mediators and perhaps point to a novel means of treatment by targeting these molecules and modulating their function during *in vivo* infection.

Another possible research avenue that remains to be explored is the potential for differential gene expression, in both the host and the bacterium, during *Shigella* infection. One example of a host gene specifically upregulated during infection with *Shigella* is IL-8 and its regulation by the eukaryotic transcription factor, NFκB, has recently been demonstrated (Philpott *et al.* 1999). However, a more comprehensive approach to identify the expression of *Shigella*-induced host genes would be the application of DNA microarray technology (reviewed in Khan *et al.* 1999). This approach will lead to the identification of gene products up- or downregulated during *Shigella* infection. Additionally, this approach could be used to attribute a particular phenotype to *Shigella* mutants that remain uncharacterized. By comparing the pattern of gene expression from wild-type versus mutant infected cells, a particular function could be ascribed for the gene product missing in these mutants. Conversely, genes expressed in the bacterium during infection of the host could also be examined. The potential to apply techniques such as signature-tagged mutagenesis (Hensel *et al.* 1995) to *Shigella* infection also remains unexplored.

It is unwise to assume that any particular, or indeed a combination of, animal models will reveal all the components involved in producing the human disease. Infections usually exhibit a restricted host range, and in the case of many important human infections such as shigellosis, the disease is essentially confined to humans. Therefore, human-specific factors that allow expression of the disease phenotype probably exist. Furthermore, in the past, bacterial infections have infected the majority of the population and caused significant mortality in children, providing the potential for skewing the surviving population towards genetic expression of factors that probably influence the host response to disease. Such factors are likely to act at the level of the innate immune response and may be represented only in humans. Identification of such factors would shed light on natural resistance and, for example, help to explain why in human *Shigella* the challenge studies, a maximum of only 70% of volunteers get the clinical disease (DuPont *et al.* 1969).

For these reasons, in shigellosis, as much as in any other bacterial infection, there is a need to develop experimental models that can more closely mimic human disease, using human cells and tissues. At present, such models remain in the developmental stage. One possibility is the further development of techniques for maintaining the viability of human tissue samples such as intestinal biopsies, which could be used to study the response of resident cells to invasion of *Shigella*. A second possibility is the refinement of techniques for grafting human mucosal tissues into SCID mice (Yan *et al.* 1993) and then repopulating the bone marrow with autologous bone marrow cells. Infection of such xenografts could then be assessed in the context of both the human intestine and immune system yet would be amenable to the manipulations achievable in the mouse. Ultimately, such studies as those described here will help form a basis of knowledge by which improved treatments and novel vaccine candidates for the prevention of shigellosis will be designed.

We thank members of the Sansonetti laboratory past and present whose work was discussed in this manuscript. We are grateful to Dr Claude Parsot for helpful discussions, Dr Michelle Rathman and Dr Maria Mavris for careful review of the manuscript. D.J.P. is supported by a Marie Curie Fellowship from the European Community; J.D.E. is supported by the Wellcome

Trust. Research from the Sansonetti Laboratory is supported by grants from the Ministere de l'Education Nationale de la Recherche et de la Technologie (Programme BIOTECH and Programme de Recherche Fondamentale en Microbiologie).

REFERENCES

Adam, T., Giry, M., Boquet, P. & Sansonetti, P. J. 1996 Rho-dependent membrane folding causes *Shigella* entry into epithelial cells. *EMBO J.* **15**, 3315–3321.

Beatty, W. L. & Sansonetti, P. J. 1997 Role of lipopolysaccharide in signaling to subepithelial polymorphonuclear leukocytes. *Infect. Immun.* **65**, 4395–4404.

Beatty, W. L., Méresse, S., Gounon, P., Davoust, J., Mounier, J., Sansonetti, P. J. & Gorvel, J.-P. 1999 Trafficking of *Shigella* lipopolysaccharide in polarized intestinal epithelial cells. *J. Cell Biol.* **145**, 689–698.

Bernardini, M. L., Mounier, J., d'Hauteville, H., Coquis-Rondon, M. & Sansonetti, P. J. 1989 Identification of *icsA*, a plasmid locus of *Shigella flexneri* that governs bacterial intra- and intercellular spread through interaction with F-actin. *Proc. Natl Acad. Sci. USA* **86**, 3867–3871.

Blocker, A., Gounon, P., Larquet, E., Niebuhr, K., Cabiaux, V., Parsot, C. & Sansonetti, P. J. 1999 The tripartite type III secreton of *Shigella flexneri* inserts IpaB and IpaC into host membranes. *J. Cell Biol.* **147**, 683–693.

Bourdet-Sicard, R., Rudiger, M., Jockusch, B. M., Gounon, P., Sansonetti, P. J. & Tran Van Nhieu, G. 1999 Binding of the *Shigella* protein IpaA to vinculin induces F-actin depolymerization. *EMBO J.* **18**, 5853–5862.

Bruce, S. R., Tack, C. J. J. & Goldstein, D. S. 1999 Microdialysis for measurement of extracellular fluid concentrations of catechols in human skeletal muscle and adipose tissue. In *Monitoring molecules in neuroscience* (ed. H. Rollema, E. Abercrombie, D. Sulzer & J. Zachheim), pp. 104–109. Newark, NJ: Rutgers.

Chen, Y., Smith, M. R., Thirumalai, K. & Zychlinsky, A. 1996 A bacterial invasin induces macrophage apoptosis by directly binding ICE. *EMBO J.* **15**, 3853–3860.

Clerc, P. L., Ryter, A., Mounier, J. & Sansonetti, P. J. 1987 Plasmid-mediated early killing of eukaryotic cells by *Shigella flexneri* as studied by infection of J774 macrophages. *Infect. Immun.* **55**, 521–527.

DuPont, H. L., Hornick, R. B., Dawkins, A. T., Snyder, M. J. & Formal, S. B. 1969 The response of man to virulent *Shigella flexneri* 2a. *J. Infect. Dis.* **119**, 296–299.

DuPont, H. L., Hornick, R. B., Snyder, M. J., Libonati, J. P., Formal, S. B. & Gangarosa, E. J. 1972 Immunity in shigellosis. II. Protection induced by oral live vaccine or primary infection. *J. Infect. Dis.* **125**, 12–16.

DuPont, H. L., Levine, M. M., Hornick, R. B. & Formal, S. B. 1989 Inoculum size in shigellosis and implications for expected mode of transmission. *J. Infect. Dis.* **159**, 1126–1128.

Fernandez-Prada, C. M., Hoover, D. L., Tall, B. D. & Venkatesan, M. M. 1997 Human monocyte-derived macrophages infected with virulent *Shigella flexneri in vitro* undergo a rapid cytolytic event similar to oncosis but not apoptosis. *Infect. Immun.* **65**, 1486–1496.

Ghayur, T. (and 13 others) 1997 Caspase-1 processes IFN-gamma-inducing factor and regulates LPS-induced IFN-gamma production. *Nature* **386**, 619–623.

Giannasca, P. J., Giannasca, K. T., Falk, P., Gordon, J. I. & Neutra, M. R. 1994 Regional differences in glycoconjugates of intestinal M cells in mice: potential targets for mucosal vaccines. *Am. J. Physiol.* **267**, G1108–G1121.

Hensel, M., Shea, J. E., Gleeson, C., Jones, M. D., Dalton, E. & Holden, D. W. 1995 Simultaneous identification of bacterial virulence genes by negative selection. *Science* **269**, 400–403.

High, N., Mounier, J., Prévost, M. C. & Sansonetti, P. J. 1992 IpaB of *Shigella flexneri* causes entry into epithelial cells and escape from the phagocytic vacuole. *EMBO J.* **11**, 1991–1999.

Hilbi, H., Chen, Y., Thirumalai, K. & Zychlinsky, A. 1997 The interleukin-1β-converting enzyme, caspase-1, is activated during *Shigella flexneri*-induced apoptosis in human monocyte-derived macrophages. *Infect. Immun.* **65**, 5165–5170.

Hilbi, H., Moss, J. E., Hersh, D., Chen, Y., Arondel, J., Banerjee, S., Flavell, R. A., Yuan, J., Sansonetti, P. J. & Zychlinsky, A. 1998 *Shigella*-induced apoptosis is dependent on caspase-1 which binds to IpaB. *J. Biol. Chem.* **273**, 32 895–32 900.

Hueck, C. J. 1998 Type III secretion systems in bacterial pathogens of animals and plants. Microbiol. *Mol. Biol. Rev.* **62**, 379–433.

Inman, L. & Cantey, J. R. 1984 Peyer's patch lymphoid follicle epithelial adherence of a rabbit enteropathogenic *Esherichia coli* (strain RDEC-1). Role of plasmid-mediated pili in initial adherence. *J. Clin. Invest.* **74**, 90–95.

Islam, M. M., Azad, A. K., Bardhan, P. K., Raqib, R. & Islam, D. 1994 Pathology of shigellosis and its complications. *Histopathology* **24**, 65–71.

Islam, D., Bardham, P. K., Lindberg, A. A. & Christensson, B. 1995 *Shigella* infection induces cellular activation of T and B cells, and distinct species-related changes in peripheral blood lymphocyte subsets during the course of the disease. *Infect. Immun.* **63**, 2941–2949.

Islam, D., Wretlind, B., Lindberg, A. A. & Christensson, B. 1996 Changes in the peripheral blood T cell receptor Vb repertoire *in vivo* and *in vitro* during shigellosis. *Infect. Immun.* **64**, 1391–1399.

Islam, D., Veress, B., Bardhan, P. K., Lindberg, A. A. & Christensson, B. 1997 *In situ* characterisation of inflammatory responses in the rectal mucosae of patients with shigellosis. *Infect. Immun.* **65**, 739–749.

Jepson, M. A., Clark, M. A., Foster, N., Mason, C. M., Bennett, M. K., Simmons, N. L. & Hirst, B. H. 1996 Targeting to intestinal M cells. *J. Anat.* **189**, 507–516.

Jung, H. C., Eckmann, L., Yang, S. K., Panja, A., Fierer, J., Morzycka-Wroblewska, E. & Kagnoff, M. F. 1995 A distinct array of proinflammatory cytokines is expressed in human epithelial cells in response to bacterial invasion. *J. Clin. Invest.* **95**, 55–62.

Khan, J., Bittner, M. L., Chen, Y., Meltzer, P. S. & Trent, J. M. 1999 DNA microarray technology: the anticipated impact on the study of human disease. *Biochim. Biophys. Acta* **1423**, M17–M28.

Kernéis, S., Bogdanova, A., Kraehenbuhl, J.-P. & Pringault, E. 1997 Conversion by Peyer's patch lymphocytes of human enterocytes into M cells that transport bacteria. *Science* **277**, 949–952.

Kotloff, K. L., Winickoff, J. P., Ivanoff, B., Clemens, J. D., Swerdlow, D. L., Sansonetti, P. J., Adak, G. K. & Levine, M. M. 1999 Global burden of *Shigella* infections: implications for vaccine development and implementation of control strategies. *WHO Bull.* **77**, 651–666.

LaBrec, E. H., Schneider, H., Magnani, T. J. & Formal, S. B. 1964 Epithelial cell penetration as an essential step in the pathogenesis of bacillary dysentery. *J. Bacteriol.* **88**, 1503–1518.

Lelouard, H., Reggio, H., Mangeat, P., Neutra, M. & Montcourrier, P. 1999 Mucin-related epitopes distinguish M cells and enterocytes in rabbit appendix and Peyer's patches. *Infect. Immun.* **67**, 357–367.

McCormick, B. A., Siber, A. M. & Maurelli, A. T. 1998 Requirement of the *Shigella flexneri* virulence plasmid in the ability to induce trafficking of neutrophils across polarised monolayers of the intestinal epithelium. *Infect. Immun.* **66**, 4237–4243.

Mallett, C., Verg, L. v. d., Collins, H. & Hale, T. 1993 Evaluation of *Shigella* vaccine safety and efficacy in an intranasal challenged mouse model. *Vaccine* **11**, 190–196.

Mandic-Muleg, I., Weiss, J. & Zychlinsky, A. 1997 *Shigella flexneri* is trapped in polymorphonuclear leukocyte vacuoles and efficiently killed. *Infect. Immun.* **65**, 110–115.

Mathan, M. M. & Mathan, V. I. 1991 Morphology of rectal mucosa of patients with shigellosis. *Rev. Infect. Dis.* **13** (Suppl. 4), S314–S318.

Ménard, R., Sansonetti, P. J. & Parsot, C. 1993 Nonpolar mutagenesis of the ipa genes defines IpaB, IpaC and IpaD as effectors of *Shigella flexneri* entry into epithelial cells. *J. Bacteriol.* **175**, 5899–5906.

Ménard, R., Sansonetti, P. J., Parsot, C. & Vasselon, T. 1994 Extracellular assoication and cytoplasmic partitioning of the IpaB and IpaC invasins of *Shigella flexneri*. *Cell* **79**, 515–525.

Ménard, R., Sansonetti, P. J. & Parsot, C. 1996 The secretion of the *Shigella flexneri* Ipa invasins is induced by the epithelial cell an controlled by IpaD. *EMBO J.* **13**, 5293–5302.

Mounier, J., Vasselon, T., Hellio, R., Lesourd, M. & Sansonetti, P. J. 1992 *Shigella flexneri* enters human colonic Caco-2 epithelial cells through the basolateral pole. *Infect. Immun.* **60**, 237–248.

Neutra, M. R., Pringault, E. & Kraehenbuhl, J.-P. 1996 Antigen sampling across epithelial barriers and induction of mucosal immune responses. *A. Rev. Immunol.* **14**, 275–300.

Nonaka, T., Kuwae, A., Sasakawa, C. & Imajoh-Ohmi, S. 1999 *Shigella flexneri* YSH6000 induces two different types of cell death, apoptosis and oncosis, in the differentiated human monoblastic cell line U937. *FEMS Microbiol. Lett.* **174**, 89–95.

Oaks, E., Wingfield, M. & Formal, S. 1985 Plaque formation by virulent *Shigella flexneri*. *Infect. Immun.* **48**, 124–129.

Owen, R. L., Pierce, N. F., Apple, R. T. & Cray Jr, W. C. 1986 M cell transport of *Vibrio cholerae* from the intestinal lumen into Peyer's patches: a mechanism for antigen sampling and for microbial transepithelial migration. *J. Infect. Dis.* **153**, 1108–1118.

Perdomo, J. J., Gounon, P. & Sansonetti, P. J. 1994a Polymorphonuclear leukocyte transmigration promotes invasion of colonic epithelial monolayer by *Shigella flexneri*. *J. Clin. Invest.* **93**, 633–643.

Perdomo, O. J., Cavaillon, J. M., Huerre, M., Ohayon, H., Gounon, P. & Sansonetti, P. J. 1994b Acute inflammation causes epithelial invasion and mucosal destruction in experimental shigellosis. *J. Exp. Med.* **180**, 1307–1319.

Phalipon, A., Kaufmann, M., Michetti, P., Cavaillon, J. M., Huerre, M. & Sansonetti, P. J. 1995 Monoclonal immunoglobulin A antibody directed against serotype-specific epitope of *Shigella flexneri* lipopolysaccharide protects against murine experimental shigellosis. *J. Exp. Med.* **182**, 769–778.

Philpott, D. J., Yamaoka, S., Isral, A. & Sansonetti, P. J. 1999 Invasive *Shigella flexneri* activates NFκB through an innate intracellular response and leads to IL-8 expression in epithelial cells. *J. Immunol.* (Submitted.)

Porcelli, S. A. & Modlin, R. L. 1999 The CD1 system: antigen presenting molecules for T cell recognition of lipids and glycolipids. *A. Rev. Immunol.* **17**, 297–329.

Sansonetti, P. J. & Phalipon, A. 1999 M cells as ports of entry for enteroinvasive pathogens: mechanisms of interaction, consequences for the disease process. *Semin. Immunol.* **11**, 193–203.

Sansonetti, P. J., Arondel, J., Fontaine, A., d'Hauteveille, H. & Bernadini, M. L. 1991 OmpB (osmo-regulation) and *icsA* (cell-to-cell spread) mutants of *Shigella flexneri*: vaccine candidates and probes to study the pathogenesis of shigellosis. *Vaccine* **9**, 416–422.

Sansonetti, P. J., Mournier, J., Prévost, M. C. & Merege, R. M. 1994 Cadherin expression is required for the spread of *Shigella flexneri* between epithelial cells. *Cell* **76**, 829–839.

Sansonetti, P. J., Arondel, A., Cavaillon, J. M. & Huerre, M. 1995 Role of IL-1 in the pathogenesis of experimental shigellosis. *J. Clin. Invest.* **96**, 884–892.

Sansonetti, P. J., Arondel, J., Cantey, R. J., Prévost, M. C. & Huerre, M. 1996 Infection of rabbit Peyer's patches by *Shigella flexneri*: effect of adhesive or invasive bacterial phenotypes on follicular-associated epithelium. *Infect. Immun.* **64**, 2752–2764.

Sansonetti, P. J., Tran Van Nhieu, G. & Egile, E. 1999a Rupture of the intestinal epithelial barrier and mucosal invasion by *Shigella flexneri*. *Clin. Infect. Dis.* **28**, 466–475.

Sansonetti, P. J., Arondel, J., Huerre, M., Harada, A. & Matsushima, K. 1999b Interleukin-8 controls bacterial transepithelial translocation at the cost of epithelial destruction in experimental shigellosis. *Infect. Immun.* **67**, 1471–1480.

Sasakawa, C., Adler, B., Tobe, T., Okada, N., Nagai, S., Komatsu, K. & Yoshikawa, M. 1988 Virulence-associated genetic regions comprising 31 kilobases of the 230-kilobase plasmid in *Shigella flexneri* 2a. *J. Bacteriol.* **170**, 2480–2484.

Serény, B. 1955 Experimental *Shigella* conjunctivitis. *Acta Microbiol. Acad. Sci. Hungary* **2**, 293–295.

Skoudy, A., Aruffo, A., Gounon, P., Sansonetti, P. J. & Tran Van Nhieu, G. 1999a CD44 binds to the *Shigella* IpaB protein and participates in bacterial invasion of epithelial cells. *Cell. Microbiol.* (In the press.)

Skoudy, A., Tran Van Nhieu, G., Mantis, N., Arpin, M., Mounier, J., Gounon, P. & Sansonetti, P. 1999b A functional role for ezrin during *Shigella flexneri* entry into epithelial cells. *J. Cell Sci.* **112**, 2059–2068.

Takahashi, K., Sasaki, T., Mammoto, A., Takaishi, K., Kameyama, T., Tsukita, S., Tsukita, S. & Takai, Y. 1997 Direct interaction of the RhoGDP dissociation inhibitor with ezrin/radixin/moesin initiates the activation of the Rho small G protein. *J. Biol. Chem.* **272**, 23 371–23 375.

Tran Van Nhieu, G., Ben-Ze'ev, A. & Sansonetti, P. J. 1997 Modulation of bacterial entry into epithelial cells by association between vinculin and the *Shigella* IpaA invasin. *EMBO J.* **16**, 2717–2729.

Tran Van Nhieu, G., Caron, E., Hall, A. & Sansonetti, P. J. 1999 IpaC induces actin polymerization and filopodia formation during *Shigella* entry into epithelial cells. *EMBO J.* **18**, 3249–3262.

Vasselon, T., Mounier, J., Hellio, R. & Sansonetti, P. J. 1992 Movement along actin filaments of the perijunctional area and *de novo* polymerization of cellular actin are required for *Shigella flexneri* colonizaion of epithelial Caco-2 cell monolayers. *Infect. Immun.* **60**, 1031–1040.

Verg, L. L. v. d., Mallett, C. P., Collins, H. H., Larsen, T., Hammack, C. & Hale, T. L. 1995 Antibody and cytokine responses in a mouse pulmonary model of *Shigella flexneri* serotype 2a infection. *Infect. Immun.* **63**, 1947–1954.

Voino-Yasenetsky, M. V. & Voino-Yasenetskaya, M. K. 1961 Experimental pneumonia caused by bacteria of the *Shigella* group. *Acta Morphol.* **XI**, 440–454.

Wassef, J., Keren, D. F. & Mailloux, J. L. 1989 Role of M cells in initial bacterial uptake and in ulcer formation in the rabbit intestinal loop model in shigellosis. *Infect. Immun.* **57**, 858–863.

Watari, M., Funato, S. & Sasakawa, C. 1996 Interaction of Ipa proteins of *Shigella flexneri* with α5β1 integrin promotes entry of the bacteria into mammalian cells. *J. Exp. Med.* **183**, 991–999.

Watari, M., Kamata, Y., Kozaki, S. & Sasakawa, C. 1997 Rho, a small GTP-binding protein, is essential for *Shigella* invasion of epithelial cells. *J. Exp. Med.* **185**, 281–292.

Way, S. S., Borczuk, A. C., Dominitz, R. & Goldberg, M. C. 1998 An essential role for gamma interferon in innate resistance to *Shigella flexneri* infection. *Infect. Immun.* **66**, 1342–1348.

Way, S. S., Borczuk, A. C. & Goldberg, M. B. 1999a Adaptive immune response to *Shigella flexneri* 2a *cydC* in immunocompetent mice and mice lacking immunoglobin A. *Infect. Immun.* **67**, 2001–2004.

Way, S. S., Borczuk, A. C. & Goldberg, M. B. 1999b Thymic independence of adaptive immunity to the intracellular pathogen *Shigella flexneri* serotype 2a. *Infect. Immun.* **67**, 3970–3979.

Yan, H.-C., Juhasz, J., Pilewski, J., Murphy, G. F., Herlyn, M. & Albeelda, S. M. 1993 Human/severe combined immunodeficient mouse chimeras. *J. Clin. Invest.* **91**, 986–996.

Zwillich, S. H., Duby, A. D. & Lipsky, P. E. 1989 T lymphocyte clones responsive to *Shigella flexneri*. *J. Clin. Microbiol.* **27**, 417–421.

Zychlinsky, A., Prévost, M. C. & Sansonetti, P. J. 1992 *Shigella flexneri* induces apoptosis in infected macrophages. *Nature* **358**, 167–169.

Zychlinsky, A., Fitting, C., Cavaillon, J. M. & Sansonetti, P. J. 1994a Interleukin-1 is released by murine macrophages during apoptosis induced by *Shigella flexneri*. *J. Clin. Invest.* **94**, 1328–1332.

Zychlinsky, A., Kenny, B., Ménard, R., Prévost, M. C., Holland, I. B. & Sansonetti, P. J. 1994b IpaB mediates macrophage apoptosis induced by *Shigella flexneri*. *Mol. Microbiol.* **11**, 619–627.

Zychlinsky, A., Thirumalai, K., Arondel, J., Cantey, J. R., Aliprantis, A. O. & Sansonetti, P. J. 1996 *In vivo* apoptosis in *Shigella flexneri* infections. *Infect. Immun.* **64**, 5357–5365.

DISCUSSION

C. W. Keevil (Centre for Applied Microbiology and Research, Porton Down, Wiltshire, UK). With respect to determining the infectious dose or LD_{50} of a gastro-intestinal pathogen, it may be worth considering the recent publication of James & Keevil (1999). This paper showed that verocytotoxigenic *Escherichia coli* 0157 attaches more avidly to enterocytes, with actin filament rearrangement, when grown microaerophilically or anaerobically rather than aerobically. Regrettably, many laboratories do not consider growing facultatively anaerobic pathogens under physiologically relevant conditions, especially low redox potential, prior to their *in vitro* or *in vivo* challenge studies. Could you comment on what anaerobic inoculum experiments have been performed by laboratories when examining the pathogenesis of *Shigella* spp.? One possible interpretation of some of your present data is that *Shigella* spp. express an anaerobic phenotype capable of enhanced attachment to epithelial cells; once they become intracellular, their phenotype will change in response to local nutrients, particularly oxygen concentration, making them fit for subsequent tissue invasion and dissemination to macrophages.

P. J. Sansonetti. I agree with Dr Keevil that not much attention has so far been paid to the effect of anaerobic growth conditions on the invasive capacity of *Shigella*. With regard to *in vitro* assays of cell invasion, growth conditions have been selected with the aim of optimizing bacterial entry into cells. It turns out that optimal conditions are the middle exponential phase of growth with aeration achieved by shacking. We are far away from anaerobiosis under such circumstances! Still, in those conditions, bacteria need to be centrifuged over the cell surface in order to achieve an intimate interaction, as no significant adherence system has ever been identified. Our preliminary evidence however, based on the sequence and annotation of the *S. flexneri* virulence plasmid, does not show any gene with a homology indicating a candidate for encoding and adhesin. This does not preclude the possibility that a pathogenicity island on the *Shigella* chromosome may encode an adherence system. In any event, the anaerobic growth conditions definitely need to be tested.

The situation seems different with regard to animal models: none of them, except infection in the macaque monkey, really reflects the situation of colonic infection that prevails in humans. In consequence, when macaque monkeys are infected intraperitoneally, the growth conditions do not really matter as the bacteria first need to survive gastric acidity, then transit through the small intestine and finally establish infection in the colon.

Under such neutral conditions, bacteria have time to adapt and express any putative specific adhesin. We believe that the best way to identify this putative adhesin will be a combination of genomics and signature-tagged mutagenesis.

DETECTION AND ANALYSIS OF GENE EXPRESSION DURING INFECTION BY *IN VIVO* EXPRESSION TECHNOLOGY

D. SCOTT MERRELL* and ANDREW CAMILLI†

Department of Molecular Biology and Microbiology,
Tufts University School of Medicine,
136 Harrison Avenue, Boston, MA 02111, USA

Many limitations associated with the use of *in vitro* models for study of bacterial pathogenesis can be overcome by the use of technologies that detect pathogen gene expression during the course of infection within an intact animal. *In vivo* expression technology (IVET) accomplishes this with versatility: it has been developed with a variety of reporter systems which allow for either *in vivo* selection or *ex vivo* screening. Selectable gene fusion systems generally allow for the complementation of a bacterial metabolic defect that is lethal *in vivo*, or for antibiotic resistance during the course of *in vivo* antibiotic challenge. In contrast, the screenable gene fusion system uses a site-specific DNA recombinase that, when expressed *in vivo*, excises a selectable gene cassette from the bacterial chromosome. Loss of this cassette can then be either screened or selected for *ex vivo*. The recombinase-based IVET can be used to detect genes that are transcriptionally induced during infection, including those expressed transiently or at low levels and, in addition, can be used to monitor the spatial and temporal expression of specific genes during the course of infection.

Keywords: IVET; RIVET; *Vibrio cholerae*; *cadA*; *ctxA*; *tcpA*

1. INTRODUCTION

During the 20th century a wealth of knowledge has been garnered concerning pathogenic microbes and the mechanisms whereby they cause disease. Empirical lines of research and epidemiological studies have yielded valuable information, which has been applied to the development of vaccines and antimicrobial compounds and the control of some microbes within their natural reservoirs. Unfortunately, an increase in the number of antibiotic-resistant strains of most pathogenic bacterial species has become a severe problem for the treatment of disease within the general population and also the prevention of disease in immunocompromised people. Due to this, the scientific community has found itself faced with the need for development

---

*(dmerrell@opal.tufts.edu)
†Author for correspondence (acamilli@opal.tufts.edu).

95

of novel antimicrobials and vaccines, and for accompanying discoveries in the basic sciences which can aid in these pursuits.

Two areas of basic research of fundamental importance are achieving a complete understanding of the physiology of pathogens during infection and the pathogenic mechanisms they employ at each stage of infection. One can imagine that distinct but partially overlapping sets of bacterial factors are required and produced at each stage of infection. For instance, the stages of infection for intestinal pathogens involve (i) entry, (ii) circumvention or survival in the face of physical barriers such as gastric acidity, (iii) primary attachment to host cells or tissue, (iv) invasion into host cells or tissues (in the case of invasive pathogens) or increased adherence by extracellular pathogens, (v) avoidance of and/or resistance to host innate immune defences, (vi) acquisition of nutrients and multiplication, and (vii) evacuation from the host to either a new host or an environmental reservoir. When considered as a whole, the prospect of understanding all of the factors produced and required by pathogens to accomplish these diverse processes becomes not only ambitious, but daunting.

Past attempts to identify and understand the function of pathogenicity factors have centred around the ability to stimulate expression of these factors *in vitro* using tissue culture cells or abiotic conditions which usually, but not always, mimic host signals. For instance, knowledge that classical biotype *Vibrio cholerae* produces cholera toxin (CT) under growth conditions of low pH and non-physiological low temperature (30°C), led to a number of elegant studies that have identified other pathogenicity factors that are co-regulated with CT (Peterson & Mekalanos 1988; Taylor *et al.* 1986) or sensory and regulatory factors that serve to regulate these pathogenicity factors (DiRita *et al.* 1991; Miller *et al.* 1987). Similar *in vitro* strategies have been employed to study many other pathogenic organisms, such as *Bordetella pertussis* (Finn *et al.* 1991) and *Salmonella typhimurium* (Valdivia & Falkow 1996). While such *in vitro* studies have been fruitful in the past, their potential use in developing a complete and accurate understanding of the factors elicited during the course of a bona fide infection is severely limited by our inability to reproduce the complex and dynamic environmental stimuli that are present *in vivo*. For instance, it is unlikely that studies done *in vitro*, including tissue culture models of infection, will reveal the identities and modes of action of pathogenicity factors that play important roles in circumventing or resisting host humoral immunity. However, development and recent advances made in *in vivo* expression technology (IVET) have provided powerful genetic tools for identifying and studying bacterial genes that are induced during infection.

IVET is designed as a promoter trap method whereby random genomic fragments are ligated in front of a promoterless reporter gene (Fig. 1). Reporter activity can then be used as an indication of transcriptional activity of the fused gene. IVET was first used to identify *in vivo*-induced (ivi) genes in a mammalian pathogen in 1993 (Mahan *et al.* 1993). Note that a similar strategy, based on antibiotic selection, was first used by Osbourn *et al.* (1987) to identify virulence genes in the plant pathogen *Xanthomonas campestris*. The initial IVET strategy as developed by Mahan and colleagues, relied upon the fact that purine auxotrophs of *S. typhimurium* are unable to survive passage through a mouse after intraperitoneal inoculation. Transcriptional fusions to a promoterless *purA–lacZY* synthetic operon were constructed and integrated into an *S. typhimurium* strain in which the chromosomal *purA* gene had been deleted (Fig. 1). Only fusions which were active during infection would result in expression of *purA*, and allow survival and multiplication of the strain in the animal. Most such strains had fusions that were also active during *in vitro* growth, i.e. these strains had constitutively active gene fusions. However, the subset of strains containing *ivi* gene fusions could easily be identified by bluewhite screening of output colonies on media containing 5-bromo-4-chloro-3-indolyl–D-galactoside due to the bicistronic nature of the *purA–lacZY* fusion. Output strains that were LacZ⁻ were said to contain gene promoters driving expression of *purA–lacZY* that were specifically active within the host environment.

The auxotroph complementation IVET strategy has now been used in a number of pathogenic species. Perhaps the most successful application involved large-scale screening of promoter fusions in *S. typhimurium* for induction in BALB/c mice and in cultured murine macrophages (Mahan *et al.* 1993; Heithoff *et al.* 1997). To date, over 100 genes have been found that are induced within these environments. While the identities of many of these genes have not been reported, those that have been reported fall into four broad categories: (i) regulators, (ii) metabolic/physiological, (iii) stress response, and (iv) unknown function. Similar screens have been conducted to identify *Pseudomonas aeruginosa* genes that are induced during infection of neutropenic mice and in response to exposure to respiratory mucus from cystic fibrosis patients, and in the plant pathogen *P. fluorescens* in response to rhizosphere colonization (Wang *et al.* 1996a, b; Rainey 1999). Each of these studies has revealed similar classes of genes and together they have proven that IVET is a powerful tool for the identification of *ivi* genes.

In the past six years, the original IVET strategy has been modified to include two additional selection strategies (Fig. 1). Each of these modifications has increased the potential application of IVET to include more diverse organisms and, theoretically, to identify a more diverse set of *ivi* genes within each organism. The first reporter modification applied to IVET involved the replacement of *purA* with *cat* (Mahan *et al.* 1995). Active promoters fused to the cat gene convey resistance to chloramphenicol (Cm). Therefore, strains containing random promoter fusions can be passaged through antibiotic-treated animals or tissue culture models to select promoter fusions that are active *in vivo*. This modification should expand the potential applications of IVET to include pathogens for which knowledge of virulence-attenuating auxotrophies is not known or is difficult to derive.

Antibiotic-based IVET strategies have been used in *S. typhimurium*, *Yersinia enterocolitica* and most recently in *Streptococcus gordonii* (Mahan *et al.* 1995; Young & Miller 1997; Kilic 1999). IVET screens conducted in *Y. enterocolitica* by Young & Miller provided the first large-scale screen for genes involved in *Yersinia* pathogenesis. In this study, *Y. enterocolitica* strains containing random transcriptional fusions to a promoterless cat gene were first selected for their ability to survive passage through a Cm-treated mouse and then screened for their inability to grow on both rich and minimal media containing Cm. In this manner, they were able to eliminate constitutively expressed genes, genes involved in nutrient uptake, and genes involved in metabolic functions. The remainder of fusions could then be said to identify virulence-associated genes that were specifically induced upon exposure to host stimuli. Once again, a broad class of genes was identified and could be subdivided into general categories: (i) stress response, (ii) response to iron starvation, (iii) cell envelope maintenance, and (iv) unknown functions.

The second reporter modification applied to IVET involved construction of a promoterless *tnpR–lacZY* reporter. The *tnpR* gene codes for the site-specific DNA resolvase from Tn$\gamma\delta$. The resolvase enzyme is able to mediate recombination between two directly repeated copies of a specific target DNA sequence, called *res1* sites. These *res1* sites are inserted, flanking a reporter gene, within the chromosome of the pathogen of interest (Fig. 1). If a particular *tnpR* fusion is transcriptionally induced during infection, even transiently and/or at a low level, the resolvase that is produced will catalyse a permanent and heritable change in the bacterium by excising the reporter from the chromosome. Resolved strains can be screened or selected after recovering the bacteria from host tissues. For instance, in the original study describing the recombinase-based IVET (RIVET), induced resolvase fusions catalysed the excision of a tetracycline-resistance (Tc^r) gene from the chromosome, resulting in conversion of the fusion strain to a Tc^s phenotype (Camilli *et al.* 1994). The strains harbouring induced fusions were then identified by replica-plating output colonies onto an agar medium supplemented with Tc.

RIVET has been applied to discover *ivi* genes in *V. cholerae* and *Staphylococcus aureus*. In the first implementation of RIVET for this purpose, over a dozen *V. cholerae* genes were identified for which the levels of transcription increased during infection of the small intestine of suckling mice (Camilli & Mekalanos 1995). Most recently, RIVET was used to identify *ivi* genes in *S. aureus* using a murine renal abscess model of disease. This study represents the first application of any IVET strategy to a Gram-positive organism, and showed its versatility as 45 staphylococcal genes were identified that were induced specifically upon infection (Lowe *et al.* 1998). Once again, a broad class of genes was identified, of which only six genes were previously known. Eleven genes were shown to have homology to non-staphylococcal genes and the others showed no homology to sequences within available databases. In addition, it was shown that mutations within seven of the identified *ivi* genes resulted in significant attenuation of pathogenicity compared with the wild-type, exemplifying the fact that IVET can identify *ivi* genes that encode essential pathogenicity factors.

Overall, IVET has proven itself as a valuable method for the identification of genes that are induced *in vivo*. In addition, it has proven applicable to a variety of different pathogenic species of bacteria. Despite their strengths, though, each of the IVET strategies has characteristics that limit the types of gene that will be identified using that strategy. For instance, for the auxotrophy complementation method, it is essential that the chosen selectable gene be required for growth in the same host compartment that the pathogenicity gene (to be identified) is expressed in. Thus, a limitation of this IVET strategy is that if the selectable gene is required throughout most of the infectious process, then preferential enrichment will occur for strains harbouring gene fusions that are constitutively and highly expressed in the host animal. Thus, pathogenicity genes that are expressed at only one stage of infection, or that are transcribed at low levels, may not be identifiable via this approach.

The antibiotic selection, while increasing the applicability of IVET, is more complicated since a suitable antibiotic must be administered and proper concentrations maintained in infected host tissues. Moreover, administration of antibiotics may disrupt the natural course of an infection and thus may not allow a completely unbiased look at gene induction. On the other hand, the former characteristic can provide some flexibility in designing a selection. For example, administering the antibiotic at only one stage of infection, or in one host compartment, should facilitate identification of pathogenicity factors that may be expressed at only one stage of infection, or in only one host compartment, respectively. In addition, it is likely that the concentration of the antibiotic can be lowered to allow identification of *ivi* genes that are transcribed at low levels *in vivo*.

RIVET-based screens suffer from two primary limitations. First, because the site-specific DNA recombinase acts efficiently at only one substrate sequence, a low level of expression of the recombinase gene fusion is sufficient to catalyse the excision event. Although this exquisite sensitivity allows detection of transient and/or low-level gene induction events, it unfortunately prohibits the identification of pathogenicity genes that have high basal levels of transcription during *in vitro* growth. This is due to the fact that a high basal level of transcription results in a strain that is unable to be constructed in the unresolved state. Second, *ex vivo* screening of strains containing active fusions has proven laborious as conversion of the strain from Tc^r to Tc^s is screened for by replica-plating. However, recent modifications have largely overcome both of these limitations, and these modifications will be discussed below along with an in-depth analysis of applications of RIVET to study the diarrhoeal pathogen *V. cholerae*.

2.   APPLICATION OF RIVET TO *V. CHOLERAE*

RIVET was first developed to study the intestinal pathogen *V. cholerae* (Camilli *et al.* 1994). *V. cholerae* is a Gram-negative facultative pathogen that is the causative agent of the epidemic and endemic diarrhoeal disease cholera. After ingestion in contaminated water or food, *V. cholerae* transits the gastric acid barrier and colonizes the relatively sterile brush border of the small intestine. At this point, *V. cholerae* resists innate immune defences, acquires nutrients, multiplies prodigiously and produces CT, the

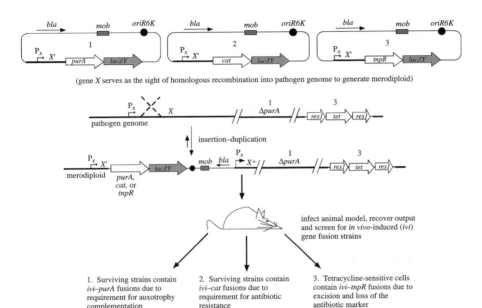

Fig. 1. Graphic depiction of three variations of IVET. As shown, auxotrophy complementation IVET selections are conducted using fusions to a promoterless *purA* gene (plasmid 1), antibiotic IVET selections are conducted using fusions to a promoterless antibiotic gene such as *cat* (plasmid 2), and recombinase-based *in vivo* expression technology (RIVET) screening is done using a promoterless *tnpR* allele (plasmid 3), which when produced, will cleave a Tc^r gene from elsewhere in the bacterial genome. Reporter gene fusion libraries are constructed by ligating random genomic fragments (designated as gene X') into the IVET vector of choice, followed by transformation into the pathogen of interest. The suicide plasmids then recombine into the chromosome by insertion–duplication creating a merodiploid. In the case of RIVET, a prescreen is required to remove strains harbouring *in vitro* active gene fusions: this is accomplished by selecting for Tc^r, LacZ$^-$ colonies. In all cases, fusion strains are passaged through an appropriate animal model of disease and collected from infected tissues and/or fluids after a period of time. In the case of the antibiotic-based IVET, the antibiotic (in this example, Cm) must be present at sufficient concentrations in animal tissues to select for *in vivo* expression of the gene fusion. Strains containing infection-induced gene fusions to *purA* and *cat* are selected in the host and are subsequently screened for lack of *in vitro* expression on LacZ indicator plates. In contrast, infection-induced gene fusions to *tnpR* are screened for, post-infection, by virtue of Tc^s and lack of expression on LacZ indicator plates.

toxic activity of which is the major cause of the profuse watery diarrhoea that ensues (for a review, see Wachsmuth *et al.* 1994). CT is an A-B-subunit toxin, which causes increased secretion of electrolytes and water into the lumen of the intestine (Sears & Kaper 1996). Additional pathogenicity factors have been identified by their coordinated *in vitro* regulation with CT. Toxin-coregulated pilus (TCP) is probably the best studied of these. This type IV bundle-forming pilus is absolutely essential for intestinal colonization in human and animal models of cholera (Herrington *et al.* 1988; Taylor *et al.* 1986), although its precise role remains unknown.

To begin to understand the intestinal physiology of *V. cholerae* and to increase our knowledge of its pathogenicity, RIVET technology has been used to identify genes that are induced within the suckling mouse and within

Table 1. *V. cholerae* infection-induced genes identified through RIVET.

Gene	Function
metabolic genes	
cysI	cysteine biosynthesis
argA	arginine biosynthesis
sucA	TCA cycle enzyme
nirT	nitrite reductase
adaptation genes	
cadA	acid tolerance
vieB	response regulator
nutrient scavenging genes	
α *hlyC*	secreted lipase
α *xds*	secreted DNase
iviVI	ABC transporter
motility genes	
α-*flaA*	flagellin antisense
α-*cheV*	chemotaxis protein antisense
iviV	chemoreceptor
genes of unknown function	
iviX	—
iviXI	—
iviXIII	—

the rabbit ligated ileal loop models of cholera (Camilli & Mekalanos 1995; Merrell & Camilli 1999). In the first of these studies, approximately 13 000 strains containing fusions that were inactive during laboratory growth on a rich agar medium (LuriaBertani (LB)) (as assessed by their LacZ$^-$ and Tc^r phenotypes, see Fig. 1) were intragastrically inoculated into suckling CD-1 mice. After 24 h, mice were killed, and bacteria were recovered from homogenates of the small intestines by plating on LB agar. Those strains containing fusions that were induced during infection were screened for by replica-plating colonies to LB supplemented with Tc (Camilli & Mekalanos 1995). A similar strategy was used for screens conducted within rabbit ligated loops (Merrell & Camilli 1999). In both studies, recovered strains that were now Tc^s could be said to contain fusions that were specifically activated *in vivo*. Both of these screens were successful and identified genes that fall into several categories: (i) metabolic, (ii) adaptive, (iii) nutrient scavenging, (iv) motility, and (v) unknown functions (Table 1).

3.   GENES INVOLVED IN METABOLISM

Identification of genes that are involved in metabolic pathways has been a common theme of every IVET search conducted thus far. This class of genes is often, unfortunately, considered uninteresting due to the fact that they are involved in so-called 'housekeeping' functions, and do not encode toxins or other factors which interact directly with the host. Perhaps this bias is due partly to the preconception that study of metabolic factors of pathogens will probably reveal nothing about the biology of the host. The fact that so many IVET searches reveal metabolic genes that are induced within the host environment, perhaps, should sound a cautionary note to the scientific community about what we consider interesting. There is, indeed, much to be learned about the growth physiology of pathogens within host environments, and these findings will in turn directly reflect upon the nature of the host compartments or tissues in which the pathogens reside. Moreover, in keeping with one of the primary goals of medical microbiological research, i.e. to develop safe and effective antimicrobials, it is worth noting that the majority of currently used antibiotics target biosynthetic and metabolic pathways (e.g. β-lactams, aminoglycosides, sulphonamides).

Among the *V. cholerae ivi*-encoded factors identified using RIVET are CysI, which catalyses an intermediate step in the biosynthesis of L-cysteine from inorganic sulphate; SucA, which is part of a multisubunit enzyme that catalyses a step in the tricarboxylic acid cycle whereby α-ketoglutarate is oxidatively decarboxylated to succinyl CoA and carbon dioxide; NirT, which codes for a tetrahaem nitrite reductase and that probably takes part in respiratory nitrate reduction; and ArgA, which catalyses the first step in the biosynthesis of L-arginine. *CysI, sucA, nirT* and *argA* were all found to be transcriptionally induced, to varying degrees, during the course of infection in the suckling mouse model, which suggested that these genes play roles in growth or survival during the course of infection. However, reduced pathogenicity, as assessed by competition assays in suckling mice, only accompanied a null mutation in *argA*, and not in the other three *ivi* genes (Camilli & Mekalanos 1995). Thus, members of this latter group of genes either do not fulfil an essential role during infection or there is redundancy of function. The identities of these *ivi* genes and knowledge of their enzymatic functions, nevertheless, provide useful information concerning the nutritional and physiological status of the mouse small intestinal environment. For example, the site(s) colonized by *V. cholerae* appears to be limiting for both L-cysteine and L-arginine. Interestingly, induction of the *Listeria monocytogenes arpJ* gene encoding a subunit of an L-arginine transporter was transcriptionally induced during infection of a tissue culture cell line (Klarsfeld *et al.* 1994). Thus, perhaps exogenous L-arginine, which is an essential amino acid supplied only through diet in humans and rodents, is of limited availability to bacterial pathogens in a variety of host tissues.

The frequent identification of *ivi* genes that appear to be non-essential for virulence points to the possible existence of overlapping or redundant pathways: these possibilities are often poorly anticipated or are overlooked. Bacteria have shown themselves to be particularly flexible and often possess a variety of back-up strategies to survive stresses and obtain essential nutrients. It is perhaps these redundant pathways that should be more frequently studied and subsequently targeted for vaccine and antimicrobial drug development.

4.   GENES INVOLVED IN ADAPTATION

The necessity to adapt to rapidly changing environments is of great importance to facultative pathogens. For instance, the natural reservoir for *V. cholerae* is aquatic, brackish water environments. This organism is capable of forming close associations with numerous planktonic microorganisms (Colwell & Huq 1994). Indeed, algal blooms have been linked to seasonal outbreaks of cholera in endemic areas of the world. Although the environmental conditions experienced by *V. cholerae* in its natural reservoir probably differ substantially from those in the human intestinal tract, *V. cholerae* has evolved the capacity to infect its human host directly from environmental reservoirs. Within minutes after ingestion, the bacteria must transit the gastric acid barrier and colonize the small intestine. Each step of this life cycle, no doubt, requires a variety of genes whose products play roles in adaptation. Two such adaptation-associated genes that are induced during infection have been identified using RIVET.

Escherichia coli CadA is an inducible lysine decarboxylase encoded by the second gene of the bicistronic *cadBA* operon (Meng & Bennett 1992). Decarboxylation of lysine produces cadaverine and carbon dioxide, and the reaction consumes a cytoplasmic proton. Cadaverine is then transported from the cell via the *cadB*-encoded lysine/cadaverine antiporter. It was hypothesized by Gale & Epps (1942) that the inducible amino-acid decarboxylases of *E. coli* play a role in surviving exposure to acidic environments. Indeed, it has subsequently been shown that CadA, Adi (an inducible arginine decarboxylase) and GadC (a putative glutamate γ-amino butyrate antiporter) play roles in a physiological adaptation process known as the acid-tolerance response (ATR) (Hersh *et al.* 1996; Lin *et al.* 1995; Park *et al.* 1996). ATR is an adaptive response whereby cells that have been exposed to a mildly acidic pH, prior to exposure to very acidic pH, show a much higher per cent survival than those that are placed directly in the very acidic pH environment (Foster & Hall 1991). The ATR of *S. typhimurium* has been extensively studied, and shown to consist of a complex network of inducible acid-survival systems that are growth-phase dependent.

The *V. cholerae cadA* homologue was identified as an *ivi* gene within both the rabbit ileal loop and within the suckling mouse models of cholera (Merrell & Camilli 1999). *In vitro* analyses revealed that *cadA* transcription could be induced in a number of ways. Growth in oxygen-limiting conditions, acidic pH, and high L-lysine concentrations were all shown to be able to increase expression of *cadA*. Determination of these parameters, combined with the fact that CadA is involved in the ATR of *S. typhimurium* led us to test and show that *V. cholerae*, which had heretofore been described

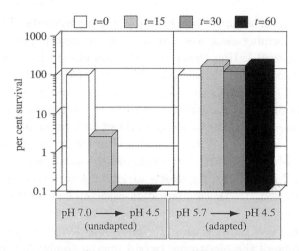

Fig. 2. The *V. cholerae* ATR to inorganic acid challenge. Wild-type *V. cholerae* was either unadapted by growth in LB pH 7.0, or adapted for 1 h by growth in LB pH 5.7 prior to acid shock in LB pH 4.5. The per cent survival at 15, 30 and 60 min was calculated by comparison with the initial numbers of colony-forming units (CFU) measured at time 0. Adapted from Merrell & Camilli (1999).

as acid sensitive, was able to mount a robust ATR (Fig. 2). Adaptation at a sublethal pH of 5.7 for 1 h endowed *V. cholerae* with resistance to killing at pH 4.5. As *V. cholerae* encounters not only low pH, but harmful organic acids during the course of its transit and colonization within the stomach and small intestine, respectively, studies were conducted that divided the ATR into two separate branches; inorganic acid (low pH) and organic acids (low pH plus the intestinal acids propionate, acetate and butyrate). CadA was shown to play a crucial role in both branches, while the major virulence regulator ToxR was shown to be necessary for organic, but not inorganic ATR (Fig. 3).

The relative importance of the *V. cholerae* ATR in the establishment and the progression of human infection is not known at present. It has been shown that mutant strains of *L. monocytogenes*, *S. typhimurium* and *Helicobacter pylori* that are compromised in their ability to survive exposure to acids are attenuated for pathogenicity. However, a *V. cholerae cadA* mutant was found to be unaffected in its ability to colonize the suckling mouse intestine. Conversely though, wild-type *V. cholerae* cells that were acid tolerized prior to infection were shown to have a reduced infectious

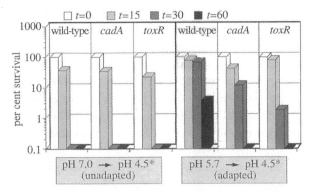

Fig. 3. The ATR of *V. cholerae* wild-type, *cadA* and *toxR* mutant strains to organic acid challenge. *V. cholerae* was either unadapted by growth in LB pH 7.0, or adapted for 1 h by growth in LB pH 5.7 containing 6.5 mM acetate, 1.9 mM butyrate and 2.8 mM propionate. Cells were then acid shocked in LB pH 4.5 plus 8.7 mM acetate, 2.5 mM butyrate and 3.7 mM propionate organic acids. The per cent survival at 15, 30 and 60 min was calculated by comparison with the initial numbers of CFUs measured at time 0. The asterisks indicate the addition of organic acids. Adapted from Merrell & Camilli (1999).

dose in suckling mice (Merrell & Camilli 1999). These findings once again point to the potential for redundancy in important functions. The fact that a *cadA* mutant is not attenuated in *vivo*, yet acid-tolerized cells are more virulent, suggests that the ATR is multifactorial and partially redundant. Indeed, it has been shown that there are more than 50 proteins induced in *S. typhimurium* upon exposure to acid, and it is reasonable to predict, *a priori*, that a subset of these may fulfil redundant functions (Foster 1991). It is also interesting to speculate that acid adaptation may decrease the infectious dose in humans, aiding in the epidemic spread of cholera. Specifically, if bacteria that are shed from the human intestine are in an acid-tolerant state, and if this state persists within contaminated waters for a period of time, subsequent infections would require a lower infectious dose thus aiding transmission of *V. cholerae*.

The importance of two-component signal transduction systems whereby a 'sensor kinase' protein monitors the environment and, upon proper stimulation, transmits a signal to a 'response regulator' protein to induce an adaptative response, is probably best highlighted by the commonality and prevalence of these systems within facultative bacterial pathogens. *VieB* (previously designated *iviVII*) was identified as an *ivi* gene in *V. cholerae* within the suckling mouse model of cholera (Camilli & Mekalanos 1995). Subsequent analysis revealed that *vieB* lies within a three gene locus that encodes a sensor kinase (VieS) and two distinct response regulator proteins (VieA and VieB). VieS contains a phosphoreceiver and two transmitter domains and is thus a member of the complex sensor kinase family exemplified by BvgS (Uhl & Miller 1996) and ArcS (Georgellis *et al.* 1997). VieA is a typical response regulator, containing an N-terminal phosphoreceiver domain and a C-terminal DNA-binding domain. In contrast, VieB is atypical because, although it contains an N-terminal phosphoreceiver domain, it lacks a C-terminal DNA-binding domain. Because *vieB* is transcribed only during infection, specifically at one particular stage of infection (after colonization of the epithelium (Lee *et al.* 1998, and see below), it has been suggested that it may play an adaptive role in this particular host niche. However, such a role has been difficult to establish experimentally, as a strain containing a null mutation in *vieB*, or one containing a deletion of all three *vie* genes, retains full virulence (Lee *et al.* 1998). Hence, either the Vie system does not play a role in adaptation, the adaptive response mediated by Vie is not required for mouse infection, or there is redundant regulation of the adaptive response. Identification of the set of genes regulated by the Vie proteins should allow specific tests of each of these possibilities to be conducted. For example, and hypothetically, if null mutations in one or more Vie-induced genes were shown to attenuate virulence, then the redundant regulation hypothesis would be supported.

During the course of infection, *V. cholerae* must encounter a variety of biochemical and nutritional parameters that are characteristic of the microenvironments through which it passes or takes up residence. Since *vieB* is induced at a particular stage of infection, its transcription must be modulated by one or more of these parameters. Despite this, extensive attempts to induce *vieB* transcription *in vitro* were unsuccessful: this failure exemplifies the notion that use of *in vitro* systems designed to mimic host parameters can be limiting due to the fact that some genes may require multiple (or cryptic) *in vivo* signals to stimulate their transcription. Hence, identification of genes like *vieB*, where no tested signal has been able to stimulate *in vitro* transcription, highlights the potential of technologies such as IVET as a means of dissecting the complexity of hostpathogen interactions.

Besides being useful for identifying *ivi* genes, RIVET has an additional functionality that other methods, including IVET, lack the ability to easily monitor the temporal and spatial patterns of *ivi* gene induction in the host. *VieB* is the first gene for which transcription during the course of infection in an intact animal was monitored (Lee *et al.* 1998). The temporal pattern of *vieB* induction was accomplished by infecting a group of animals with a *vieB::tnpR* fusion strain and then recovering bacteria from the small intestine at different times. The percentage of bacterial cells in each sample that had resolved was measured by scoring for loss of the marker gene (*tet*) within the excisable cassette. It was determined that *vieB* transcription was induced approximately 3–4 h post-inoculation (Fig. 4).

Once the time of *vieB* induction was known, the spatial pattern of induction in the small intestine at this time could be determined. This procedure involved removing and dissecting the small intestine and caecum into ten segments of equal length at 3.5 h post-inoculation, recovering bacteria from each segment, and replica-plating colonies to determine the extent of resolution. Resolved (Tc^s) cells were found throughout the small intestine, but with a gradient from high to low levels of resolution corresponding to the

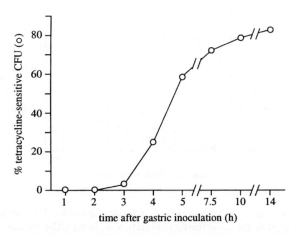

Fig. 4. Kinetics of transcriptional induction of *vieB* during infection. *V. cholerae vieB::tnpR* was used to intragastrically inoculate suckling CD-1 mice. At the post-inoculation times indicated on the *x*-axis, the small intestines were removed and homogenized. The per cent Tcs CFUs per intestine, shown on the *y*-axis, was determined by replica-plating colonies. Adapted from Lee *et al.* (1999).

proximal to distal segments (Fig. 5). In this experimental system the bolus of the bacterial inoculum travels down the lumen of the small intestine during the first 3–4 h with only a minority of cells adhering to the intestinal wall (Fig. 5, solid circles). Three conclusions were drawn from these data. First, because the majority of bacteria that had flowed down the lumen to the ileum by 3.5 h remained unresolved, *vieB* induction was not occurring in the lumen of the stomach or duodenum. Second, the duodenum must contain a *vieB*-inducing environment, and it remains possible that more downstream segments do as well. Third, because the highest percentage of resolved cells was observed for the minority of bacteria that had colonized the duodenal epithelium, either *vieB* induction occurs during the act of colonization or soon thereafter.

The exact host signals necessary for transcriptional induction of *vieB* have not been determined, but it has subsequently been shown that strains that are blocked for the ability to colonize (via a mutation in *tcpA*) never induce *vieB* transcription *in vivo*. This result suggests that

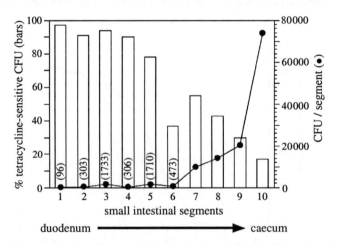

Fig. 5. Spatial pattern of *vieB* transcription induction in the small intestine. *V. cholerae vieB::tnpR* was used to intragastrically inoculate a suckling CD-1 mouse. At 3.5 h post-inoculation the small intestine and caecum were removed and the former was dissected into nine segments of equal length. The caecum was considered the tenth segment. For each segment, the total CFUs (shown on the right axis) and per cent Tc^s CFU (shown on the left axis) were determined. The total CFU for segments 1–6 are shown in parentheses above each data point. Adapted from Lee *et al.* (1999).

a colonization-specific signal is sent after attachment to the host mucosa (S. Lee and A. Camilli, unpublished data). The flexibility and usefulness of RIVET for determining the sites, times and requirements for gene induction *in vivo* is exemplified by these *vieB* studies. With advancements recently made in RIVET (discussed in §8), virtually any gene that is transcriptionally induced *in vivo*, no matter how high its basal level of *in vitro* transcription, can be analysed during the course of infection.

5.   GENES INVOLVED IN NUTRIENT SCAVENGING

Once a pathogen enters its host it must not only survive the onslaught of innate immune mechanisms, but must also be able to acquire nutrients that may be in limited quantity or, in some cases, quite different in composition from the previous environment. It is therefore not surprising that pathogenic species of bacteria have evolved the ability to acquire nutrients in diverse host environments. One would expect, for this reason, that one prominent subclass of RIVET-identified genes will encode proteins that play roles in the acquisition of nutrients during the course of infection.

HlyC, *iviVI*, and *xds* were all identified as *ivi* genes in *V. cholerae*, and are predicted to play diverse roles in the acquisition of nutrients during the course of infection. *HlyC* encodes a secreted protein that shows a high degree of similarity to the *P. aeruginosa* LipA lipase (Camilli & Mekalanos 1995). A strain with a null mutation in *hlyC* was unable to hydrolyse emulsified tributyrin thus confirming that HlyC is a secreted triacylglycerol lipase. The exact role that HlyC plays during the course of an infection has not been determined, but it is possible that this lipase is involved in the breakdown of host cell membranes to fatty-acid chains that are subsequently metabolized. An alternative possibility is that HlyC plays a role in defence against host immune cells by damaging their membranes in a way that disrupts cell function.

Xds was identified as an *ivi* gene that is induced in both the rabbit ileal loop and suckling mouse models of cholera (D. S. Merrell, M. Angelichio and A. Camilli, unpublished data). The *xds* gene encodes a secreted extracellular DNase (Focareta & Manning 1991). Though the exact role of Xds during the course of infection is not known, there are several possibilities. First, it may function in nutrient acquisition by breakdown of extracellular DNA into nucleotides that could then be taken up and used by *V. cholerae*.

Second, Xds may degrade the DNA constituent of the mucus gel that covers the intestinal epithelium and that probably serves as a viscous barrier to attachment by the *V. cholerae* cells. Therefore, breakdown of this material by Xds would facilitate colonization. An experimental determination of the importance of Xds for virulence is complicated by the fact that *V. cholerae* produces a second, distinct extracellular DNase that may function redundantly with Xds. Focareta & Manning (1991) examined a strain in which both DNase structural genes had been disrupted, but found no significant reduction in the LD_{50} for this strain in the suckling mouse model of disease. It remains possible, however, that this double mutant strain may have a more subtle defect in colonization that is not revealed by LD_{50} determinations, but which may be revealed by a more sensitive test such as a competition assay in suckling mice.

IviVI is predicted to encode a polypeptide that has a high degree of similarity to the ATP-binding cassette (ABC) family of permeases (Camilli & Mekalanos 1995). It contains a canonical Walker A box motif that is highly conserved in members of this transporter family (Ames *et al.* 1990). ABC transporters have been well characterized in pathogens such as *S. typhimurium* and *Streptococcus pneumoniae* (Hiles *et al.* 1987; Alloing *et al.* 1990), and are often involved in the uptake of nutrients such as short peptide fragments and carbohydrates. Additionally, it has been shown that the substrate-binding components of ABC transporters are important for pathogenicity in some bacterial species in that they facilitate adhesion to host cells (Andersen 1994; Jenkinson 1992; Pei & Blaser 1993). Currently, neither the molecule that is transported by IviVI nor the role of IviVI in *V. cholerae* pathogenesis has been determined.

The initial RIVET screen for *V. cholerae ivi* genes was not comprehensive by any means, and thus it is expected that dozens, perhaps hundreds, more *ivi* genes involved in biosynthesis and nutrient acquisition await identification. Knowledge of the members of these two classes of *ivi* genes will reveal much about the nutritional status of the intestinal sites colonized by *V. cholerae*, and this body of information may apply to other pathogens that colonize the small intestine or use it as a site of entry into deeper tissues. It is possible that this information could then be used to target certain biosynthetic or nutrient uptake pathways for antimicrobial drug screening or development.

6. GENES INVOLVED IN MOTILITY

The importance of motility in the pathogenic life cycle of *V. cholerae* is a complex issue. Some data indicate that motility is required for full pathogenicity of *V. cholerae* in various adult animal models of cholera (Freter *et al.* 1981; Richardson 1991). Other data however, indicate that motility can be detrimental in the suckling mouse model. In fact, non-chemotactic mutant strains survive in greater number than wild-type vibrios and produce a more rapid and severe disease (Freter & O'Brien 1981). That motility is a detriment within the suckling mouse is supported by the following two recent findings made using RIVET. First, out of three *V. cholerae ivi* genes identified that are hypothesized to play roles in motility and chemotaxis (Table 1), two of these appear to encode antisense transcripts that may be involved in inhibition of these processes. Specifically, α-*flaA* and α-*cheV* were both identified as transcriptional fusions to *tnpR* that were induced during the course of infection of suckling mice. While the biological significance of these antisense transcripts is not known, it is tempting to speculate that they are directly involved in downregulating motility and chemotaxis, respectively, during the course of infection. Second, a strain containing a null mutation in *cheV*, which encodes a putative chemotaxis protein, colonized suckling mice twice as well as the wild-type parental strain (Camilli & Mekalanos 1995). Downregulation of motility and chemotaxis could be useful for formation and maintenance of microcolonies on intestinal epithelia or, alternatively, could prevent *V. cholerae* from colonizing the wrong compartment in the small intestine: either would act as a means of increasing the efficiency of the infection. These suppositions are also supported by data that indicate that motility and virulence gene expression are reciprocally regulated in *V. cholerae* (Gardel & Mekalanos 1996; Harkey *et al.* 1994).

7. GENES OF UNKNOWN FUNCTION

The last class of *V. cholerae* genes identified by RIVET are those that are predicted to encode polypeptides with no similarities to database sequences or that are similar to hypothetical polypeptides of unknown function. This intriguing class of genes has been a major class identified by all previous IVET screens. This fact points to our still-fledgling knowledge of the roles that many gene products play in the growth and infectious process of pathogenic microbes and points to the necessity for studies designed to elucidate the roles of these many genes. Although genome sequencing projects are dramatically increasing the number of hypothetical polypeptides to which some Ivi polypeptides have striking similarity, future research must begin to focus on the elucidation of the actual functions of these many factors and their potential roles in pathogenicity.

8.   THE 'NEW AND IMPROVED' RIVET

The original RIVET suffered from two limitations. First, it was somewhat laborious due to the fact that strains recovered from the host animal had to be screened via replica-plating to determine those that had resolved and were now Tc^s. Second, RIVET was exquisitely sensitive, in that only fusions that were transcribed at very low levels *in vitro* could be constructed and maintained in the unresolved state. This limited the scope of *ivi* genes that could be identified to include only those that were transcriptionally silent *in vitro*, but then were subsequently upregulated *in vivo*. This restriction excluded all of the known pathogenicity genes of *V. cholerae*, and thus, probably, some unknown genes as well. Modifications to RIVET, in the forms of a modified excisable cassette that allows selection of resolved strains and a system for reduced sensitivity, have dealt with each of these limitations and increased the range of studies that can now be conducted.

The most obvious way to decrease the labour required for performing RIVET screens was to modify the excisable cassette in such a manner as to allow direct selection of resolved strains after recovery from infected host tissues. This modification has been accomplished by the incorporation of a counter-selectable marker, *sacB*, into the sequence flanked by the res1 sites. *SacB* encodes levansucrase, which is an enzyme that results in the metabolism of sucrose into a by-product that is toxic to many Gram-negative bacterial species. *V. cholerae* strains that carry the gene and are grown in the presence of sucrose are inviable (Butterton *et al.* 1993). Therefore, resolved strains, which have lost the *sacB*-containing cassette, can be directly selected for by plating small intestinal homogenates on sucrose-containing media. This facilitates screening large numbers of output cells for those containing fusions that were activated *in vivo*.

Use of the original RIVET strategy was greatly limited by the extreme sensitivity of the system. This sensitivity was a function of the fact that the produced resolvase need act at only one target sight per genome, and thus very low levels of transcription of a *tnpR* fusion would result in resolution. Recently, however, this limitation has been largely overcome by the creation of a 'tunable' RIVET whereby the sensitivity is lowered in user-defined increments (Lee *et al.* 1999). Specifically, several alleles of *tnpR* were generated that contain down-mutations in the ribosome-binding sequence (RBS). Accordingly, these alleles show a range of decreased translational efficiencies and, therefore, decreased productions of resolvase for any given level of transcription.

The mutant RBS alleles of *tnpR* were generated by polymerase chain reaction using a partially degenerate forward primer that randomized three critical bases in the RBS. These mutant alleles were transcriptionally fused to an iron-repressible promoter, P_{irgA}, within the genome of a *V. cholerae* strain containing the *res1–tet–res1* cassette. Finally, the resulting fusion strains were individually screened for those that showed decreased levels of resolution after growth at an iron concentration low enough to result in 100% resolution of the wild-type *tnpR* fusion strain. Three *tnpR* alleles were isolated that, upon further study using the P_{irgA} fusion strain, were found to require two- to fourfold lower concentrations of iron to produce 50% resolution. One of the three *tnpR* alleles was subsequently shown to serve as a very useful reporter gene for detecting the *in vivo* induction of several *V. cholerae* virulence genes (see §9). Use of these 'reduced sensitivity' *tnpR* alleles in new RIVET screens should allow identification of a broader class of *ivi* genes: specifically, ones having low to moderate basal levels of expression during *in vitro* growth, but that are induced to yet higher levels during infection.

Use of the enhanced RIVET to screen for *ivi* genes thus proceeds by first generating several gene fusion libraries, one for each mutant RBS *tnpR* allele. For example, three libraries could be generated; one that uses the wild-type *tnpR* gene, a second that uses a mutant *tnpR* allele ($tnpR^{mut168}$) that is translated approximately two-fold less efficiently, and a third that uses a *tnpR* allele ($tnpR^{mut135}$) that is translated approximately fourfold less efficiently. Each gene fusion library is pre-screened to collect unresolved strains, which are then passaged through the animal (Fig. 6). Finally, the bacteria are collected from infected tissue after the infection has run its

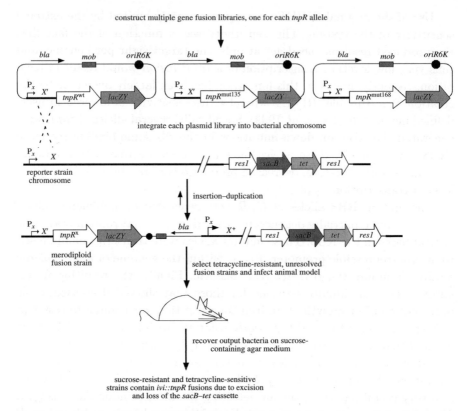

Fig. 6. Graphic depiction of a RIVET screen for infection-induced genes. As shown, the screen is conducted using three different promoterless *tnpR* alleles that vary in their translational efficiencies due to different RBSs. Each *tnpR* allele will therefore yield a different class of infection-induced genes: specifically, classes of genes that differ in their basal levels of transcription during *in vitro* growth. Construction of the reporter gene fusion libraries and their subsequent screening is as in the legend to Fig. 1, with one notable exception: in the present scheme, resolved strains are directly selected from intestinal homogenates through their ability to grow on sucrose-containing agar media.

course, and are plated on a sucrose-containing agar medium to select re-solved strains. The sucrose-resistant strains contain *ivi* genes fused to *tnpR*, which were induced at some point during infection to mediate resolution. Each of the *tnpR* reporter alleles should result in the identification of distinct sets of *ivi* genes, but which will probably overlap to some extent. Specifically, the wild-type *tnpR* allele should identify *ivi* genes that are transcriptionally silent or nearly so during *in vitro* growth. The $tnpR^{mut168}$ allele should identify some of the same *ivi* genes as the wild-type allele, but in addition those that are transcribed at low levels *in vitro*. Finally, the $tnpR^{mut135}$ allele should identify some of the same genes as the other two alleles, but in addition those that are transcribed at still higher levels during *in vitro* growth.

Clearly, the comprehensiveness of the RIVET screen can be increased (or decreased) by altering two parameters. First, the number of unique gene fusion strains that are screened in animals can be increased, and this of course has a direct bearing on the comprehensiveness of the screen. Second, the number of different *tnpR* alleles used to construct the gene fusion libraries, as well as the breadth of their combined range of translational efficiencies, can be increased. This would have the effect of increasing the types of *ivi* genes that could be identified with respect to their levels of transcription *in vitro*. Theoretically, any gene whose level of transcription increases at some point during the infectious process can be identified using the enhanced RIVET.

9. USE OF THE ENHANCED RIVET TO MONITOR *IVI* GENE INDUCTION DURING INFECTION

As described above (see §4 on *vieB*), RIVET has an additional application; monitoring the transcriptional induction of *ivi* genes during infection of a host animal. Initial attempts to monitor the spatio-temporal pattern of induction of known pathogenicity genes of *V. cholerae* using the original RIVET were unsuccessful due to the moderate basal levels of transcription of these genes *in vitro*. This was first demonstrated by the construction of transcriptional fusions of the wild-type *tnpR* gene to *tcpA* and *ctxA*. *TcpA* codes for the pilin subunit of the TCP, and *ctxA* codes for the enzymatic subunit of CT. The basal levels of expression of these genes resulted in immediate excision of the *res1–tet–res1* cassette from the chromosome, thus eliminating the possibility of monitoring expression patterns of either gene *in vivo* (Lee *et al.* 1999). However, it was found that when either *tcpA* or *ctxA* was fused to the $tnpR^{mut135}$ allele, which is translated fourfold less efficiently, resolution did not occur during *in vitro* growth but did occur *in vivo* (see first set of columns in Fig. 7(a), (b)). In addition, growth of the fusion strains in AKI broth, which is the only known *in vitro* condition that induces *tcpA* and *ctxA* expression (Iwanga *et al.* 1986), also resulted in resolution.

The use of a mutant RBS *tnpR* allele to measure increases in the *in vivo* transcription of *tcpA* and *ctxA* exemplifies what has been termed the 'tunable' RIVET. This methodology proceeds by, first, isolating a set of mutant alleles of *tnpR* that exhibit a wide range of translational efficiencies. Next, the set of *tnpR* alleles are screened for the one allele that 'tunes' the gene of interest to the appropriate level of resolvase expression, i.e. a level of expression that mediates resolution only upon an increase of gene transcription during infection. This tunable RIVET should, theoretically, make possible the analysis of any gene that shows an increase in transcription within a particular environment, such as during infection of host tissues.

The above results were the first direct experimental data confirming the long-held hypothesis that *tcpA* and *ctxA* virulence genes are transcriptionally induced during infection. The *tcpA* and *ctxA* fusions to the *tnpR*[mut135] allele were subsequently used to determine the effect of various regulatory gene mutations on *tcpA* and *ctxA* expression, respectively. *ToxR* and *TcpP* are homologous inner membrane proteins that sense certain environmental parameters such as pH, osmolarity and some amino acids, and in turn regulate transcription of *toxT*, *ctxA*, and other genes (Carroll *et al.* 1997; Hase & Mekalanos 1998; Skorupski & Taylor 1997). ToxR and TcpP each associate with another inner membrane protein, ToxS and TcpH, respectively, that are required for proper signalling (Carroll *et al.* 1997; DiRita & Mekalanos 1991; Hase & Mekalanos 1998). ToxT is a cytoplasmic transcription factor that is autoregulatory and, importantly, is responsible for activating transcription of a number of virulence genes such as *tcpA* and *ctxA* (DiRita *et al.* 1991). *TcpA* expression in AKI broth was found to require *toxR*, *tcpPH* and *toxT* (Fig. 7(a)). Interestingly though, *toxR* and *tcpPH* were not required for induction of *tcpA* during the course of a suckling mouse infection. However, mutations in all three of these regulatory genes or, a single mutation in *toxT*, negated this induction. Hence, induction of *tcpA* during infection is accomplished in a partially redundant manner by ToxR and TcpPH, in a ToxT-dependent manner. Perhaps most surprisingly, *tcpA* transcription was found to be dependent on prior TcpA expression. This was shown by finding that a *tcpA::tnpR*[mut135] fusion strain harbouring a null mutation in *tcpA* only resolved to low levels during infection. These data are in contrast to complete resolution of the same strain *in vitro* in AKI broth: this shows

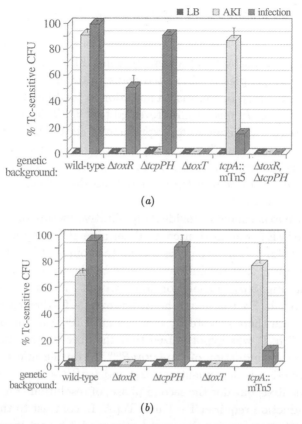

Fig. 7. Regulation of *tcpA* and *ctxA* transcription during growth *in vitro* and during infection. (*a*) Extent of resolution of *V. cholerae* strains containing a *tcpA::tnpR*^{mut135} transcriptional fusion and null mutations in the genes shown on the *x*-axis after 10 h of growth in LB broth, AKI broth or in the small intestine of suckling CD-1 mice. (*b*) Extent of resolution of *V. cholerae* strains containing a *ctxA::tnpR*^{mut135} transcriptional fusion and null mutations in the genes shown on the *x*-axis. *V. cholerae* was recovered from the *in vitro* and *in vivo* environments and the per cent Tc^s CFUs determined. The means and standard deviations from multiple experiments are shown. Adapted from Lee *et al.* (1999).

that *tcpA* autoregulation is an in vivo-specific phenomenon. It is likely that a TCP-mediated interaction with either the host or other vibrios, is what is required for full induction of the *tcpA* gene within the intestinal tract. However, this interaction appears to only require the low basal level of *tcpA* expression that is observed prior to, and during, the initial stage of infection (see below).

Investigation of *ctxA* expression also revealed novel aspects of the regulation of this important pathogenicity gene. Mutations in either *toxR*, *tcpPH* or *toxT*, completely abrogated transcriptional induction of *ctxA* in AKI broth (Fig. 7(*b*)). In contrast, a mutation in *tcpA* had no effect on *ctxA* expression *in vitro*. *In vivo* studies revealed that *toxR*, *toxT*, and *tcpA*, but not *tcpPH*, were each required for expression of *ctxA*. Thus, as was true for *tcpA*, there are concrete differences between the regulatory requirements for *ctxA* expression during *in vitro* growth versus during infection. Moreover, ToxR-mediated regulation of *tcpA* differs from that of *ctxA in vivo*, which is an unexpected result.

The above experiments provided a qualitative picture of *tcpA* and *ctxA* induction. RIVET was next used to determine the temporal patterns of expression of these virulence genes during the course of infection. Interestingly, it was found that resolution mediated by the *tcpA::tnpR*^{mut135} fusion was biphasic (Fig. 8(*a*)). Resolution starts within 1 h of infection, and levels off from 2 to 3 h. By the fourth hour, however, a further round of resolution begins, and there is a steady increase until 100% of the bacteria have resolved by 7 h. It was hypothesized that the observed resolution kinetics reflect a biphasic induction of *tcpA::tnpR*^{mut135} during infection. The first phase of induction is dependent on *toxR* and *tcpPH* as mutations in either eliminate the first, but not the second phase, of resolution (Fig. 8(*a*)). Both phases of induction required ToxT and TcpA. In contrast to this pattern of expression, *ctxA* expression does not begin until 3 h post-inoculation, and resolution (mediated by the *ctxA::tnpR*^{mut135} fusion) increases to 100% by 6 h (Fig. 8(*b*)). The induction of *ctxA* was abolished by mutations in *toxR*, *toxT* or *tcpA*, but not by a mutation in *tcpPH*. Because *ctxA* induction is dependent upon prior *tcpA* expression, it can be postulated that CT is not produced until after TCP-mediated colonization has occurred. This latter model is gratifying from a standpoint of efficiency and potency of virulence factor expression, in that CT is not produced until after the bacteria have colonized the epithelium, which contains the target cells for CT.

Interestingly, a widely held model of virulence factor regulation in *V. cholerae* asserts that *tcpA* and *ctxA* are co-regulated (DiRita *et al.* 1991). This model is based on the coordinate regulation of genes in the ToxR-regulon, such as *tcpA* and *ctxA*, seen during *in vitro* growth. Unfortunately,

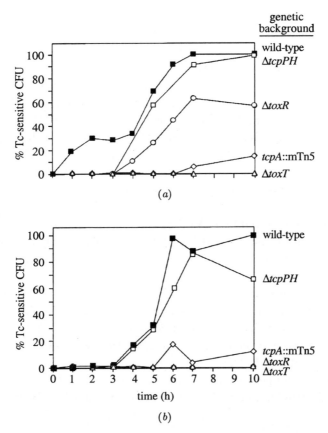

Fig. 8. Kinetics of *tcpA* and *ctxA* transcriptional induction during infection. (*a*) Resolution kinetics of the same strains as in Fig. 7(*a*), in suckling mice. (*b*) Resolution kinetics of the same strains as in Fig. 7(*b*), in suckling mice. *V. cholerae* was recovered at the times indicated on the *x*-axis, and the per cent Tc^s CFUs was determined from at least 150 colonies. Genetic backgrounds are shown to the right of each curve. Adapted from Lee *et al.* (1999).

due to various technical problems, it has never been possible to investigate the validity of this model *in vivo*. Use of RIVET has provided the first evidence that *in vitro* and *in vivo* regulation and expression of these two major pathogenicity factors are different. Moreover, these data represent the first indication that transcriptional induction of *tcpA* during infection is complex, wherein two phases of induction occur in response to two potentially different signals. The first signal is received soon after infection, resulting in low levels of expression of *tcpA* (Lee *et al.* 1999). Later, after colonization has occurred, we see a further increase of *tcpA* expression, presumably in response to a second, colonization-dependent signal.

All of these findings point to the extreme importance of developing an understanding of events as they occur during the course of infection in a whole animal. By using RIVET one can not only identify those genes that are induced specifically during the course of an infection, but can conduct important spatio-temporal analyses to discern the time and site of induction of each *ivi* gene. In this way, important information concerning host–pathogen interactions, such as what host compartments contain inducing signals that trigger pathogenicity gene expression, can be determined. In addition, the new 'tunable' RIVET is a powerful tool that can be used to derive an accurate picture of the regulatory hierarchy of *ivi* gene expression during infection. Specifically, the effects of mutations within regulatory and effector genes on the *in vivo* pattern of *ivi* gene expression can be easily assessed. Future use of IVET, RIVET and other *in vivo* technologies not discussed here should increase our knowledge of the physiology and pathogenicity of human pathogens, and should aid in the discovery and development of novel vaccines and treatments that will help us in our effort to control infectious diseases.

D.S.M. was supported by US National Institutes of Health (NIH) training grant AI 07422, and A.C. by NIH AI 40262 and Pew Scholars Award P0168SC.

REFERENCES

Alloing, G., Trombe, M. C. & Claverys, J. P. 1990 The *ami* locus of the gram-positive bacterium *Streptococcus pneumoniae* is similar to binding protein-dependent transport operons of gram-negative bacteria. *Mol. Microbiol.* **4**, 633–644.

Ames, G. F., Mimura, C. S. & Shyamala, V. 1990 Bacterial periplasmic permeases belong to a family of transport proteins operating from *Escherichia coli* to human: traffic ATPases. *FEMS Microbiol. Rev.* **6**, 429–446.

Andersen, R. N., Ganeshkumar, N. & Kolenbrander, P. E. 1994 Sequence, of *Streptococcus gordonii* PK488 scaA coaggregation adhesin gene, and ATP-binding cassette system. In *Abstracts 94th Annual Meeting of the American Society for Microbiology* D-120, p. 117. Washington, DC: American Society for Microbiology.

Butterton, J. R., Boyko, S. A. & Calderwood, S. B. 1993 Use of the *Vibrio cholerae irgA* gene as a locus for insertion and expression of heterologous antigens in cholera vaccine strains. *Vaccine* **11**, 1327–1335.

Camilli, A. & Mekalanos, J. J. 1995 Use of recombinase gene fusions to identify *Vibrio cholerae* genes induced during infection. *Mol. Microbiol.* **18**, 671–683.

Camilli, A., Beattie, D. & Mekalanos, J. 1994 Use of genetic recombination as a reporter of gene expression. *Proc. Natl Acad. Sci. USA* **91**, 2634–2638.

Carroll, P. A., Tashima, K. T., Rogers, M. B., DiRita, V. J. & Calderwood, S. B. 1997 Phase variation in *tcpH* modulates expression of the *ToxR* regulon in *Vibrio cholerae*. *Mol. Microbiol.* **25**, 1099–1111.

Colwell, R. R. & Huq, A. 1994 Environmental reservoir of *Vibrio cholerae*. The causative agent of cholera. *Ann. NY Acad. Sci.* **740**, 44–54.

DiRita, V. J. & Mekalanos, J. J. 1991 Periplasmic interaction between two membrane regulatory proteins, ToxR and ToxS, results in signal transduction and transcriptional activation. *Cell* **64**, 29–37.

DiRita, V. J., Parsot, C., Jander, G. & Mekalanos, J. J. 1991 Regulatory cascade controls virulence in *Vibrio cholerae*. *Proc. Natl Acad. Sci. USA* **88**, 5403–5407.

Finn, T. M., Shahin, R. & Mekalanos, J. J. 1991 Characterization of vir-activated TnphoA gene fusions in *Bordetella pertussis*. *Infect. Immun.* **59**, 3273–3279.

Focareta, T. & Manning, P. A. 1991 Distinguishing between the extracellular DNases of *Vibrio cholerae* and development of a transformation system. *Mol. Microbiol.* **5**, 2547–2555.

Foster, J. W. 1991 Salmonella acid shock proteins are required for the adaptive acid tolerance response. *J. Bacteriol.* **173**, 6896–6902.

Foster, J. W. & Hall, H. K. 1991 Inducible pH homeostasis and the acid tolerance response of *Salmonella typhimurium*. *J. Bacteriol.* **173**, 5129–5135.

Freter, R. & O'Brien, P. C. 1981 Role of chemotaxis in the association of motile bacteria with intestinal mucosa: chemotactic responses of *Vibrio cholerae* and description of motile nonchemotactic mutants. *Infect. Immun.* **34**, 215–221.

Freter, R., Allweiss, B., O'Brien, P. C., Halstead, S. A. & Macsai, M. S. 1981 Role of chemotaxis in the association of motile bacteria with intestinal mucosa: *in vitro* studies. *Infect. Immun.* **34**, 241–249.

Gale, E. F. & Epps, H. M. R. 1942 The effect of the pH of the medium during growth on the enzymic activities of bacteria (*E. coli* and *M. lysodiekticus*) and the biological significance of the changes produced. *Biochemistry* **36**, 600–619.

Gardel, C. L. & Mekalonos, J. J. 1996 Alterations in *Vibrio cholerae* motility phenotypes correlate with changes in virulence factor expression. *Infect. Immun.* **64**, 2246–2255.

Georgellis, D., Lynch, A. S. & Lin, E. C. C. 1997 *In vitro* phosphorylation study of the Arc two-component signal transduction system of *Escherichia coli*. *J. Bacteriol.* **179**, 5429–5435.

Harkey, C. W., Everiss, K. D. & Peterson, K. M. 1994 The *Vibrio cholerae* toxincoregulated-pilus gene *tcpI* encodes a homolog of methyl-accepting chemotaxis proteins. *Infect. Immun.* **62**, 2669–2678.

Hase, C. C. & Mekalanos, J. J. 1998 TcpP protein is a positive regulator of virulence gene expression in *Vibrio cholerae*. *Proc. Natl Acad. Sci. USA* **95**, 730–734.

Heithoff, D. M., Conner, C. P., Hanna, P. C., Julio, S. M., Hentschel, U. & Mahan, M. J. 1997 Bacterial infection as assessed by *in vivo* gene expression. *Proc. Natl Acad. Sci. USA* **94**, 934–939.

Herrington, D. A., Hall, R. H., Losonsky, G., Mekalanos, J. J., Taylor, R. K. & Levine, M. M. 1988 Toxin, toxin-coregulated pili, and the *toxR* regulon are essential for *Vibrio cholerae* pathogenesis in humans. *J. Exp. Med.* **168**, 1487–1492.

Hersh, B. M., Farooq, F. T., Barstad, D. N., Blankenhorn, D. L. & Slonczewski, J. L. 1996 A glutamate-dependent acid resistance gene in *Escherichia coli*. *J. Bacteriol.* **178**, 3978–3981.

Hiles, I. D., Gallagher, M. P., Jamieson, D. J. & Higgins, C. F. 1987 Molecular characterization of the oligopeptide permease of *Salmonella typhimurium*. *J. Mol. Biol.* **195**, 125–142.

Iwanga, M., Yamamoto, K., Higa, N., Ichinose, Y., Nakasone, N. & Tanabe, M. 1986 Culture conditions for stimulating cholera toxin production by *Vibrio cholerae* O1 El Tor. *Microbiol. Immunol.* **30**, 1075–1083.

Jenkinson, H. F. 1992 Adherence, coaggregation, and hydrophobicity of *Streptococcus gordonii* associated with expression of cell surface lipoproteins. *Infect. Immun.* **60**, 1225–1228.

Kilic, A. O., Herzberg, M. C., Meyer, M. W., Zhau, X. & Tao, L. 1999 Streptococcal reporter gene-fusion vector for identification on *in vivo* expressed genes. *Plasmid* **42**, 67–72.

Klarsfeld, A. D., Goossens, P. L. & Cossart, P. 1994 Five *Listeria monocytogenes* genes preferentially expressed in infected mammalian cells: *plcA, purH, purD, pyrE* and an arginine ABC transporter gene, *arpJ. Mol. Microbiol.* **13**, 585–597.

Lee, S. H., Angelichio, M. J., Mekalanos, J. J. & Camilli, A. 1998 Nucleotide sequence and spatiotemporal expression of the *Vibrio cholerae vieSAB* genes during infection. *J. Bacteriol.* **180**, 2298–2305.

Lee, S. H., Hava, D. L., Waldor, M. K. & Camilli, A. 1999 Regulation and temporal expression patterns of *Vibrio cholerae* virulence genes during infection. *Cell* **99**, 625–634.

Lin, J., Lee, I. S., Frej, J., Slonczewski, J. L. & Foster, J. W. 1995 Comparative analysis of extreme acid survival in *Salmonella typhimurium, Shigella flexneri* and *Escherichia coli*. *J. Bacteriol.* **177**, 4097–4104.

Lowe, A. M., Beattie, D. T. & Deresiewicz, R. L. 1998 Identification of novel staphylococcal virulence genes by *in vivo* expression technology. *Mol. Microbiol.* **27**, 967–976.

Mahan, M. J., Slauch, J. M. & Mekalanos, J. J. 1993 Selection of bacterial virulence genes that are specifically induced in host tissues. *Science* **259**, 686–688.

Mahan, M. J., Tobias, J. W., Slauch, J. M., Hanna, P. C., Collier, R. J. & Mekalanos, J. J. 1995 Antibiotic-based IVET selection for bacterial virulence genes that are specifically induced during infection of a host. *Proc. Natl Acad. Sci. USA* **92**, 669–673.

Meng, S. Y. & Bennett, G. N. 1992 Nucleotide sequence of the *Escherichia coli* cad operon: a system for neutralization of low extracellular pH. *J. Bacteriol.* **174**, 2659–2669.

Merrell, D. S. & Camilli, A. 1999 The *cadA* gene of *Vibrio cholerae* is induced during infection and plays a role in acid tolerance. *Mol. Microbiol.* **34**, 836–849.

Miller, V. L., Taylor, R. K. & Mekalanos, J. J. 1987 Choleratoxin transcriptional activator ToxR is a transmembrane DNA binding protein. *Cell* **48**, 271–279.

Osbourn, A. E., Barber, C. E. & Daniels, M. J. 1987 Identification of plant-induced genes of the bacterial pathogen *Xanthomonas campestris* pathovar campestris using a promoter-probe plasmid. *EMBO J.* **6**, 23–28.

Park, Y. K., Bearson, B., Bang, S. H., Bang, I. S. & Foster, J. W. 1996 Internal pH crisis, lysine decarboxylase and the acid tolerance response of *Salmonella typhimurium*. *Mol. Microbiol.* **20**, 605–611.

Pei, Z. & Blaser, M. J. 1993 PEB1, the major cell-binding factor of *Campylobacter jejuni*, is a homolog of the binding component in gram-negative nutrient transport systems. *J. Biol. Chem.* **268**, 18 717–18 725.

Peterson, K. M. & Mekalanos, J. J. 1988 Characterization of the *Vibrio cholerae* ToxR regulon: identification of novel genes involved in intestinal colonization. *Infect. Immun.* **56**, 2822–2829. (Erratum *in Infect. Immun.* 1989 **57**, 660.)

Rainey 1999 Adaptation of *Pseudomonas fluorescens* to the plant rhizosphere. *Environ. Microbiol.* **1**, 243–257.

Richardson, K. 1991 Roles of motility and flagellar structure in pathogenicity of *Vibrio cholerae*: analysis of motility mutants in three animal models. *Infect. Immun.* **59**, 2727–2736.

Sears, C. L. & Kaper, J. B. 1996 Enteric bacterial toxins: mechanisms of action and linkage to intestinal secretion. *Microbiol. Rev.* **60**, 167–215.

Skorupski, K. & Taylor, R. K. 1997 Control of the ToxR virulence regulon *in Vibrio cholerae* by environmental stimuli. *Mol. Microbiol.* **25**, 1003–1009.

Taylor, R. K., Miller, V. L., Furlong, D. B. & Mekalanos, J. J. 1986 Identification of a pilus colonization factor that is coordinately regulated with cholera toxin. *Annali Sclavo Collana Monografica* **3**, 51–61.

Uhl, M. A. & Miller, J. F. 1996 Integration of multiple domains in a two-component sensor protein: the *Bordetella pertussis* BvgAS phosphorelay. *EMBO J.* **15**, 1028–1036.

Valdivia, R. H. & Falkow, S. 1996 Bacterial genetics by flow cytometry: rapid isolation of *Salmonella typhimurium* acid-inducible promoters by differential fluorescence induction. *Mol. Microbiol.* **22**, 367–378.

Wachsmuth, I. K., Blake, P. A. & Olsvik, O. 1994 Vibrio cholerae *and cholera: molecular to global perspectives*. Washington, DC: American Society for Microbiology.

Wang, J., Lory, S., Ramphal, R. & Jin, S. 1996a Isolation and characterization of *Pseudomonas aeruginosa* genes inducible by respiratory mucus derived from cystic fibrosis patients. *Mol. Microbiol.* **22**, 1005–1012.

Wang, J., Mushegian, A., Lory, S. & Jin, S. 1996b Large-scale isolation of candidate virulence genes of *Pseudomonas aeruginosa* by *in vivo* selection. *Proc. Natl Acad. Sci. USA* **93**, 10 434–10 439.

Young, G. M. & Miller, V. L. 1997 Identification of novel chromosomal loci affecting *Yersinia enterocolitica* pathogenesis. *Mol. Microbiol.* **25**, 319–328.

MEASUREMENT OF BACTERIAL GENE
EXPRESSION *IN VIVO*

ISABELLE HAUTEFORT* and JAY C. D. HINTON†

*Institute of Food Research, Norwich Research Park,
Norwich NR4 7UA, UK*

The complexities of bacterial gene expression during mammalian infection cannot be addressed by *in vitro* experiments. We know that the infected host represents a complex and dynamic environment, which is modified during the infection process, presenting a variety of stimuli to which the pathogen must respond if it is to be successful. This response involves hundreds of *ivi* (*in vivo*-induced) genes which have recently been identified in animal and cell culture models using a variety of technologies including *in vivo* expression technology, differential fluorescence induction, subtractive hybridization and differential display. Proteomic analysis is beginning to be used to identify IVI proteins, and has benefited from the availability of genome sequences for increasing numbers of bacterial pathogens. The patterns of bacterial gene expression during infection remain to be investigated. Are *ivi* genes expressed in an organ-specific or cell-type-specific fashion? New approaches are required to answer these questions. The uses of the immunologically based *in vivo* antigen technology system, *in situ* PCR and DNA microarray analysis are considered. This review considers existing methods for examining bacterial gene expression *in vivo*, and describes emerging approaches that should further our understanding in the future.

Keywords: green fluorescent protein (GFP); *in vivo* expression technology; *in vivo* induced fluorescence (IVIF); fluorescence-activated cell sorting; *in vivo* antigen technology; Lux

1. INTRODUCTION

Understanding the expression of genes involved in infection remains the holy grail for research into bacterial virulence. Over the past three decades, research has focused on bacteria grown in monoculture in laboratory media to characterize the responses to many environmental stimuli. The simultaneous development of molecular biological technology and molecular genetic approaches has facilitated the dissection of a plethora of regulatory systems (Mekalanos 1992). The scene has been set for the investigation of patterns of gene expression in more complex systems and the consideration of the bacterial response to the host environment *in situ*. It is likely that this will

---

* (isabelle.hautefort@bbsrc.ac.uk)
† Author for correspondence. (jay.hinton@bbsrc.ac.uk).

reveal fascinating and novel bacterial responses, as well as new hierarchies of genetic regulation (Hinton 1997). In this review we describe the current tools for the analysis of bacterial gene expression *in vivo*, their limitations and some promising new approaches.

2.   APPROACHES FOR ASSESSING PATTERNS OF *IN VIVO* GENE EXPRESSION

(a) *Reporter systems for indirect monitoring of bacterial mRNA levels* in vivo

Bacterial gene expression is assessed by the direct or indirect measurement of mRNA levels. Reporter genes are used to monitor transcription indirectly by putting genes that encode an assayable protein under the control of promoters of interest. Various reporter gene products have been used for molecular genetic analyses, and they all share the ability to glow, to fluoresce or to be assayed colorimetrically (Miller 1992; Sala-Newby *et al.* 1999). The increased use of reporter systems that do not naturally occur in most bacterial or mammalian cells has improved the sensitivity of monitoring bacterial gene expression *in situ*. These recent developments now allow gene expression to be studied not only in large bacterial populations but also in individual bacterial cells.

(i) *β-galactosidase*

The reliable *β*-galactosidase reporter system has been extensively used to monitor gene activity in response to various environmental conditions (Casadaban & Cohen 1979; Jacob & Monod 1961; Silhavy & Beckwith 1985). The chromogenic substrate nitrophenyl-*β*-D-galactopyranoside has been used for simple and accurate quantification of *β*-galactosidase activity for cultures grown *in vitro*. Increasingly, *β*-galactosidase is being used for analysis of gene expression *in vivo*; for example, expression of the *spvB* virulence gene of *Salmonella typhimurium* was monitored with a *β*-galactosidase reporter system in cultured macrophages and non-phagocytic cells (Fierer *et al.* 1993). The activation of *Salmonella* PhoP-regulated genes was assessed in phagosomes within macrophages with *β*-galactosidase transcriptional fusions (Aranda *et al.* 1992). However, a major limitation of the *β*-galactosidase reporter system is that it is invasive, requiring permeabilization of target bacterial cells. Because the efficiency of cell permeabilization varies, observed differences in β-galactosidase activity can reflect altered substrate uptake rather than changing levels of gene expression, preventing accurate analysis of individual bacterial cells (Nwoguh *et al.* 1995). Furthermore, studies involving animal and cell models require the isolation of the bacteria prior to β-galactosidase assay. Despite the development of sensitive fluorogenic substrates for the quantification of gene expression (Garcia-del-Portillo *et al.* 1992; Handfield *et al.* 1998; Heithoff *et al.* 1997; Slauch *et al.* 1994), and a new electrochemical approach for online monitoring of β-galactosidase activity (Biran *et al.* 1999), robust assessment of the kinetics of gene expression *in vivo* requires other approaches.

(ii) *Luciferases*

Luciferases isolated from fireflies or bioluminescent bacteria (LuxAB from *Vibrio* spp.) have been popular as reporters for gene expression in a variety of organisms including Gram-negative and Gram-positive bacteria, yeast, plant cells and transgenic plants (Kirchner *et al.* 1989; Langridge *et al.* 1991(*a, b*); Meighen 1991; Schauer 1988; Stewart & Williams 1992).

Several reasons have been used to justify the choice of luciferase as a reporter for gene expression. (i) LuxAB has a short half-life, ensuring that photon production reflects real-time gene expression. This can be advantageous, but does make the system sensitive to experimental pertubations. (ii) The signal produced by luciferase is ideal for gene expression studies because low levels of light can be measured and linearly quantified over several orders of magnitude. For example, a comparison of the LuxAB and β-galactosidase reporter systems to assess *spvB* gene expression from *S. typhimurium* inside mammalian cells showed that light emission mediated by Lux was a faster and more easily detectable signal than β-galactosidase activity (Pfeifer & Finlay 1995). (iii) As luciferase enzymes are not widespread in bacterial pathogens, endogenous background does not present a problem.

Three examples demonstrate the value of the lux reporter; the mini-Mu derivative, Mudlux was used to assess the response of *S. typhimurium* to adhesion to cultured macrophages (Francis & Gallagher 1993). In *Yersinia pseudotuberculosis*, *luxAB* fusions showed that *yopE* and *yopH* expression was highest during early stages of colonization of either Peyer's patches or the spleen of infected mice (Forsberg & Rosqvist 1994) and a *yopE:lux* fusion was used to demonstrate that *yopE* was induced by contact with HeLa cells (Pettersson *et al.* 1996). However, the toxicity of the aldehyde substrate of LuxAB and of the intracellular expression of luciferase in enterobacteria (Gonzalez-Flecha & Demple 1994) and the variation of Lux activity with oxygen concentration has led the reliability of Lux for measurement of bacterial gene expression *in vivo* to be questioned, particularly in low oxygen environments (Camilli 1996). A further limitation of Lux as a reporter gene is that the photon yield of luciferase does not permit visualization of gene expression in individual bacterial cells.

Lux has been used for the detection of bacterial pathogens such as *Mycobacterium smegmatis* during infection (Sarkis *et al.* 1995). Contag *et al.* (1995) used the Lux operon derived from *Photorhabdus*, which does not require an exogenous substrate, to develop a method for non-invasive optical monitoring of bacterial infections in whole animals. *Salmonella* bacteria that expressed high levels of Lux were visualized within a live mouse with an intensified charge-coupled-device (CCD) camera. Despite photon absorption by mouse organs and skin, bacteria were seen in many infected tissues, including the caecum, spleen and liver. This promising approach has not yet been applied to direct monitoring of gene expression in whole animals.

(iii) *Green fluorescent protein*

Experiments addressing real-time or *in situ* measurements of gene expression tend to use fluorescence-emitting reporter systems, which have the greatest sensitivity. All of these require exogenous substrates, except the 238 residue green fluorescent protein (GFP) protein from *Aequoria* jellyfish, which naturally emits green fluorescence after excitation by blue light (Cubitt *et al.* 1995; Prasher *et al.* 1992; Ward 1998). Because investigation of gene expression *in vivo* requires non-invasive reporters, GFP-based systems have several advantages, particularly when compared to β-galactosidase and luciferase (Jaiwal *et al.* 1998). GFP does not require the addition of any substrate or co-factor. It is a small protein remaining fluorescent even when fused to proteins of interest, making it ideal for studying protein translocation. GFP allows detection and quantification of bacterial gene expression in tissue culture models and promises real-time visualization of gene expression in different pathogens. Expression of GFP does not cause attenuation or reduce bacterial invasion of cultivated cell lines (Valdivia & Falkow 1997a; Zhao *et al.* 1998).

Initially, the use of *gfp* as a reporter gene of bacterial promoter activity was hampered by technical limitations such as weak fluorescence intensity and formation of inclusion bodies following over-expression (Valdivia *et al.* 1998). Brighter, more soluble and blue-shifted mutants of GFP obtained by DNA shuffling or oligo-directed mutagenesis (Cormack *et al.* 1996; Crameri *et al.* 1995; Heim *et al.* 1994; Heim & Tsien 1995; Siemering *et al.* 1996) are suitable for *in situ* analyses (Chalfie *et al.* 1994).

GFP has been used for a range of different purposes including use as a transfection marker in eukaryotic cells, such as *Schizosaccharomyces pombe* (Atkins & Izant 1995). Microbial ecological applications include the use of GFP as a bacterial identification marker (Eberl *et al.* 1997). GFP has had considerable impact in localizing proteins within bacteria or after translocation into the cytosol of the host cell (Chen *et al.* 1999; Jacobi *et al.* 1998; Weiss *et al.* 1999). Increasing use is being made of GFP as a reporter of gene activity in various *in vivo* models, such as cultivated cell lines (*S. typhimurium*, Valdivia *et al.* 1996; Falkow 1997; *Y. pseudotuberculosis*, Valdivia *et al.* 1996; *Mycobacterium* spp., Parker & Bermudez 1997; Dhandayuthapani *et al.* 1995; Kremer *et al.* 1995) or in an animal model (*Mycobacterium marinum*, Barker *et al.* 1998). GFP reporters have also been used to investigate regulation of gene expression in *S. typhimurium* and *Candida albicans* (Cirillo *et al.* 1998; Morschäuser *et al.* 1998). The limitations of GFP as a reporter of gene expression are discussed in §2(c)(ii).

(b) *Analytical techniques for measuring gene expression in vivo*

The use of fluorescent and light-emitting reporters to monitor virulence gene expression and protein localization requires accurate visualization *in vivo*. This has been facilitated by significant enhancements of microscopic and flow cytometric technologies, detailed in §2(c)(ii).

(i) *Microscopic approaches*

Image capture and digital analysis

The capture of images from fluorescent microscopes originally involved relatively insensitive video grabbing devices. CCD cameras offer significant improvements because they are composed of arrays of light-sensitive units (pixels) that allow imaging at higher resolution (Cinelli 1998; Entwistle 1998; Fung & Theriot 1998). This powerful tool can accurately detect weak signals associated with bacteria (Hinnebusch & Bendich 1997; Lewis *et al.* 1994). Improvement of image quality requires reliable background subtraction to improve contrast. To monitor bacterial gene expression *in vivo*, accurate quantification of signals is required. The development of digital imaging analysis has resolved this problem, allowing improved quality of both two-dimensional (2D) and three-dimensional (3D) images, and simplifying data storage (Shotton 1998).

Confocal microscopy

Confocal microscopy is an important development of light microscopy, which allows 3D analysis of relatively thick tissue sections. The confocal optical system resembles a moveable pinhole that admits light from a single focal plane only (Sheppard & Shotton 1997). This technique allows detection of fluorescing signal throughout tissue sections and is helpful for identifying the cell type in which bacteria reside. The localization of *S. typhimurium* in liver sections of mice was elegantly shown by confocal microscopy of immunohistochemical-stained sections (Richter-Dahlfors *et al.* 1997).

Image deconvolution

This software-based approach assembles 3D images by deblurring a library of images acquired with a conventional microscope to restore out-of-focus photons to their focus plane (Shaw 1995). The improved image contrast allows localization of the proteins in bacterial cells, and produces images that rival those obtained with a confocal microscope (Glaser *et al.* 1997).

Multiphoton microscopy

Confocal laser scanning microscopy has been limited by the lack of reliable fluorophores that could be excited by the ultraviolet part of the spectrum. One alternative is the two-photon-based molecular excitation system in which the sample is submitted to a stream of pulsed infrared light with a pulse frequency that makes one dye molecule likely to simultaneously absorb two long-photons in the focal plane. By absorbing two photons, the dye molecule combines their energy to reach its excited state and to fluoresce

(Denk *et al.* 1990). This two-photon excitation can be used in conjunction with confocal microscopy to provide improved resolution of 3D images. Photobleaching, a problem with conventional fluorescent microscopy, is confined to the vicinity of the focal plane. As two-photon excitation allows sharp localization of fluorescence emission, this technique is useful for the observation of intracellular protein trafficking.

Microscopic analyses of reporter gene expression *in vivo* have several significant limitations: (i) high quality instruments (microscopes and CCD cameras) remain relatively expensive; (ii) animal tissues often give intense auto-fluorescence that overlays and masks weaker signals of interest; (iii) immunohistochemical techniques often require permeabilization of the specimen to allow reliable detection of epitopes; certain permeabilization solvents alter the fluorescence of reporter proteins such as GFP; (iv) deconvolution software is extremely processor intensive, making image processing lengthy.

Consequently, microscopic approaches are ideal for detecting bacteria and for determining the mammalian cell-type specificity of infecting bacteria. However, microscopy of tissue sections does not give a reliable measurement of bacterial gene expression but is limited to determining whether a gene is simply 'on' or 'off'. Accurate analyses of the expression of bacterial reporter genes in vivo require the use of flow cytometry and fluorescence-activated cell sorting (FACS).

(ii) *Flow cytometry*

Flow cytometry has been used extensively to study eukaryotic cells (Simons 1999) and its use for prokaryotic gene expression measurements is developing; it is a powerful technique for the assessment of intracellular fluorescence in individual bacterial cells. In combination with reporter gene technology, FACS can be used to analyse large populations of cells and to fractionate bacteria according to their level of gene expression. During FACS analysis, diffraction of argon laser light by bacteria gives information relating to size. This 488 nm light excites certain fluorophores causing emission of bacterial fluorescence which is amplified and detected. Signals are collected, digitized and stored for further analysis. Flow cytometry can measure the dynamics of gene induction in individual cells in response to environmental stimuli as discussed in §2(c)(ii). Flow cytometric analysis of some bacterial species is complicated by aggregation, which can cause artefacts; *Mycobac-teria* spp. require detergent and sonication treatment prior to cell sorting (Kremer *et al.* 1995; L. Ramakrishnan, R. Valdivia and S. Falkow, personal communication).

(c) *Techniques for identification of* in vivo-*induced genes*

Reporter genes have been used as the basis for selection procedures designed to isolate *in vivo*-induced (*ivi*) genes as outlined below. Several reviews have recently considered these techniques (Amann & Kïhl 1998; Camilli 1996; Falkow 1997; Handfield & Levesque 1999; Relman & Wright 1998; Valdivia & Falkow 1997b, 1998). Analysis of *ivi* genes has already elucidated aspects of *in vivo* regulation of virulence gene expression and pathogen adaptation (Cotter & Miller 1998; Foster 1999; Guiney 1997; Mahan *et al.* 1996; VanBogelen *et al.* 1999).

(i) In vivo *expression technology*

The genetic approach of *in vivo* expression technology (IVET) allows the identification of host-induced genes during infection (Chiang *et al.* 1999; Handfield & Levesque 1999; Mahan *et al.* 1993). IVET relies on the inser-tion of random chromosomal fragments upstream of a selectable gene that is required for survival in an animal model. The only bacteria that survive passage through the host must carry a fusion that expresses an active pro-moter. An *in vitro* pre-screening procedure ensures that only true *ivi* genes are identified.

IVET has been used to identify hundreds of *ivi* genes (Merrell & Camilli, this volume; Chiang *et al.* 1999; Handfield *et al.* 1998; Heithoff *et al.* 1997). The role of these genes in virulence is now being studied, and is complicated by the fact that between 25 and 50% of *ivi* genes do not resemble any other sequences in the database making it difficult to guess at their role; they are designated as having no known function (FUN) (Handfield & Levesque 1999; Hinton 1997). Auxotrophic and antibiotic resistance-based IVET se-lections are rather stringent and tend to identify genes that are highly expressed *in vivo*. These selection systems could also miss genes that are transiently expressed during infection. However, an improved recombinase fusion approach has been used to adapt IVET for the identification of genes expressed transiently during infection, which should allow the identification of organ-specific *ivi* genes (Camilli *et al.* 1994; Camilli & Mekalanos 1995; Merrell & Camilli, this issue).

(ii) *Differential fluorescence induction*

Differential fluorescence induction (DFI) is an alternative method for the positive selection of *ivi* genes during infection of cultured cells (Valdivia & Falkow 1997*a*). This system uses optimized GFP mutant proteins, which express up to 35-fold greater fluorescence than wild-type GFP and were selected by FACS (Cormack *et al.* 1996). These 'enhanced GFPs' (EGFP) allow individual bacteria bearing inducible gene fusions to be isolated by cell sorting (Valdivia *et al.* 1998). DFI involves the insertion of random genomic DNA upstream of the *gfp* reporter gene on a plasmid, followed by introduction to the bacterial pathogen. Bacteria harbouring *gfp* fusions are pooled and used for infection of cultured mammalian cells. FACS is used to isolate GFP-expressing bacteria released from lysed cells. These bacteria are cultivated on laboratory medium and a second FACS step is used to isolate bacteria with low *in vitro* fluorescence carrying *ivi* fusions that were only induced *in vivo*. Subsequent cell infection and bacteria sorting is used to enrich the *in vivo*-induced population. DFI was first applied to *S. typhimurium* to identify genes induced under acid conditions or within macrophages or epithelial cells (Valdivia *et al.* 1996; Valdivia & Falkow 1997*a*; Cirillo *et al.* 1998). A similar *gfp*-based approach has been used recently to identify macrophage-induced genes in *M. marinum* (Barker *et al.* 1998). In *M. smegmatis* and BCG, *gfp* transcriptional fusions to one heat-shock promoter *hsp60* have successfully been detected in animal tissue or infected macrophages using FACS, fluorescence and laser confocal microscopy (Dhandayuthapani *et al.* 1995; Kremer *et al.* 1995). DFI offers advantages to IVET for the identification of promoters that are only weakly and transiently induced during infection and offers accurate relative quantification of the level of gene expression in individual bacteria.

The analysis of bacterial GFP gene expression with flow cytometry does present some problems. Because sample exposure time is in the order of milliseconds in flow cytometric analysis, promoter strength is best detected and quantified by driving *gfp* expression from multicopy plasmids (Valdivia & Falkow 1997*a*). These limitations of detection explain why the current published examples describe plasmid-borne GFP reporter systems. Single-copy

GFP expression has not been intense enough for accurate measurement. Our current reliance on plasmid-based GFP fusions prevents the detection of context- or topological-dependent effects of gene regulation.

(iii) *Limitations of* ivi *gene identification*

The IVET and DFI approaches identify *ivi* genes but do not address their role in virulence. The complementary approach of signature-tagged mutagenesis (STM) is a direct screen that allows the identification of *in vivo* survival or virulence-associated genes (Hensel *et al.* 1995). This technology is based on transposon mutagenesis and allows negative selection of genes whose expression is required for survival *in vivo* (Chiang & Mekalanos 1998; Hensel 1998; Unsworth & Holden, this issue).

The three approaches of IVET, DFI and STM can be used to recognize different classes of *ivi* or virulence-associated genes by testing the same gene library or mutant pool in different animal models. For example, STM has been used to screen for Tn917 signature-tagged mutants of *Staphylococcus aureus* in the three models of mouse abscess, bacteraemia and wound infection (Coulter *et al.* 1998; Mei *et al.* 1997; Schwan *et al.* 1998).

IVET will not allow the identification of avirulence genes for which expression must be reduced *in vivo* to cause virulence (Galan 1998), but these could be recognized by a modified DFI approach.

The IVET and DFI *in vivo* expression technologies are based on transcriptional fusions and by definition will not detect genes that are post-transcriptionally regulated *in vivo*. Complementary approaches for identifying post-transcriptional induction *in vivo* are described below.

(d) *Direct measurement of bacterial mRNA levels* in vivo

The reporter-based systems described above have been used to follow mRNA expression indirectly. These systems are generally reliable, and much of the gene expression data obtained with reporter genes has subsequently been confirmed by direct monitoring of mRNA levels. However, reporter genes can occasionally give misleading data, which is only apparent when different systems for measuring gene expression are compared, as has been reported for *hns* and *proU* expression in *E. coli* and *Salmonella* (Forsberg *et al.* 1994; Free & Dorman 1995).

A number of techniques have been developed for the direct measurement of bacterial mRNA expression, as described below.

(i) *Subtractive hybridization*

A lack of genetic tools has prevented the application of reporter gene technology to many bacterial pathogens, and has led to the development of alternative techniques for the identification of *ivi* genes. Initially developed in eukaryotic cells, subtractive hybridization is now being used to study prokaryotic systems, such as the infection of cultivated macrophages by *M. avium* (Plum & Clark-Curtiss 1994). In this study, mycobacterial mRNA was converted to cDNA by reverse transcription. Biotin-labelled cDNA prepared from *M. avium* grown in broth was used to subtract constitutively expressed housekeeping genes from the cDNA of macrophage-derived *M. avium* using streptavidin-coated paramagnetic beads.

Unfortunately, the construction of representative cDNA libraries for subtractive hybridization is hampered by the instability of bacterial mRNA and the problems of isolating sufficient high-quality mRNA from small populations of bacteria growing *in vivo*. In addition, transiently expressed genes may not be well represented in the cDNA library. A PCR-based modification may prove to be a crucial innovation for subtractive hybridization, as it allows representation of low-abundance mRNA and requires a smaller number of bacteria (Sharma *et al.* 1993).

As well as being used to analyse gene expression, subtractive hybridization has also been applied to the study of genetic acquisition and bacterial evolution. Genomic subtractive hybridization was used for identifying virulence genes from strains of *Pseudomonas aeruginosa* isolated from cystic fibrosis patients (Schmidt *et al.* 1998) and genes that were specific to *S. typhimurium* and absent from *S. typhi* (Emmerth *et al.* 1999).

(ii) *Differential display*

Differential display was developed to identify eukaryotic genes that are induced under particular conditions (Kozian & Kirschbaum 1999), and is beginning to be applied to the identification of bacterial *ivi* genes.

This 'black-box' approach involves the construction of cDNA libraries from the mRNA of bacteria grown under different conditions. It consists of six steps: (i) isolation of RNA; (ii) reverse transcription; (iii) PCR amplification of cDNA species; (iv) electrophoretic separation of the resulting fragments; (v) reamplification of the fragments that vary between conditions, cloning and sequencing; and (vi) confirmation of the differential gene expression (Handfield & Levesque 1999).

Differential display has been used to identify virulence genes induced by adherence to red blood cells of uropathogenic *E. coli* to host cells (Zhang & Normark 1996) and to identify the novel *eml* locus in *Legionella pneumophila*, which is induced during early stages of macrophage infection (Abu Kwaik & Pederson 1996).

The differential display approach has the advantage of being applicable to more than two conditions at the same time, lending itself to multiplex analyses. Unlike the subtractive hybridization technique, differential display can detect both up- and downregulation of *in vivo*-induced genes. However, the combination of PCR with differential display commonly generates false-positive results that do not relate to *ivi* genes.

(e) *Proteomic approaches and* in vivo *expression*

The patterns of gene expression and the changes in mRNA levels shown by the above techniques simply reflect regulation at the transcriptional level which cannot be directly related to protein expression. Reporters such as alkaline phosphatase have been used to generate translational gene fusions to address this issue (Manoil & Beckwith 1985). However, the stability of such fusion proteins can be aberrant, giving misleading data concerning protein expression. To determine the role of post-transcriptional, translational or post-translational regulation *in vivo*, the entire bacterial protein complement (proteome) must be studied during infection. Approaches for proteome analysis are constantly being refined and improved, and are currently based on the 2D polyacrylamide gel electrophoresis technique (O'Farrell 1975). Separation from *in vitro*- and *in vivo*-grown bacteria, first by their isoelectric point and second by their molecular weight, allows detection of IVI proteins. Recent advances in mass spectrometry combined with increasing availability of genomic sequence data should allow IVI proteins to be identified for many pathogens (Humphery-Smith *et al.* 1997; O'Connor *et al.* 1998).

Two-dimensional gel electrophoresis has been used to characterize acid- and alkali-induced proteins during aerobic and anaerobic growth of *E. coli* (Blankenhorn *et al.* 1999) as well as Salmonella IVI proteins induced within host macrophages (Abshire & Neidhardt 1993; Buchmeier & Heffron 1990). IVI proteins of *S. typhimurium* that are expressed within Henle-407 cells were identified by using a labelled lysine precursor, which cannot be used by mammalian cells (Burns-Keliher *et al.* 1997). Two-dimensional gels showed that *S. typhimurium* exhibits cell-type- and species-specific gene expression in cultured cells (Burns-Keliher *et al.* 1998). However, this approach is limited because only a restricted set of proteins can be visualized on 2D gels. Furthermore, this technology has only been used in cultured cells and is unlikely to prove applicable to whole animal models. Unfortunately, the proteomic approach cannot assess protein expression in individual bacterial cells because of the quantities of protein currently required for 2D gel analysis.

3.   FUTURE APPROACHES

The technologies outlined above are summarized in Table 1, and have been used to identify hundreds of bacterial genes that are induced *in vivo*. Our current challenge is to determine the stage of pathogenesis at which these bacterial genes are expressed. Do *ivi* genes show organ-specific patterns of expression? Are *ivi* genes only switched on in certain cell types in the host? Are *ivi* genes switched on at early or late stages of infection? Do *ivi* genes show a temporal cascade of gene expression following induction? These questions require new approaches for monitoring virulence gene expression *in situ*.

(a) *Immunological-based detection of bacterial gene expression*

IVI virulence determinants can elicit a long-lived immune response in the host. Antibodies contained in sera from an infected host can subsequently be used to screen for bacterial IVI proteins. Comparative immunological detection of *Borrelia burgdorferi* proteins expressed *in vivo* or *in vitro* was achieved by screening a genomic expression library with sera obtained from infected or immunized mice after injection of live or killed pathogens (Suk *et al.* 1995). These sera effectively discriminated between *in vivo*- and *in vitro*-expressed proteins.

Alternatively, IVI proteins can be recognized following immunization of individual mice with individual proteins encoded by each open reading frame (ORF) of a bacterial pathogen and screening for protection against subsequent challenge with wild-type bacteria. Sykes & Johnston (1999) described an elegant procedure involving the insertion of the promoterless ORF studied between a cytomegalovirus promoter and a eukaryotic terminator. These linear expression elements (LEE) generate transcriptionally active units, which were used individually to immunize mice. Specific immunization of the animals was checked on immunoblots with pathogen whole-cell lysates. This technique was used successfully to identify IVI proteins of *M. tuberculosis* (Felgner & Liang 1999). However, this complex strategy does require a complete genome sequence and careful optimization of the inoculum administration required to achieve a protective immune response.

An alternative method for monitoring protein expression involves the creation of protein fusions to specific epitope tags that can be identified immunologically. Induction of *spv* gene expression in *Salmonella* has been observed in murine macrophages by using transcriptional fusions of *spv* gene promoters and the fimbrial cluster of *E. coli* KS71A. Expression of *spv* genes was measured by agglutination of the bacteria with anti-fimbrial antibodies (Rhen *et al.* 1993).

(b) In vivo-*induced antigen technology*

A novel strategy to identify IVI proteins based on natural infection in humans has recently been described (M. Handfield, personal communication; Fig. 1). IVIAT was developed using sera from patients infected with *Actinobacillus actinomyce-temcomitans*, the aetiological agent of localized juvenile periodontitis, and offers several advantages to existing approaches. First and foremost, it is based on actual human infection and does not rely on potentially misleading animal models of disease. No particular genetic construct is required, allowing IVIAT to be used for any pathogen that can be grown *in vitro* (Handfield *et al.* 1999). This approach is rapid, simple and involves well-established methodologies, which facilitates the screening of a whole genome. Following growth of the pathogen *in vitro*, whole bacterial cells and total protein extracts are used to subtract antibodies contained in human patient sera that are directed against *in vitro*-expressed bacterial proteins. In parallel, a genomic expression plasmid library of the pathogen is constructed in *E. coli*. The remaining antibacterial antibodies

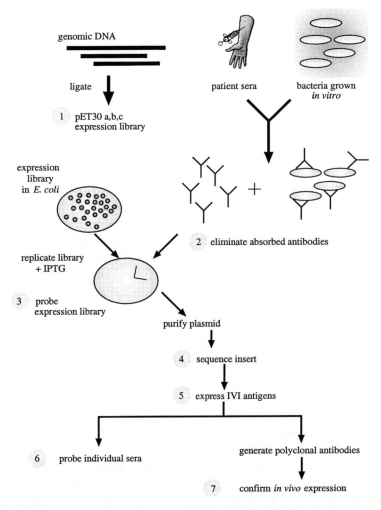

Fig. 1. Diagram showing the principle of IVIAT (M. Handfield, personal communication). IPTG, isopropylthio-beta-D-galactoside.

in human sera are specific for IVI proteins and are subsequently used to identify these proteins from the expression library. A large amount of these specific antibodies can be purified and used to probe biological samples isolated from infected patients demonstrating *in vivo* protein induction via a direct immunofluorescence technique. IVIAT has been successfully used to study bacterial (*A. actinomyce-temcomitans* and *P. aeruginosa*) and yeast (*Candida* spp.) infections.

(c) In vivo *induction of fluorescence*

The limitations of luciferase and β-galactosidase reporter technology described earlier mean that a new approach was required for the monitoring of gene expression in individual bacterial cells *in vivo*. DFI relied on FACS to measure bacterial gene expression either *in vitro* or following isolation of bacteria from infected macrophages, epithelial cells or other mammalian cells. Our interest in organ-specific patterns of gene expression has led to the development of an improved system for assessing the transcriptional response of *Salmonella* virulence genes during murine infection (J. C. D. Hinton and I. Hautefort, unpublished data). *In vivo* induction of fluorescence (IVIF) uses a plasmid-based EGFP reporter system to monitor gene expression in tissues of infected mice. EGFP fluorescence can be detected and quantified in individual bacteria by fluorescent microscopic and flow cytometric analysis.

The IVIF system does not perturb the invasion, spread or multiplication of *S. typhimurium* in mice. EGFP expression gives easily detectable fluorescence for both microscopic observations of organ sections and flow cytometric analysis on bacteria released from infected organs. The stability of EGFP (half-life 424 h) makes it ideal for the detection of transient gene induction *in vivo*.

Cultivated cell lines have already been used to determine gene induction kinetics (Valdivia & Falkow 1997*a*), but such cell-culture-based approaches cannot mimic the complex and changing environment that occurs at the site of infection. IVIF allows the identification of the pattern of *ivi* gene expression by sacrificing mice at different time-points after infection and sampling from several sites. An example of the data that can be generated by IVIF (Fig. 2) shows the expression of a *S. typhimurium ssrA:gfp* fusion in an infected mouse spleen.

(d) In situ *PCR*

Although this technology has not yet been applied to the analysis of gene expression *in vivo*, a method for the measurement of intracellular mRNA

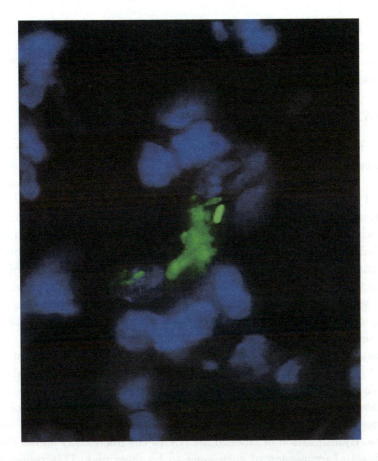

Fig. 2. Fluorescence micrograph showing expression of a plasmid-borne transcriptional *ssrA:gfp* fusion in *S. typhimurium*. A DAPI-stained spleen section is shown from an intraperitoneally inoculated BALB/c mouse, with cell nuclei appearing as blue. Individual *S. typhimurium* 12023 bacteria expressing GFP appear as green (I. Hautefort, J. M. Sidebotham and J. C. D. Hinton, unpublished data).

levels is available. Hodson *et al.* (1995) outlined a method for *in situ* PCR of bacterial cells which was applied to microbial identification. More recently, Tolker-Nielsen *et al.* (1997) used *in situ* PCR on *Salmonella* to detect as little as one *lac* mRNA molecule per bacterial cell. The same method was used to demonstrate that *groEL* is cell-cycle regulated and that *lac* expression in *Salmonella* exhibited two distinct subpopulations (Holmstrom *et al.* 1999;

Tolker-Nielsen *et al.* 1998). This *in situ* PCR approach has great promise for assessing gene expression in bacteria grown *in vitro*, and could potentially be applied *in vivo* to visualize organ-specific or cell-type-specific gene expression.

(e) *DNA microarrays*

A great many bacterial genome sequences are either now available or will be completed soon. These represent bacteria of academic, medical or industrial interest. Virulent *E. coli* and *Salmonella* genome sequences will shortly be released and the genomes of pathogenic organisms such as *Campylobacter* and *Neisseria* are already being exploited. The new 'functional genomic' approach involves concerted efforts to define gene function through analysis of global gene expression (transcriptome), protein expression (proteome) and genetics (construction of knockout strains). As an example of what is expected to be achieved in bacterial systems, DNA microarray analysis of the yeast genome has already identified hundreds of genes regulated during sporulation and yielded thousands of knockout strains with well-characterized phenotypes (Chu *et al.* 1998; Winzeler *et al.* 1999).

The availability of genome sequences has facilitated an important development of nucleic acid hybridization, the DNA microarray or chip. Microarrays are generally manufactured on glass microscope slides, and consist of an ordered grid of thousands of DNA spots. These can either be oligonucleotides or PCR products corresponding to every predicted ORF on the genome. Hybridization of labelled mRNA to the microarray allows the relative level of expression of each gene to be measured. This represents a significant breakthrough as it is now possible to get information for the expression of all bacterial genes from a single experiment.

The first bacterial application was described by De Saizeu *et al.* (1998) using an oligonucleotide-based microarray to look at expression of a subset of 100 genes from Streptococcus pneumoniae. Richmond *et al.* (1999) compared the reliability of DNA arrays by comparing arraying and labelling methods for *E. coli*. They concluded that DNA microarrays hybridized with fluorescently labelled samples gave the most reliable and sensitive data, and showed that this technology can be effectively applied to bacterial gene regulation. Tao *et al.* (1999) used an *E. coli* array to identify genes induced by growth on minimal media, and to recognize a new role for *rpoS* expression in exponentially growing cells.

Table 1. Approaches for monitoring bacterial gene expression *in vivo*.

Techniques	Organisms	Cell type or animal	References
in vivo expression technology (IVET)	*Salmonella typhimurium*	mouse	Mahan *et al.* 1993
differential fluorescence induction (DFI)	*S. typhimurium*	epithelial cells, macrophages, dendritic cells	Valdivia & Falkow 1997
signature-tagged mutagenesis (STM)	*S. typhimurium*	mouse	Hensel *et al.* 1995
subtractive hybridization	*Mycobacterium avium*	macrophages	Plum & Clark-Curtiss 1994
differential display	uropathogenic *E. coli*	red blood cells	Zhang & Normark 1996
proteomic analysis	*S. typhimurium*	epithelial cells, macrophages	Buchmeier & Heffron 1990; Abshire & Neidhardt 1993
DNA microarrays	*Streptococcus pneumoniae* *Escherichia coli* *Mycobacterium tuberculosis*	none	De Saizeu *et al.* 1998; Richmond *et al.* 1999; Tao *et al.* 1999
in situ PCR	*S. typhimurium*	none	Tolker-Nielsen *et al.* 1997

DNA microarrays have the potential to revolutionize the study of *ivi* gene expression since they can yield global information concerning temporal and spatial aspects of gene expression. However, the limiting stage of microarray technology for *ivi* gene expression is the isolation of sufficient high-quality bacterial mRNA from infected tissue. Improved mRNA purification coupled with linear RNA amplification technology (Eberwine *et al.* 1992) could make microarrays the method of choice for *ivi* gene expression analysis in the future.

We thank Gordon Dougan for critical appraisal of the manuscript, Carmen Beuzon, David Holden and Margaret Jones for their assistance, and Paul Pople for preparing Fig. 1. J.C.D.H. acknowledges support from the Wellcome Trust (ref. 045490), and from the Biotechnology and Biological Sciences Research Council. I.H. is supported by a Training and Mobility of Researchers fellowship from the European Union (contract number ERBFMRXCT9).

REFERENCES

Abshire, K. Z. & Neidhardt, F. C. 1993 Analysis of proteins synthesized by *Salmonella typhimurium* during growth within host macrophage. *J. Bacteriol.* **175**, 3734–3743.

Abu Kwaik, Y. & Pederson, L. L. 1996 The use of differential display-PCR to isolate and characterise a *Legionella pneumophila* locus induced during the intracellular infection of macrophages. *Mol. Microbiol.* **21**, 543–556.

Amann, R. & Kïhl, M. 1998 *In situ* methods for assessment of microorganisms and their activities. *Curr. Opin. Microbiol.* **1**, 352–358.

Aranda, C. M. A., Swandon, J. A., Loomis, W. P. & Miller, S. J. 1992 *Salmonella typhimurium* activates virulence gene transcription within acidified macrophages phagosomes. *Proc. Natl Acad. Sci. USA* **89**, 10 079–10 083.

Atkins, D. & Izant, J. G. 1995 Expression and analysis of the green fluorescent protein gene in the fission yeast *Schizosaccharomyces pombe*. *Curr. Genet.* **28**, 585–588.

Barker, L. P., Brooks, D. M. & Small, P. L. C. 1998 The identification of *Mycobacterium marinum* genes differentially expressed in macrophage phagosomes using promoter fusions to green fluorescent protein. *Mol. Microbiol.* **29**, 1167–1177.

Biran, I., Klimentiy, L., Hengge-Aronis, R., Eliora, Z., Rishpon, R. & Rishpon, J. 1999 On-line monitoring of gene expression. *Microbiology* **145**, 2129–2133.

Blankenhorn, D., Phillips, J. & Slonczewski, J. L. 1999 Acid- and base-induced proteins during aerobic and anaerobic growth of *Escherichia coli* revealed by two-dimensional gel electrophoresis. *J. Bacteriol.* **181**, 2209–2216.

Buchmeier, N. A. & Heffron, F. 1990 Induction of *Salmonella* stress proteins upon infection of macrophages. *Science* **248**, 730–732.

Burns-Keliher, L., Portteus, A. & Curtiss III, R. 1997 Specific detection of *Salmonella typhimurium* proteins synthesized intracellularly. *J. Bacteriol.* **179**, 3604–3612.

Burns-Keliher, L., Nickerson, C. A., Morrow, B. J. & Curtiss III, R. 1998 Cell-specific proteins synthesized by *Salmonella typhimurium*. *Infect. Immun.* **66**, 856–861.

Camilli, A. 1996 Noninvasive techniques for studying pathogenic bacteria in the whole animal. *Trends Microbiol.* **4**, 295–296.

Camilli, A. & Makalanos, J. J. 1995 Use of recombinase gene fusions to identify *Vibrio cholerae* genes induced during infection. *Mol. Microbiol.* **18**, 671–683.

Camilli, A., Beattie, D. T. & Mekalanos, J. J. 1994 Use of genetic recombination as a reporter of gene expression. *Proc. Natl Acad. Sci. USA* **91**, 2634–2638.

Casadaban, M. J. & Cohen, S. N. 1979 Lactose genes fused to exogenous promoters in one step using Mu-*lac* bacteriophage: *in vivo* probe for transcriptional control sequences. *Proc. Natl Acad. Sci. USA* **76**, 4530–4533.

Chalfie, M., Tu, Y., Euskirchen, G., Ward, W. W. & Prasher, D. C. C. 1994 Green fluorescent protein as marker of gene expression. *Science* **263**, 802–805.

Chen, J. C., Weiss, D. S., Ghigo, J.-M. & Beckwith, J. 1999 Septal localization of FtsQ, an essential cell division protein in *Escherichia coli*. *J. Bacteriol.* **181**, 521–530.

Chiang, S. L. & Mekalanos, J. J. 1998 Use of signature-tagged mutagenesis to identify *Vibrio cholerae* genes critical for colonization. *Mol. Microbiol.* **30**, 175–188.

Chiang, S. L., Mekalanos, J. J. & Holden D. W. 1999 *In vivo* genetic analysis of bacterial virulence. *A. Rev. Microbiol.* **53**, 129–154.

Chu, S., DeRisi, J., Eisen, M., Mulholland, J., Botstein, D., Brown, P. O. & Herskowitz, I. 1998 The transcriptional program of sporulation in budding yeast. *Science* **282**, 699–705.

Cinelli, A. R. 1998 Flexible method to obtain high sensitivity, low-cost CCD cameras for video microscopy. *J. Neurosci. Meth.* **85**, 33–43.

Cirillo, D. M., Valdivia, R. H., Monack, D. & Falkow, S. 1998 Macrophage-dependent induction of the *Salmonella* pathogenicity island 2 type III secretion system and its role in intracellular survival. *Mol. Microbiol.* **30**, 175–188.

Contag, C. H., Contag, P. R., Mullins, J. I., Spilman, S. D., Stevenson, D. K. & Benaron, D. A. 1995 Photonic detection of bacterial pathogen in living hosts. *Mol. Microbiol.* **189**, 593–603.

Cormack, B. P., Valdivia, R. H. & Falkow, S. 1996 FACS-optimized mutants of the green fluorescent protein (GFP). *Gene* **173**, 33–38.

Cotter, P. A. & Miller, J. F. 1998 *In vivo* and *ex vivo* regulation of bacterial virulence gene expression. *Curr. Opin. Microbiol.* **1**, 17–26.

Coulter, S. N., Schwan, W. R., Ng, E. Y. W., Langhorne, M. H., Ritchie, H. D., Westbrock-Wadman, S., Hufnagle, W. O., Folger, K. R., Bayer, A. S. & Stover, C. K. 1998 *Staphylococcus aureus* genetic loci impacting growth and survival in multiple infection environments. *Mol. Microbiol.* **30**, 393–404.

Crameri, A., Whitehorn, E. A., Tate, E. & Stemmer, W. P. C. 1995 Improved green fluorescent protein by molecular evolution using DNA shuffling. *Nature Biotech.* **14**, 315–319.

Cubitt, A. B., Heim, R., Adams, S. R., Boyd, A. E., Gross, L. A. & Tsien, R. Y. 1995 Understanding, improving and using green fluorescent proteins. *Trends Biochem. Sci.* **20**, 448–455.

De Saizieu, A., Certa, U., Warrington, J., Gray, C., Keck, W. & Mous, J. 1998 Bacterial transcript imaging by hybridisation of total RNA to oligonucleotide arrays. *Nature Biotech.* **16**, 45–48.

Denk, W., Strickler, J. H. & Webb, W. W. 1990 Two-photon laser scanning fluorescence microscopy. *Science* **248**, 73–76.

Dhandayuthapani, S., Via, L. E., Thomas, C. A., Horowitz, P. M., Deretic, D. & Deretic, V. 1995 Green fluorescent protein as a marker for gene expression and cell biology of mycobacterial interactions with macrophages. *Mol. Microbiol.* **17**, 901–912.

Eberl, L., Schulze, R., Ammendola, A., Geisenberger, O., Erhart, R., Sternberg, C., Molin, S. & Amann, R. 1997 Use of green fluorescent protein as a marker for ecological studies of activated sludge communities. *FEMS Microbiol. Lett.* **149**, 77–83.

Eberwine, J., Yeh, H., Miyashiro, K., Cao, Y., Nair, S., Finnell, R., Zettel, M. & Coleman, P. 1992 Analysis of gene expression in single live neurons. *Proc. Natl Acad. Sci. USA* **89**, 3010–3014.

Emmerth, M., Goebel, W., Miller, S. I. & Hueck, C. J. 1999 Genomic subtraction identifies *Salmonella typhimurium* prophages, F-related plasmid sequences, and a novel fimbrial operon, *stf*, which are absent in *Salmonella typhi. J. Bacteriol.* **181**, 5652–5661.

Entwistle, A. 1998 A comparison between the use of a high-resolution CCD camera and 35 mm film for obtaining coloured micrographs. *J. Microsc.* **192**, 81–89.

Falkow, S. 1997 Invasion and intracellular sorting of bacteria: searching for bacterial genes expressed during host/pathogen interactions. In Perspectives Series: host/pathogen interactions. *J. Clin. Invest.* **100**, 239–243.

Felgner, P. L. & Liang, X. 1999 Debugging expression screening. *Nature Biotech.* **17**, 329–330.

Fierer, J., Eckmann, L., Fang, F. Pfeifer, C., Finlay, B. B. & Guiney, D. 1993 Expression of the *Salmonella* virulence plasmid gene *spvB* in cultured macrophages and non-phagocytic cells. *Infect. Immun.* **61**, 5231–5236.

Forsberg, J. & Rosqvist, R. 1994 *In vivo* expression of virulence genes of *Yersinia pseudotuberculosis. Infect. Agents Dis.* **2**, 275–278.

Forsberg, J., Pavitt, G. D. & Higgins, C. F. 1994 Use of transcriptional fusions to monitor gene expression: a cautionary tale. *J. Bacteriol.* **176**, 2128–2132.

Foster, J. W. 1999 When protons attack: microbial strategies of acid adaptation. *Curr. Opin. Microbiol.* **2**, 170–174.

Francis, K. P. & Gallagher, M. P. 1993 Light emission from a mudlux transcriptional fusion in *Salmonella typhimurium* is stimulated by hydrogen peroxide and by interaction with the mouse macrophage cell line J774.2. *Infect. Immun.* **61**, 640–649.

Free, A. & Dorman, C. J. 1995 Coupling of *Escherichia coli hns* mRNA levels to DNA synthesis by autoregulation: implications for growth phase control. *Mol. Microbiol.* **18**, 101–113.

Fung, D. C. & Theriot, J. A. 1998 Imaging techniques in microbiology. *Curr. Opin. Microbiol.* **1**, 346–351.

Galan, J. E. 1998 'Avirulence genes' in animal pathogens? *Trends Microbiol.* **6**, 3–6.

Garcia-del-Portillo, F., Foster, J. W., Maguire, M. E. & Finlay, B. B. 1992 Characterization of the micro-environment of *Salmonella typhimurium*-containing vacuoles within MDCK epithelial cells. *Mol. Microbiol.* **6**, 3289–3297.

Glaser, P., Sharpe, M. E., Raether, B., Perego, M., Ohlsen, K. & Errington, J. 1997 Dynamic, mitotic-like behavior of a bacterial protein required for accurate chromosome partitioning. *Genes Dev.* **11**, 1160–1168.

Gonzalez-Flecha, B. & Demple, B. 1994 Intracellular generation of superoxide as a by-product of *Vibrio harveyi* luciferase expressed in *E. coli. J. Bacteriol.* **176**, 2293–2299.

Guiney, D. G. 1997 Regulation of bacterial virulence gene expression by host environment. In Perspectives Series: host/pathogen interactions. *J. Clin. Invest.* **100**, S7–S11.

Handfield, M. & Levesque, R. C. 1999 Strategies for isolation of *in vivo* expressed genes from bacteria. *FEMS Microbiol. Rev.* **23**, 69–91.

Handfield, M., Schweizer, H. P., Mahan, M. J., Sanschagrin, F., Hoang, T. & Levesque, R. C. 1998 Asd-GFP vectors for *in vivo* expression technology in *Pseudomonas aeruginosa* and other Gram-positive bacteria. *Biotechniques* **24**, 261–264.

Handfield, M., Brady, L. J., Progulske-Fox, A. & Hillman, J. D. 2000 *In vivo* induced antigen technology (IVIAT): a novel method to identify microbial genes expressed specifically during human infections. *Trends Microbiol.* (In the press.)

Heim, R. & Tsien, R. Y. 1995 Engineering green fluorescent protein for improved brightness, longer wavelengths and fluorescence resonance energy transfer. *Curr. Biol.* **6**, 178–182.

Heim, R., Prasher, D. C. & Tsien, R. Y. 1994 Wavelength mutations and post-translational autoxidation of green fluorescent protein. *Proc. Natl Acad. Sci. USA* **91**, 12 501–12 504.

Heithoff, D. M., Conner, C. P., Hanna, P. C., Julio, S. M., Hentschel, U. & Mahan, M. J. 1997 Bacterial infection as assessed by *in vivo* gene expression. *Proc. Natl Acad. Sci. USA* **94**, 934–939.

Hensel, M. 1998 Whole genome scan for habitat-specific genes by signature-tagged mutagenesis. *Electrophoresis* **19**, 608–612.

Hensel, M., Shea, J. E., Gleeson, C., Jones, M. D., Dalton, E. & Holden, D. W. 1995 Simultaneous identification of bacterial virulence genes by negative selection. *Science* **269**, 400–403.

Hinnebusch, B. J. & Bendich, A. J. 1997 The bacterial nucleoid visualized by fluorescent microscopy of cells lyzed within agarose: comparison of *Escherichia coli* and spirochetes of the genus *Borrelia*. *J. Bacteriol.* **179**, 2228–2237.

Hinton, J. C. D. 1997 The *Escherichia coli* genome sequence: the end of an era or the start of FUN? *Mol. Microbiol.* **26**, 417–422.

Hodson, R. E., Dustman, W. A., Garg, R. P. & Moran, M. A. 1995 *In situ* PCR for visualization of microscale distribution of specific genes and gene products in prokaryotic communities. *Appl. Environ. Microbiol.* **61**, 4074–4082.

Holmstrom, K., Tolker-Nielsen, T. & Molin, S. 1999 Physiological states of individual *Salmonella typhimurium* cells monitored by *in situ* reverse transcription PCR. *J. Bacteriol.* **181**, 1733–1738.

Humphery-Smith, I., Cordwell, S. J. & Blackstock, W. P. 1997 Proteome research: complementarity and limitations with respect to the RNA and DNA worlds. *Electrophoresis* **18**, 1217–1242.

Jacob, F. & Monod, J. 1961 Genetic regulatory mechanisms in the synthesis of proteins. *J. Mol. Biol.* **3**, 318–356.

Jacobi, C. A., Rogenkamp, A., Zumbihl, R., Leitritz, L. & Heesemann, J. 1998 *In vitro* and *in vivo* expression studies of *yopE* from *Yersinia enterolitica* using the *gfp* reporter gene. *Mol. Microbiol.* **30**, 865–882.

Jaiwal, S. P. K., Ahad, A. & Sahoo, L. 1998 Green fluorescent protein: a novel reporter gene. *Curr. Sci.* **74**, 402–405.

Kirchner, G., Roberts, J. L., Gustafson, G. D. & Ingolia, T. D. 1989 Active bacterial luciferase from a used gene: expression of *Vibrio harveyi luxAB* translational fusion in bacteria, yeast and plant cells. *Gene* **81**, 349–354.

Kozian, D. H. & Kirschbaum, B. J. 1999 Comparative gene-expression analysis. *Trends Biotech.* **17**, 73–78.

Kremer, L., Baulard, A., Estaquier, J., Poulain-Godefroy, O. & Locht, C. 1995 Green fluorescent protein as a new expression marker in mycobacteria. *Mol. Microbiol.* **17**, 913–922.

Langridge, W. H. R., Escher, A., Koncz, C., Schell, J. & Szalay, A. A. 1991a Bacterial luciferase genes: a light emitting reporter system for *in vivo* measurement of gene expression. *Technique J. Meth. Cell Mol. Biol.* **3**, 91–97.

Langridge, W. H. R., Escher, A. & Szalay, A. A. 1991b Measurement of bacterial luciferase as a reporter enzyme *in vivo* in transformed bacteria, yeast, plant cells and in transgenic plants. *Technique J. Meth. Cell Mol. Biol.* **3**, 99–108.

Lewis, P. J., Nwoguh, C. E., Barer, M. R., Harwood, C. R. & Errington, J. 1994 Use of digitized video microscopy with a fluorogenic enzyme substrate to demonstrate cell- and compartment-specific gene expression in *Salmonella enteritidis* and *Bacillus subtilis*. *Mol. Microbiol.* **13**, 655–662.

Mahan, M. J., Slauch, J. M. & Mekalanos, J. J. 1993 Selection of bacterial virulence genes that are specifically induced in host tissues. *Science* **259**, 686–688.

Mahan, M. J., Slauch, J. M. & Mekalanos, J. J. 1996 Environmental regulation of virulence gene expression in *Escherichia coli*, *Salmonella*, and *Shigella* spp. In Escherichia coli *and* Salmonella: *cellular and molecular biology* (ed. F. C. Neidhardt), pp. 2803–2815. Washington, DC: American Society of Microbiology.

Manoil, C. & Beckwith, J. 1985 *TnphoA*: a transposon probe for protein export signals. *Proc. Natl Acad. Sci. USA* **82**, 8129–8133.

Mei, J. M., Nourbakhsh, F., Ford, C. W. & Holden, D. W. 1997 Identification of *Staphylococcus aureus* virulence genes in a murine model of bacteraemia using signature-tagged mutagenesis. *Mol. Microbiol.* **26**, 399–407.

Meighen, E. A. 1991 Molecular biology of bacterial bioluminescence. *Microbiol. Rev.* **55**, 123–142.

Mekalanos, J. J. 1992 Environmental signals controlling expression of virulence determinants in bacteria. *J. Bacteriol.* **174**, 1–7.

Miller, J. H. 1992 Additional systems used to monitor gene expression. In *A short course in bacterial genetics*, pp. 63–67. Cold Spring Harbor, NY: Cold Spring Harbor Laboratory Press.

Morschhäuser, J., Michel, S. & Hacker, J. 1998 Expression of a chromosomally integrated, single copy *gfp* gene in *Candida albicans*, and its use as a reporter of gene regulation. *Mol. Gen. Genet.* **257**, 412–420.

Nwoguh, C. E., Harwood, C. R. & Barer, M. R. 1995 Detection of induced β-galactosidase activity in individual non-culturable cells of pathogenic bacteria by quantitative cytological assay. *Mol. Microbiol.* **17**, 545–554.

O'Connor, C. D., Farris, M., Hunt, L. G. & Neville, J. 1998 The proteome approach. *Meth. Microbiol.* **27**, 191–204.

O'Farrell, P. H. 1975 High-resolution two-dimensional electrophoresis of proteins. *J. Biol. Chem.* **25**, 4007–4021.

Parker, A. E. & Bermudez, L. E. 1997 Expression of the green fluorescent protein (GFP) in *Mycobacterium avium* as a tool to study the interaction between mycobacteria and host cells. *Microb. Pathogen.* **22**, 193–198.

Pettersson, J., Nordfelth, R., Dubinina, E., Bergman, T., Gustafsson, M., Magnusson, K. E. & Wolf-Watz, H. 1996 Modulation of virulence factor expression by pathogen target cell contact. *Science* **273**, 1231–1233.

Pfeifer, C. G. & Finlay, B. B. 1995 Monitoring gene expression of Salmonella inside mammalian cells: comparison of luciferase and β-galactosidase fusion systems. *J. Microbiol. Meth.* **24**, 155–164.

Plum, G. & Clark-Curtiss, J. E. 1994 Infection of *Mycobacterium avium* gene expression following phagocytosis by human macrophages. *Infect. Immun.* **62**, 476–483.

Prasher, D. C., Eckenrode, V. K., Ward, W. W., Prendergast, F. G. & Cormier, M. J. 1992 Primary structure of *Aequoria victoria* green-fluorescent protein. *Gene* **111**, 229–233.

Relman, D. A. & Wright, A. 1998 Molecular and cellular microbiology: new tools of the trade in techniques. *Curr. Opin. Microbiol.* **1**, 337–339.

Rhen, M., Riikonen, P. & Taira, S. 1993 Transcriptional regulation of *Salmonella enterica* virulence plasmid genes in cultured macrophages. *Mol. Microbiol.* **10**, 45–56.

Richmond, C. S., Glasner, J. D., Mau, R., Jin, H. & Blattner, F. R. 1999 Genome-wide expression profiling in *Escherichia coli* K-12. *Nucl. Acids Res.* **27**, 3821–3835.

Richter-Dahlfors, A., Buchan, A. M. J. & Finlay, B. B. 1997 Murine salmonellosis studied by confocal microscopy: *Salmonella typhimurium* resides intracellularly inside macrophages and exerts a cytotoxic effect on phagocytes *in vivo*. *J. Exp. Med.* **186**, 569–580.

Sala-Newby, G. B., Kendall, J. M., Jones, H. E., Taylor, K. M., Badminton, M. N., Llewellyn, D. H. & Campbell, A. K. 1999 Bioluminescent and chemiluminescent indications for molecular signalling and function in living cells. In *Fluorescent and luminescent probes for biological activity. Biological Techniques Series*, 2nd edn (ed. W. T. Mason), pp. 251–272. San Diego, CA: Academic Press.

Sarkis, G. J., Jacobs, W. R. & Hatfull, G. F. 1995 L5 luciferase reporter mycobacteriophages: a sensitive toll for the detection and assay in mycobacteria. *Mol. Microbiol.* **15**, 1055–1067.

Schauer, A. T. 1988 Visualizing gene expression with luciferase fusions. *Trends Biotechnol.* **6**, 23–27.

Schmidt, K. D., Schmidt-Rose, T., Romlig, U. & Tummler, B. 1998 Differential genome analysis of bacteria by genomic subtractive hybridization and pulsed field gel electrophoresis. *Electrophoresis* **19**, 509–514.

Schwan, W. R., Coulter, S. N., Ng, E. Y. W., Langhorne, M. H., Ritchie, H. D., Brody, L. L., Westbrock-Wadman, S., Bayer, A. S., Folger, K. R. & Stover, C. K. 1998 Identification and characterization of the PutP proline permease that contributes to *in vivo* survival of *Staphylococcus aureus* in animal models. *Infect. Immun.* **66**, 567–572.

Sharma, P., Lonnenborg, A. & Stougaard, J. J. 1993 PCR-based construction of subtractive cDNA library using magnetic beads. *BioTechniques* **15**, 610–611.

Shaw, P. J. 1995 Comparison of wide-field/deconvolution and confocal microscopy for 3D imaging. In *Handbook of biological confocal microscopy*, 2nd edn (ed. J. B. Pawley), pp. 373–387. New York: Plenum Press.

Sheppard, C. R. J. & Shotton, D. M. 1997 *Confocal laser scanning microscopy*. Oxford, UK: Bios Scientific.

Shotton, D. M. 1998 Image resolution and digital image processing in electron light microscopy. In *Cell biology, a laboratory handbook*, vol. 3, 2nd edn (ed. J. E. Celis), pp. 85–98. San Diego, CA: Academic Press.

Siemering, K. R., Golbik, R., Sever, R. & Haseloff, J. 1996 Mutations that suppress the thermosensitivity of green fluorescent protein. *Curr. Biol.* **6**, 1653–1663.

Silhavy, T. J. & Beckwith, J. R. 1985 Uses of *lac* fusions for the study of biological problems. *Microbiol. Rev.* **49**, 398–418.

Simons, E. R. 1999 Flow cytometry: use of multiparameter kinetics to evaluate several activation parameters simultaneously in individual living cells. In *Fluorescent and luminescent probes for biological activity. Biological Techniques Series*, 2nd edn (ed. W. T. Mason), pp. 527–539. San Diego, CA: Academic Press.

Slauch, J. M., Mahan, M. J. & Mekalanos, J. J. 1994 Assay of β-galactosidase using a fluorescent substrate. *Biotechniques* **16**, 642–644.

Stewart, G. S. A. B. & Williams, P. 1992 lux genes and the applications of bacterial bioluminescence. *J. Gen. Microbiol.* **138**, 1289–1300.

Suk, K., Das, S., Sun, W., Jwang, B., Barthold, S. W., Flavell, R. A. & Fikrig, E. 1995 *Borrelia burgdorferi* genes selectively expressed in the infected host. *Proc. Natl Acad. Sci. USA* **92**, 4269–4273.

Sykes, K. F. & Johnston, S. A. 1999 Linear expression elements: a rapid, *in vivo*, method to screen for gene functions. *Nature Biotech.* **17**, 355–359.

Tao, H., Bausch, C., Richmond, C., Blattner, F. R. & Conway, T. 1999 Functional genomics: expression analysis of *Escherichia coli* growing on minimal and rich media. *J. Bacteriol.* **181**, 6425–6440.

Tolker-Nielsen, T., Holmstrom, K. & Molin, S. 1997 Visual-isation of specific gene expression in individual *Salmonella typhimurium* cells by *in situ* PCR. *Appl. Environ. Microbiol.* **63**, 4196–4203.

Tolker-Nielsen, T., Holmstrom, K., Boe, L. & Molin, S. 1998 Non-genetic population heterogeneity studied by *in situ* polymerase chain reaction. *Mol. Microbiol.* **27**, 1099–1105.

Valdivia, R. H. & Falkow, S. 1997a Fluorescent-based isolation of bacterial genes expressed within host cells. *Science* **277**, 2007–2011.

Valdivia, R. H. & Falkow, S. 1997b Probing bacterial gene expression within host cells. *Trends Microbiol.* **5**, 360–363.

Valdivia, R. H. & Falkow, S. 1998 Flow cytometry and bacterial pathogenesis. *Curr. Opin. Microbiol.* **1**, 359–363.

Valdivia, R. H., Hromockyj, A. E., Monack, D., Ramakrishnan, L. & Falkow, S. 1996 Applications for green fluorescent protein (GFP) in the study of hostpathogen interactions. *Gene* **173**, 47–52.

Valdivia, R. H., Cormack, B. P. & Falkow, S. 1998 The uses of green fluorescent protein in prokaryotes. In GFP properties, applications and protocols (ed. M. Chalfie and S. Kain), pp. 121–138. New York: Wiley-Liss.

VanBogelen, R. A., Greis, K. D., Blumenthal, R. M., Tani, T. H. & Matthews, R. G. 1999 Mapping regulatory networks in microbial cells. *Trends Microbiol.* **7**, 320–328.

Ward, W. W. 1998 Biochemical and physical properties of green fluorescent protein. In *GFP: properties, applications and protocols* (ed. M. Chalfie and S. Kain), pp. 45–75. New York: Wiley-Liss.

Weiss, D. S., Chen, J. C., Ghigo, J.-M., Boyd, D. & Beckwith, J. 1999 Localization of FtsI (PBP3) to the septal ring requires its membrane anchor, the Z ring, FtsA, FtsQ and FtsL. *J. Bacteriol.* **181**, 508–520.

Winzeler, E. A. (and 51 others) 1999 Functional characterization of the *S. cerevisiae* genome by gene deletion and parallel analysis. *Science* **285**, 901–906.

Zhang, J. P. & Normark, S. 1996 Induction of gene expression in *Escherichia coli* after pilus-mediated adherence. *Science* **273**, 1234–1236.

Zhao, H., Thompson, R. B., Lockatell, V., Johnson, D. E. & Mobley, H. L. T. 1998 Use of green fluorescent protein to assess urease gene expression by uropathogenic *Proteus mirabilis* during experimental ascending tract infection. *Infect. Immun.* **66**, 330–335.

IDENTIFICATION AND ANALYSIS OF BACTERIAL VIRULENCE GENES *IN VIVO*

KATE E. UNSWORTH and DAVID W. HOLDEN*

Department of Infectious Diseases, Imperial College School of Medicine, Hammersmith Hospital, Du Cane Road, London W12 ONN, UK

We dedicate this paper to our friend and colleague Geoffrey Banks (1939–1999)

Signature-tagged mutagenesis is a mutation-based screening method for the identification of virulence genes of microbial pathogens. Genes isolated by this approach fall into three classes: those with known biochemical function, those of suspected function and some whose functions cannot be predicted from database searches. A variety of *in vitro* and *in vivo* methods are available to elucidate the function of genes of the second and third classes. We describe the use of some of these approaches to study the function of the *Salmonella* pathogenicity island 2 type III secretion system of *Salmonella typhimurium*. This virulence determinant is required for intracellular survival. Secretion by this system is induced by an acidic pH, and its function may be to alter trafficking of the *Salmonella*-containing vacuole. Use of a temperature-sensitive non-replicating plasmid and competitive index tests with other genes show that *in vivo* phenotypes do not always correspond to those predicted from *in vitro* studies.

Keywords: signature-tagged mutagenesis; SPI-2; type III secretion; competitive index

1. INTRODUCTION

The study of bacterial mutants *in vivo* can provide important insights into the diverse mechanisms of virulence employed by pathogenic organisms. Signature-tagged mutagenesis (STM) is one such approach, enabling the identification of virulence genes from a variety of pathogens. The original STM screen identified the *Salmonella* pathogenicity island 2 (SPI-2) type III secretion system (TTSS). The attenuated virulence of SPI-2 mutants has been the starting point for further analysis of the secretion system by our group and others. The function of SPI-2 was not obvious from its sequence, and various *in vitro* assays have been carried out to investigate the TTSS in greater detail. In this paper we discuss the results of those experiments and others that have addressed the role of SPI-2 *in vivo*.

---

*Author for correspondence (d.holden@ic.ac.uk).

2. SIGNATURE-TAGGED MUTAGENESIS

STM was developed as an *in vivo* genetic screen for the identification of microbial virulence genes. The intention was to allow a relatively rapid unbiased search for virulence genes using an animal host to select against strains carrying mutations in genes affecting virulence, among a mixed population of mutants. Searches have been carried out by screening individual bacterial mutants in an appropriate host (Miller *et al.* 1989; Bowe *et al.* 1998), but this is a relatively labour intensive and expensive approach if large numbers of strains are to be screened. By tagging each mutant with a different DNA 'signature', STM allows large numbers of different strains to be screened at the same time in the same animal host.

STM is a comparative hybridization technique that employs as mutagens a collection of transposons or other recombinogenic transforming DNA molecules, each modified by the incorporation of a different DNA sequence tag. The tags were originally designed as short DNA segments containing a 40 bp variable central region flanked by invariant 'arms' that facilitate the co-amplification and labelling of the central portions by polymerase chain reaction (PCR). When the tagged mutagens integrate into the genome of an organism, each individual mutant can in theory be distinguished from every other mutant based on the different tags carried by the strains.

In STM, mutagenized strains are stored individually in arrays (usually in the wells of microtitre dishes), and colony or dot blots are made from these arrays, or from plasmids carrying the tags or from the tags themselves. Pools of mutants are then inoculated into an appropriate animal host, and PCR is used to prepare labelled probes representing the tags present in the inoculum (input) and recovered from the host (output). Hybridization of the tags from the input and output pools to the colony or dot blots permits the identification of mutants that fail to grow *in vivo*, because the tags carried by these mutants will not be present in the output pools. These strains can then be recovered from the original arrays and the nucleotide sequence of DNA flanking the mutation site can be determined (Fig. 1).

The STM approach has a number of inherent limitations. First, it relies on the ability of the pathogen in question to replicate *in vivo* as a mixed population, and can only be expected to identify virulence genes whose mutant phenotypes cannot be transcomplemented by other virulent strains present in the same inoculum. The types of virulence genes

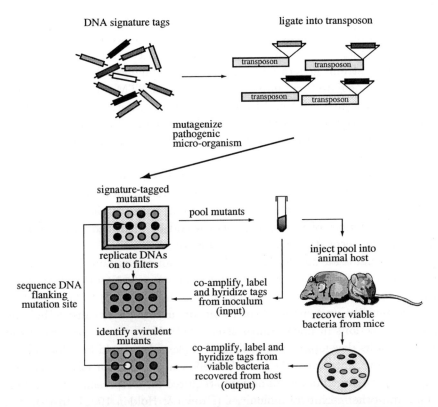

Fig. 1. Principles of STM. A transposon is modified by incorporation of DNA signature tags. A collection of different insertional mutants of a bacterial pathogen, each carrying a different tag, is assembled in a microtitre dish. The mutants are pooled and used as the inoculum for an appropriate animal model of infection. After a period of time in which virulent bacterial strains have multiplied, bacteria are recovered from the host. Signature tags representing strains in the inoculum and strains recovered from the animal are separately amplified using primers that anneal to invariant sequences flanking the tags, then labelled and used to probe membranes carrying DNA from the mutants in the microtitre dish. An avirulent mutant is identified by its failure to yield a signal on the membrane hybridized with the tags recovered from the animal (Tang & Holden 1999).

that could fail to be recovered as a result of mixed infection include those which encode secreted toxins, and factors which interfere with a systemic immune response (by altering cytokine production, for example). Second, STM usually involves an animal model of a human disease, and these models cannot be expected to faithfully reproduce all of the features of the bacterial–human interaction. Third, as with any genetic screen involving single mutations, any gene whose loss can be compensated for by another (genetic redundancy) is unlikely to be identified. Fourth, it is self evident that mutant strains generated by insertional mutagenesis have to be viable in order to inoculate them into an animal host, so STM (along with any insertional mutagenesis system) cannot provide any information about genes that are essential for life (with the proviso that insertions into 3'-ends of genes may result in truncated proteins which lose part but not all of an essential function).

Therefore, even if the mutant bank to be screened is large, and the genome of the pathogen is saturated with mutations, not all virulence genes of interest are likely to be recovered using STM; however, the hope was that a sufficient number of interesting genes would be identified to provide the basis for long-term studies into their functions.

The stimulus for developing STM came from work in our laboratory on *Aspergillus fumigatus*, and in particular the demonstration that two fungal products thought (on the basis of circumstantial evidence) to be important in pathogenicity, were in fact dispensable for this fungus to cause lethal pulmonary infections in mice (Smith *et al.* 1994). However, it was obvious at the time of its development that STM of *Aspergillus* would be a formidable task, not least because of the lack of an efficient method for insertional mutagenesis, the large number of mutants required for a representative mutant bank, and other technical difficulties (Brown & Holden 1998). Instead the *Salmonella typhimurium*–mouse interaction was chosen as a model system in which to assess the feasibility of the approach. This was for three reasons. First, excellent and straightforward molecular genetic methods were available for insertional mutagenesis using modified transposons (de Lorenzo & Timmis 1994). Second, the mouse model of systemic infection had been used extensively by many groups, is considered a good model of human typhoid fever and was known to support growth of a mixed population of different strains simultaneously (Mahan *et al.* 1993). Third, many virulence genes had already been identified in *S. typhimurium* (Groisman & Ochman 1994), and the successful re-isolation of these would help to validate the method.

In the original method, tags were incorporated into a mini Tn5 transposon and their suitability was checked prior to use by hybridization of amplified, labelled tags to DNA colony blots from the *S. typhimurium* mutants used to generate the probes. Mutants whose tags failed to yield clear signals on autoradiograms were discarded, and those that gave good signals were reassembled into new pools for screening in animals (Hensel *et al.* 1995). Pools of 96 strains were inoculated via the intraperitoneal route into mice. Viable bacteria were recovered from the spleens of animals three days post-inoculation, and the proportions of different mutants assessed by hybridization analysis of tags. Approximately 4% of the mutants tested were provisionally identified as attenuated in virulence. DNA regions flanking the transposon insertion sites of attenuated mutants were then recovered for sequencing either by restriction digestion and plasmid cloning, or by inverse PCR. The insertion points of 16 of the attenuated strains mapped to a 40 kb pathogenicity island at 30.7 centisomes on the chromosome (Hensel *et al.* 1995; Shea *et al.* 1996). The locus was named SPI-2 to distinguish it from the well-studied *Salmonella* pathogenicity island 1 (SPI-1) at 63 centisomes. It should be noted that SPI-2 genes were independently identified by a genome comparison approach (Ochman *et al.* 1996) and by differential fluorescence induction (Valdivia & Falkow 1997).

3.   VARIATIONS AND IMPROVEMENTS

The STM method was subsequently modified to avoid the pre-screening process (Mei *et al.* 1997). In this version, a series of tagged Tn*917* transposons were selected prior to mutagenesis of *Staphylococcus aureus* on the basis of efficient tag amplification and labelling, and lack of cross-hybridization to other tags. These modified transposons were then used separately to generate a large number of *S. aureus* mutants that were arrayed according to the tags they carry. As the same tags can be used to generate an infinite number of mutants, this obviates the need to pre-screen mutant strains for the suitability of tags. A second advantage is that since the identity of the tag in each mutant is known, hybridization analysis can be done using plasmid or tag DNA dot blots rather than colony blots. This increases the sensitivity of the assay and allows the use of non-radioactive detection methods (Mei *et al.* 1997). In a further modification of the technique, Cormack *et al.* (1999) first introduced each of 96 tags at a disrupted *URA3* locus of *Candida glabrata*. Then for each tagged strain, many different mutants were created by insertional mutagenesis, using a plasmid that simultaneously complemented the *URA3* mutation. Although in most cases the tags are not physically closely linked to the mutations, the DNA flanking the mutations could be cloned easily by plasmid rescue in *Escherichia coli*. STM or variations of it has now been applied successfully in several microbial pathogens, including *S. typhimurium* (Hensel *et al.* 1995), *S. aureus* (Mei *et al.* 1997; Coulter *et al.* 1998; Schwan *et al.* 1998), *Streptococcus pneumoniae* (Polissi *et al.* 1998), Vibrio cholerae (Chiang & Mekalanos 1998), Proteus mirabilis (Zhao *et al.* 1999), *Yersinia enterocolitica* (Darwin & Miller 1999), *Legionella pneumophila* (Edelstein *et al.* 1999) and *C. glabrata* (Cormack *et al.* 1999).

For successful application of STM, it is essential to consider a number of parameters in order to obtain reproducible hybridization patterns from different animals inoculated with the same pool of mutants. As the number of different mutant strains in a pool is increased, so is the probability that fully virulent mutants will not be recovered in sufficient numbers to yield hybridization signals, and this could result in false identification of attenuated mutants. When *S. typhimurium* was inoculated into the peritoneal cavities of mice, pools of 96 different mutants gave reproducible hybridization signals after three days of infection, whereas pools of 192 did not (Hensel *et al.* 1995). With *V. cholerae* in a colonization model in mice, it was necessary to reduce the complexity of the orally inoculated pool to 48 strains to give reproducible results (Chiang & Mekalanos 1998).

In some cases, it may be difficult to achieve reproducible output signals even if the inoculum dose is varied. At low doses, there may not be enough cells of any one virulent mutant to initiate a successful infection, but at high doses a failure to recover the same mutants from different animals probably reflects a bottleneck in the infection process that selects individual cells stochastically, and these then grow out as the infection proceeds. Attempting to solve this problem by further increasing the size of inoculum may overwhelm the animal's immune defences, resulting in rapid death of the host or the growth of mutant strains that would otherwise be attenuated.

The route of administration of the bacterial inoculum obviously plays a major role in determining the numbers of bacterial strains that reach the target organ(s) and tissues, and therefore the reproducibility of tag hybridization signals. For example, when 10^5 *S. typhimurium* cells representing a pool of 96 mutants were inoculated by the intraperitoneal route, reproducible hybridization signals were obtained for the vast majority of strains recovered from the spleens of infected animals. When the same inoculum was given orally, however, only a small percentage of mutants were subsequently found in the spleens, and the identity of these varied from animal to animal (D. W. Holden, unpublished data).

Another important aspect of the STM screening process is the time-point at which bacteria are recovered to prepare tags for hybridization analysis. If the incubation period is short, virulent cells may have had insufficient time to outgrow the attenuated strains to a degree that is reflected in a clear difference in hybridization signal intensity of tags on the blots. On the other hand, if the period is too long, then there may be a risk that some virulent strains may simply outgrow other virulent strains stochastically.

The parameters described above are obviously interrelated and must be optimized empirically for each pathogenhost interaction in order to obtain reproducible hybridization patterns using tags recovered from at least two animals infected with the same pool of mutants.

In the STM study of *S. aureus* by Mei *et al.* (1997), 1248 Tn*917* mutants were tested in a murine model of bacteraemia. The majority of loci from 50 mutants that were identified as attenuated were predicted by sequence similarity to be involved in cell surface metabolism (e.g. peptidoglycan cross-linking and transport functions), nutrient biosynthesis and cellular repair processes, but most of the remainder had no known function. A slightly larger signature-tagged mutant bank was constructed using the same transposon and tested in models of bacteraemia, abscess and wound formation and endocarditis (Coulter *et al.* 1998). This enabled the identification of various genes affecting growth and virulence in specific disease states, as well as 18 that are important in at least three of the infection models. Many of these genes appear to be involved in the same kinds of processes as those identified in the earlier study; indeed, seven of the genes identified by Mei *et al.* (1997) were also found by Coulter *et al.* (1998).

It is curious that whereas many of the genes identified by STM in Gram-negative bacteria are clearly 'virulence determinants' in the classical sense of the term, many of the *S. aureus* genes identified by STM appear to have more fundamental roles in bacterial metabolism. Genes for known virulence determinants, such as toxins, extracellular matrix-binding proteins and their regulators, such as *agr* (Lowy 1998), have not yet been identified by these screens (Mei *et al.* 1997; Coulter *et al.* 1998; Lowe *et al.* 1998). In the case of STM, transposon mutagenesis is not fully random and may have favoured mutation of certain areas of the chromosome over others. Second, mutation of genes encoding toxins may result in transcomplementable phenotypes. Third, the screens carried out to date are probably not saturating with respect to mutations. Fourth, the failure to identify genes for extracellular matrix-binding proteins for example, may be explained by genetic redundancy for these proteins (Greene *et al.* 1995). Also, STM may not identify mutations causing small reductions in survival.

4.   FURTHER ANALYSIS OF MUTANT STRAINS

When a mutant strain of interest has been identified, it is of course very important to obtain as much DNA sequence information around the site of the insertional mutation as is possible. With the rapid completion of many bacterial genome sequencing projects this task should become increasingly easy. It is essential to confirm that the tagged mutation is the cause of virulence attenuation, either by transduction of a transposon, recreation of the mutation by directed mutagenesis, outcrossing, and/or complementation with a functional allele, to show that the mutation and the virulence phenotype are linked. The DNA sequence might also indicate if the gene is part of an operon, and if so the phenotype could be due to a polar effect on a downstream gene. Directed mutagenesis of downstream genes is then necessary to establish which gene(s) are the cause of the attenuation.

Genes identified by an STM screen can be classified into three groups: first those whose products have a known biochemical function or whose function can be predicted on the basis of sequence similarities; second, genes with unknown specific function and relationship to virulence, but whose sequence gives some indication of functional class, such as an ATP-binding cassette transporter or a two-component regulatory system; and third, genes whose predicted amino-acid sequences give no clue to function. Several STM screens have identified genes of the first class. These are typically genes involved in biosynthesis of nutrients which are in limiting supply *in vivo*, such as amino acids and nucleotides (Mei *et al.* 1997). Knowledge that strains which lack the ability to synthesize a particular metabolite are attenuated in virulence is important because it can provide valuable information on the nutritional restrictions of different tissues *in vivo*. Furthermore, pathogens can sense the reduced availability of metals such as magnesium and iron in their extracellular environment and respond not only by synthesizing appropriate uptake systems but also by regulating other virulence factors (Garcia Vescovi *et al.* 1996; Dussurget & Smith 1998).

To study the functions of genes which fall into the second and third classes, mutant strains can be subjected to a range of physiological assays, ranging from the general to the specific. Knowledge that the strain has a growth defect *in vivo* is a useful starting point. The types of assays which are most appropriate will vary from system to system, but some general questions that can be asked are listed in Table 1. First the degree of virulence attenuation can be described in greater detail, using LD_{50} and the more sensitive and increasingly popular competitive index (CI) tests. *In vivo* studies can then be carried out with mutant strains (either as single or mixed infections with the wild-type parent strain) to describe the *in vivo* growth kinetics of the mutant and to identify the time and body site where the virulence defect is first apparent (Shea *et al.* 1999). The mutant strains can be tested for growth defects *in vitro*, in rich and minimal media, to establish if the phenotype is *in vivo* specific, and to identify an auxotrophy not apparent from the DNA sequence of the gene. Careful microscopic examination of the mutant strain may identify morphological abnormalities, such as a defect in motility or cell shape. A series of more specific physiological tests, using knowledge of the hostpathogen interaction, can also be carried out. Another approach that we have begun to use in our laboratory is the construction of double mutants and *in vivo* analysis of these to determine if the gene in question is functionally connected with another known virulence gene. An example of this approach is given below.

Table 1. Analysis of virulence genes of unknown function.

---

virulence attenuation (LD_{50} and CI)

growth kinetics *in vivo*

starvation survival and growth rates in different media (auxotrophy?)

morphological abnormalities

physiological assays exploiting knowledge of pathogenhost interaction stress resistance (osmotic, pH, reactive N and O inter-mediates, antimicrobial peptides, etc.) resistance to complement and neutrophil killing adherence to extracellular matrix proteins or host cells invasion of host cells (if intracellular)

genetics: study of interactions with other genes by construction and analysis of double mutants *in vitro* and *in vivo*

In addition to these tests, molecular studies can be carried out using the genes themselves, such as construction of fusions with reporters to learn more about the timing and control of their expression, and the generation of antibodies to localize the protein within the bacterial cell and infected tissue, and possibly for immunoprecipitation experiments if the protein is suspected to interact with other bacterial or host proteins.

The remainder of this paper discusses the functions of a group of genes which fall into the second class, encoding the SPI-2 TTSS, with particular emphasis on the use of *in vivo* studies to help elucidate their function and relationship to other virulence factors required for systemic pathogenesis.

5.   ORGANIZATION OF SPI-2 TTSS GENES

The discovery of the SPI-2 genes (Hensel *et al.* 1995) was surprising because it was already known that *S. typhimurium* has a TTSS (Inv/Spa) that is involved in epithelial cell entry. Indeed, *Salmonella enterica* appears to be unique among pathogenic bacteria in having two functionally distinct TTSS.

Comparison of the boundaries of SPI-2 with the corresponding region of the *E. coli* chromosome showed that SPI-2 is a 40 kb region of *Salmonella*-specific DNA inserted into a tRNA^{val} gene, which is occupied in *E. coli* by 9 kb of *E. coli*-specific sequence (Hensel *et al.* 1997a). Hybridization analysis has demonstrated the presence of SPI-2 in the majority of serovars of *S. enterica*, but sequences hybridizing to the TTSS are absent from the ancestral species *Salmonella bongori* (Ochman & Groisman 1996; Hensel *et al.* 1997a). This is in contrast to the Inv/Spa TTSS of SPI-1, which is present in both species of *Salmonella* and was therefore probably acquired prior to SPI-2. A GC content of 41.4%, which is considerably lower than the *Salmonella* genome average of 52%, suggests that SPI-2 was acquired by horizontal transfer from an unknown source. The pathogenicity island has a mosaic structure, with the TTSS encoded within the centisome 31 region. The centisome 30 region, which is not required for virulence, encodes genes involved in tetrathionate respiration (Hensel *et al.* 1999).

Four operons of SPI-2 appear to encode the secretion system; these have been designated 'regulatory', 'structural I', 'secretory' and 'structural II' (Hensel *et al.* 1997b; Cirillo *et al.* 1998). However, this nomenclature requires revision in the light of the recent finding that the *spiC* (also called *ssaB*) gene of the structural I operon encodes a secreted protein (Uchiya *et al.* 1999). Collectively, the products of 13 genes of the structural II operon are most similar to their homologues in the locus of enterocyte effacement of enteropathogenic *E. coli*, and are no more similar to the corresponding proteins from Inv/Spa than they are to those of the TTSS of the plant pathogen *Erwinia* (Elliot *et al.* 1998).

The principal proteins regulating SPI-2 gene expression form the two-component system SsrAB, and their genes comprise the regulatory operon. The products of *ssaL*, *ssaM* and *ssaP* of the structural II operon do not have significant similarity to products of other TTSS, and might be important for the specific function of the SPI-2 TTSS (Hensel *et al.* 1997*b*). The predicted protein SsaV is similar to members the LcrD family of inner membrane transport proteins (Galan *et al.* 1992; Hensel *et al.* 1997*b*), and probably forms a protein-conducting membrane channel of the secreton. The product of *ssaN* has similarity to YscN, a cytoplasmic protein that energizes the Yersinia TTSS through its ATPase activity (Woestyn *et al.* 1994). SsaC, encoded by the structural I operon, is similar to InvG (Ochman *et al.* 1996; Shea *et al.* 1996), a member of the PulD family of translocases (Kaniga *et al.* 1994), which is thought to form an outer membrane pore (Crago & Koronakis 1998). Most of the products of the structural I and II operons, together with some proteins encoded by the secretory operon, probably form a secreton that spans the inner and outer membranes of the bacterial cell.

The fourth, secretory operon encodes proteins classed as putative secreted proteins and their cytoplasmic chaperones. The type III secretion apparatus secretes different classes of proteins. The translocon, which is formed by YopB/D in *Yersinia* (Neyt & Cornelis 1999) and IpaB/C in *Shigella* (Blocker *et al.* 1999), forms a pore in the eukaryotic membrane through which further proteins are translocated into the target cell. In the case of SPI-2 this probably involves SseC, which is 24% identical and 48% similar to EspD of EPEC and YopB and has three predicted hydrophobic membrane-spanning domains (Hensel *et al.* 1998). Second, in some secretion systems a protein, such as EspA of EPEC, forms a link between the bacterium and target cell (Ebel *et al.* 1998; Knutton *et al.* 1998), and could act as a channel for the delivery of proteins into the host cell (Frankel *et al.* 1998). SseB of SPI-2 is 25% identical and 47% similar to EspA and its SPI-2-dependent secretion onto the bacterial cell surface has recently been demonstrated (Beuzon *et al.* 1999), making it a likely candidate for this role. Although the functions of several of the structural components of TTSS are conserved in a wide variety of bacteria, those of the effector molecules are quite different. For example the *Yersinia* effectors (YOPS) have a cytotoxic function, whereas the effectors of SPI-1 are involved in epithelial cell invasion (Hueck 1998). The distinction between effector and translocator may, however, be arbitrary: for example IpaC of *Shigella* is a component of the pore formed by the bacterium in the host cell membrane (Blocker *et al.* 1999), but is also involved in activating cytosolic actin polymerization (Tran Van Nhieu 1999). The sequence of SPI-2 does not reveal any candidate effectors such as kinases or phosphatases. However, a gene of the structural I operon, *spiC*, encodes a protein that is translocated by SPI-2 into host cells where it appears to interfere with vacuolar fusion events (Uchiya *et al.* 1999). With the exception of SpiC no other effector proteins of SPI-2 have been identified. It is possible that, as is the case for Inv/Spa, further effector proteins may be encoded outside the pathogenicity island.

Chaperone proteins of the TTSS bind and stabilize secreted proteins in the bacterial cytoplasm. They are small, usually with an acidic pI, and are often encoded upstream of the gene encoding their target. Two predicted proteins of SPI-2 are similar to chaperones of other TTSS: SscA, which is 26% identical to LcrH of *Yersinia pestis*, and SscB, which is 23% identical to IppI of *Shigella* (Hensel *et al.* 1998).

6.   REGULATION OF GENE EXPRESSION

Several environmental factors affect SPI-2 gene expression *in vitro*. Magnesium and phosphate starvation leads to transcriptional activation of a *luc* fusion to *sseA*, the first gene of the secretory operon (Deiwick *et al.* 1999). However, the effect is dependent on growth phase, with maximal activity occurring several hours after these ions are removed. It is therefore likely that phosphate and magnesium starvation are not the expression stimuli *per se* and that other factors present at stationary phase are required for SPI-2 gene expression.

Expression of genes within the TTSS is dependent upon the two-component regulatory system SsrAB. Transcription of an *sseA::luc* fusion was reduced 250–300 times in an *ssrA⁻* or *ssrB⁻* mutant background when bacteria were grown in minimal media (Deiwick *et al.* 1999). The dependence of SPI-2 transcription upon *ssrAB* expression has been confirmed within host cells; transcription of an *ssaH::gfp* fusion identified by its 400-fold intracellular induction was not induced within macrophages in strains carrying *ssrA* disruptions (Valdivia & Falkow 1997). Expression of *ssrA* is induced eightfold within mammalian cells soon after entry, and reaches a maximum after 2 h. Although *ssrAB* is able to induce its own expression, it is also affected by other unidentified regulators, because an *ssrA::gfp* fusion was induced fourfold in an *ssrA⁻* background. Treatment of macrophages with the vacuolar proton ATPase inhibitor bafilomycin A resulted in two- to threefold reduced induction of *ssrAB* expression. An acidic vacuolar pH is therefore required for optimal SPI-2 gene expression in host cells (Cirillo *et al.* 1998).

Although *ssaH* was identified as a macrophage-activated gene, further experiments showed that expression of the *gfp* reporter was induced in a variety of cell types. The *in vivo* relevance of these *in vitro* studies has been confirmed by demonstration of SPI-2 gene transcription in mouse spleno-cytes and hepatocytes three days post-inoculation (Cirillo *et al.* 1998).

7. FUNCTIONAL ANALYSIS OF THE SPI-2 TTSS

SPI-2 TTSS mutant strains are attenuated by several orders of magnitude when administered orally, intraperitoneally or intravenously to Ity[S] or Ity[R] mice (Shea *et al.* 1996, 1999). This makes SPI-2 mutants among the most attenuated that have been described in *S. typhimurium*. The attenuation of SPI-2 mutants after intraperitoneal inoculation indicates that the role of the secretion system is at a stage of pathogenesis subsequent to epithelial cell entry. This is in contrast to SPI-1 mutants, which are defective in invasion of the gut epithelium and are thus only attenuated after oral inoculation. After oral infection, SPI-2 mutant strains reach the Peyer's patches but do not persist there and cannot reach the mesenteric lymph nodes, spleen or liver in significant numbers (Cirillo *et al.* 1998). The notion that SPI-2 is important for the establishment or maintenance of systemic infection is supported by the recent work of Tsolis *et al.* (1999), who found that, whereas SPI-1 mutants are unable to cause disease in the calf diarrhoea model, SPI-2 mutants are not significantly attenuated.

Although SPI-2 mutant strains have no obvious morphological defects and have similar growth rates to wild-type strains in minimal and rich media, work from several laboratories has established that they fail to accumulate in a variety of host cells, including RAW and J774 macrophages (Cirillo *et al.* 1998; Hensel *et al.* 1998; Ochman *et al.* 1996; Uchiya *et al.* 1999), elicited peritoneal macrophages (Hensel *et al.* 1998), and Hep-2 and HeLa epithelial cells (Cirillo *et al.* 1998). SPI-2 mutants are recovered in lower numbers than wild-type from the mesenteric lymph nodes, liver and spleen during systemic infection (Cirillo *et al.* 1998; Shea *et al.* 1999). However, since the primary sites of *S. typhimurium* replication *in vivo* are splenic macrophages and liver Kupffer cells (Richter-Dahlfors *et al.* 1997),

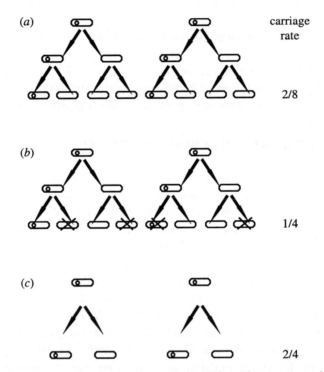

Fig. 2. Use of a temperature-sensitive non-replicating plasmid to measure relative growth and killing rates of *S. typhimurium* in the mouse. (*a*) The dilution of the plasmid in the wild-type population is proportional to the number of rounds of bacterial cell division. (*b*) If a mutant strain has increased sensitivity to host killing mechanisms there will be lower bacterial numbers and fewer plasmid-carrying cells than for the wild-type strain. (*c*) If a mutant strain has a decreased ability to divide normally in the host, there will be lower bacterial numbers but the same number of plasmid-carrying mutant cells as in the wild-type strain. Adapted from Gulig & Doyle (1993).

these are most probably the cells in which the SPI-2 secretion system mediates its effect *in vivo*. *In vivo* time-course experiments in BALB/c mice reveal an approximate tenfold increase in the numbers of SPI-2 mutants between 4 and 24 h following intraperitoneal inoculation. The bacterial load then remains relatively static for several days before being cleared. This failure of mutant strains to accumulate further could be the result of either a reduced ability of the bacteria to replicate or an increased susceptibility to the bactericidal defences of the host, or both. It is possible to distinguish between these two possibilities *in vivo* by measuring the dilution of a temperature-sensitive plasmid (pHSG422) within the bacterial population during infection (Benjamin *et al.* 1990; Gulig & Doyle 1993). This plasmid cannot replicate at 37°C and becomes diluted in the population as the bacteria divide. The total number of bacteria and the proportion of these carrying the plasmid are counted by plating the inoculum and spleen homogenates from infected mice with the appropriate antibiotics. If the difference between the total numbers of wild-type and mutant strains is due to an increased susceptibility of the mutants to killing, the number of mutant cells carrying the plasmid will be lower than the number of plasmid-carrying wild-type cells. If, however, there is no significant difference in sensitivity to killing but the mutants are unable to replicate efficiently, the numbers of each strain carrying the plasmid should be approximately equal (Fig. 2). After a 16 h infection of wild-type and SPI-2 mutant strains in mice, the numbers of each strain carrying the non-replicating plasmid were approximately equal in spleen homogenates, although there were in total tenfold more wild-type than mutant cells (Shea *et al.* 1999). This indicates that SPI-2 enables growth of the bacterial population rather than conferring resistance to host killing mechanisms. It also showed that by 16 h, wild-type cells of the inoculum had undergone five or six divisions whereas the SPI-2 mutant cells had undergone approximately half as many divisions. Recent work suggests that the failure of mutants to accumulate intracellularly is connected with an inability to modify trafficking of the *S. typhimurium* phagosome (Uchiya *et al.* 1999).

A characteristic of *S. typhimurium* growth in host cells is its ability to replicate in a specialized phagosome, the *Salmonella*-containing vacuole (SCV), which avoids interacting with lysosomal compartments. SPI-2 could secrete and translocate factors into the cytosol to alter the trafficking pathways of the host cell (Fig. 3). Uchiya *et al.* (1999) found that the

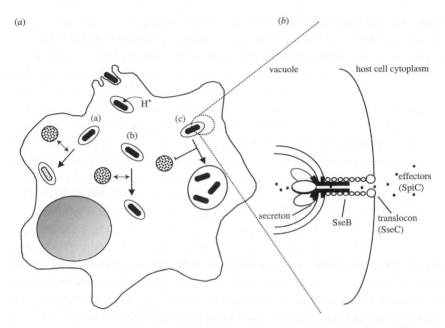

Fig. 3. (*a*) Putative trafficking of *S. typhimurium* within the macrophage. After entry, the SCV undergoes acidification (Rathman *et al.* 1996). (a) Phagosomes containing heat-killed bacteria interact with lysosomal compartments and are ultimately degraded (Rathman *et al.* 1997). (b) SCVs containing SPI-2 mutant strains interact with lysosomal compartments (Uchiya *et al.* 1999). *In vivo* data suggest that this does not result in an increased rate of killing but the bacteria are unable to replicate to the levels of wild-type cells (Shea *et al.* 1999). (c) A proportion of wild-type *S. typhimurium* SCVs do not interact with lysosomal compartments (Rathman *et al.* 1997). They form a specialized vacuole in which bacterial replication occurs. (*b*) Putative structure of the SPI-2 TTSS. A link between the bacterium and host vacuolar membrane is formed by SseB. SseC forms a pore in the vacuolar membrane for translocation of effectors, including SpiC (Uchiya *et al.* 1999), into the host cell cytoplasm.

SPI-2-encoded protein SpiC is secreted into the cytosol of J774 macrophages 1–6 h after bacterial uptake, but only by strains of *S. typhimurium* with a functional SPI-2 secretion system. Infection with wild-type *S. typhimurium* reduces the overall rate of phagosomelysosome fusion within the cell, as demonstrated by co-localization of bovine serum albumin (BSA)-gold with lysosomal markers. *spiC* mutants do not have this characteristic, and trafficking of BSA-gold in macrophages infected with these strains is similar to that of uninfected J774 cells.

If the function of SPI-2 is to restrict interactions between the SCV and toxic lysosomal compartments, we might expect to see increased killing of SPI-2 mutants compared with wild-type cells *in vivo*. However, the experiments with the non-replicating plasmid *in vivo* described above show that there is little difference in the degree of killing sustained by the two strains and that the defect of SPI-2 mutants is related to their inability to divide at a normal rate. Perhaps SpiC is therefore involved in altering SCV trafficking which enables intracellular *S. typhimurium* to obtain nutrients or other factors necessary for growth and replication inside the vacuole, rather than avoiding killing by lysosomal enzymes or other host cell mechanisms.

8.   REGULATION OF SECRETION

The SCV undergoes acidification to pH 4.0–5.0 within 1 h after bacterial uptake (Rathman *et al.* 1996). Beuzon *et al.* (1999) investigated whether secretion of a SPI-2 protein *in vitro* could be induced by growth in media simulating the conditions within the vacuole. The secreted protein studied, SseB, is similar to EspA of EPEC, which is a component of a filamentous surface appendage forming a contact between the bacterium and the eukaryotic cell (Ebel *et al.* 1998; Knutton *et al.* 1998). SseB is secreted onto the surface of the cell, from which it can be removed by extraction with hydrophobic agents that allow recovery of proteins weakly associated with the bacterial cell surface but do not remove known membrane proteins. This secretion was shown to be dependent on a functional TTSS, and occurred only at pH 4.5–5.0 after growth to stationary phase in minimal media. The induction of secretion was independent of expression, as there was no difference in intracellular levels of SseB at pH 5.0 and 7.5, and SseB was not secreted at pH 7.5 when expressed from a constitutive promoter (Beuzon *et al.* 1999). In *Yersinia*, regulation of secretion involves the protein YopN, which appears to act as a plug blocking the secretion pore until it is itself secreted. SPI-2 does not contain a homologue of yopN but some further control mechanism must exist because SseB is expressed but not secreted in minimal media at neutral pH, and is secreted shortly after a shift to acidic media, indicating that the secreton is probably assembled before the pH shift.

Little is known about the regulation and timing of secretion by SPI-2 *in vivo*. However, SpiC was detected inside infected macrophages within 1 h of entry (Uchiya *et al.* 1999). This is difficult to reconcile with the findings of Cirillo *et al.* (1998), who showed that expression of the SPI-2 regulatory and secretory apparatus genes does not increase greatly until 2 h post-uptake, and reaches a maximum after 6 h. Although SseB is immediately secreted onto the cell surface after acidification of the culture media, this is presumably conditional on the presence of a preformed secretion apparatus. Expression of SPI-2 genes *in vitro* requires low magnesium and phosphate and occurs as cells enter the stationary phase of growth. However, it is not known at what stage of *in vivo* infection the secretion apparatus is assembled. Further work is therefore needed to determine the kinetics of SPI-2 secretion within host cells *in vivo*.

9.   RELATIONSHIPS BETWEEN VIRULENCE FACTORS *IN VIVO*

In a mixed infection of BALB/c mice, the ability of two *S. typhimurium* strains to colonize the spleen provides a measure of their relative virulence. The CI, defined as the output ratio of mutant to wild-type bacteria divided by the input ratio of mutant to wild-type bacteria, is a quantitative value for the degree of attenuation of a mutant strain, with the CI of a wild-type strain versus a fully virulent derivative being approximately 1.0.

A further development of this method in our laboratory is its use to distinguish whether two genes have the same biochemical function *in vivo*. This is achieved by creating a double mutant strain, which is combined in a mixed inoculum with a strain carrying only one of the mutations. The attenuation caused by the mutation present in both strains will be equivalent, so the CI calculated from this experiment will reflect the effect of the second mutation in the absence of the first gene. If the functions of the two genes are totally independent, the CI will be equal to the CI of a strain carrying only the second mutation versus the wild-type strain, and the CI of the double mutant against the wild-type strain should reflect the effects of the two mutations in an additive manner. If the two genes are involved in the same biochemical function, no additional attenuation should be conferred on a single mutant strain by the acquisition of the second mutation, and the predicted CI of the double mutant versus the single will be approximately 1.0. In a situation where the functions of the two genes partially overlap the CI may be intermediate. However, because the function of one virulence gene may be dependent on the function of another biochemically unrelated gene, it is difficult to interpret a phenotype that is not completely additive. The lack of further attenuation by the presence of a second mutation may not necessarily mean that the functions of two genes are directly related, but simply that the attenuation caused by one mutation prevents the infection process reaching the stage when the second gene acts. Manipulation of the animal model (by varying the inoculum dose or route of infection, for example) may allow the effects of the second mutation to be observed. If introduction of a second mutation increases attenuation in a completely additive fashion, it can be concluded that the genes in question are not related *in vivo*.

It has been suggested that the *spv* genes, encoded on the *S. typhimurium* virulence plasmid, may be related to SPI-2 because of the striking similarities of their mutant phenotypes (Shea *et al.* 1999). SPI-2 and *spv* mutant strains have a similar level of attenuation in the mouse, which cannot be transcomplemented *in vivo* by the presence of virulent strains. Both sets of genes are transcribed within host cells, and mutants are defective for intracellular survival, the defect being related to bacterial growth

Table 2. CI values for mixed infections of *S. typhimurium* strains in mice. (Mice were inoculated intraperitoneally with approximately 10^5 colony-forming units (CFU) of a mixed inoculum. Mouse spleens were harvested after 48 h for enumeration of bacterial CFU. The strains used were differentiated on the basis of antibiotic sensitivity. The CI was calculated as the output ratio of strain 2 to strain 1 divided by the input ratio of strain 2 to strain 1. Adapted from Shea *et al.* (1999).)

Strain 1	Strain 2	CI
12023	$ssaV^-$	0.058
$ssaV^-$	$ssaV^-$	1.14
$ssaV^-$	$ssaV^-$, $sseC^-$	1.28[a]
12023	$purD^-$	0.005
$ssaV^-$	$ssaV^-$, $purD^-$	0.008[b]
12023	$spvA^-$	0.068
$ssaV^-$	$ssaV^-$, $spvA^-$	0.052[b]

[a]Not significantly different ($p > 0.5$) from the CI obtained for the mixed infection $ssaV^-$ versus $ssaV^-$.
[b]Significantly different ($p > 0.05$) from the CI obtained for the mixed infection $ssaV^-$ versus $ssaV^-$.

rate, not an increased susceptibility to killing. However, the ability of an $ssaV^-$, $spvA^-$ strain to colonize the spleen in competition with an $ssaV^-$ strain was not statistically different from the ability of an $spvA^-$ strain to colonize the spleen in competition with a wild-type strain (Table 2) (Shea *et al.* 1999). The attenuation conferred by these two mutations is therefore entirely additive, indicating that they most probably have completely independent functions.

Many aspects of pathogenesis and survival in the host cannot be simulated in a test tube or cell culture model, and artificial factors that do not exist in a natural infection may complicate the interpretation of *in vitro* results. As a result of these limitations, a mutant phenotype observed in an *in vitro* assay does not necessarily mean that this is relevant *in vivo*. Where different *in vitro* techniques have produced conflicting results, it is especially important to try to obtain *in vivo* evidence to clarify the situation. A case in point is the contribution of the wide domain regulator PhoPQ to regulation of SPI-2 gene expression. An *ssaH::gfp* fusion did not require a functional *phoP* gene for intramacrophage induction (Valdivia & Falkow 1997). Furthermore, Cirillo *et al.* (1998) found that intramacrophage SPI-2 gene expression was reduced by bafilomycin, whereas PhoPQ-dependent expression is not affected by this drug (Rathman 1996), and it was concluded from this that SPI-2 expression is independent of PhoPQ. However, Deiwick *et al.* (1999) reported that levels of *in vitro* SscA expression were greatly reduced in a *phoP^-* strain. In our laboratory, *in vivo* competition experiments involving a *phoPQ* mutant and a *phoPQ*, SPI-2 double mutant have failed to provide any evidence that these two systems interact in the host, because the CI of this combination was similar to that of a SPI-2 mutant against the wild-type strain, and the CI of the double mutant against the wild-type strain reflected a completely additive effect on attenuation of the two mutations (C. Beuzon, unpublished data).

10. CONCLUDING REMARKS

A variety of genetic approaches are now available to study the behaviour of bacterial pathogens *in vivo*. STM allows a mutation-based screen for genes that contribute to pathogenesis, which has been applied successfully to a variety of pathogens including *S. typhimurium*. Where the functions of genes identified by this and other approaches are not clear from analysis of their nucleotide sequence, further analysis of the mutant strains *in vitro*, and the use of the CI test and plasmid segregation studies, can provide important information about the functions and interactions of these genes *in vivo*. It is worth noting that, in the case of *S. typhimurium* SPI-2 TTSS, the results of some *in vitro* studies are not in accord with the phenotypes of mutant strains *in vivo*, and should therefore be interpreted with caution, until more sophisticated methods are available to study gene function *in vivo*.

Work in the laboratory of D.W.H. was supported by the UK Medical Research Council. We thank C. Beuzon, S. Garvis and C. Tang for critical reviews.

REFERENCES

Benjamin Jr, W. H., Hall, P., Roberts, S. J. & Briles, D. E. 1990 The primary effect of the *ity* locus is on the rate of growth of *Salmonella typhimurium* that are relatively protected from killing. *J. Immunol.* **144**, 3143–3151.

Beuzon, C. R., Banks, G., Deiwick, J., Hensel, M. & Holden, D. W. 1999 pH-dependent secretion of SseB, a product of the SPI-2 type III secretion system of *Salmonella typhimurium*. *Mol. Microbiol.* **33**, 806–816.

Blocker, A., Gounon, P., Larquet, E., Niebuhr, K., Cabiaux, V., Parsot, C. & Sansonetti, P. 1999 The tripartite type III secreton of *Shigella flexneri* inserts IpaB and IpaC into host membranes. *J. Cell Biol.* **147**, 683–693.

Bowe, F., Lipps, C. J., Tsolis, R. M., Groisman, E., Heffron, F. & Kusters, J. G. 1998 At least four percent of the *Salmonella typhimurium* genome is required for fatal infection of mice. *Infect. Immun.* **66**, 3372–3377.

Brown, J. S. & Holden, D. W. 1998 Insertional mutagenesis of pathogenic fungi. *Curr. Opin. Microbiol.* **1**, 390–394.

Chiang, S. L. & Mekalanos, J. J. 1998 Use of signature-tagged transposon mutagenesis to identify *Vibrio cholerae* genes critical for colonization. *Mol. Microbiol.* **27**, 797–806.

Cirillo, D. M., Valdivia, R. H., Monack, D. M. & Falkow, S. 1998 Macrophage-dependent induction of the *Salmonella* pathogenicity island 2 type III secretion system and its role in intracellular survival. *Mol. Microbiol.* **30**, 175–188.

Cormack, B. P., Ghori, N. & Falkow, S. 1999 An adhesin of the yeast pathogen *Candida glabrata* mediating adherence to human epithelial cells. *Science* **285**, 578–582.

Coulter, S. N., Schwan, W. R., Ng, E. Y. W., Langhorne, M. H., Ritchie, H. D., Westbrock-Wadman, S., Hufnagle, W. O., Folger, K. R., Bayer, A. S. & Stover, C. K. 1998 *Staphylococcus aureus* genetic loci impacting growth and survival in multiple infection environments. *Mol. Microbiol.* **30**, 393–404.

Crago, A. M. & Koronakis, V. 1998 *Salmonella* InvG forms a ring-like multimer that requires the InvH lipoprotein for outer membrane localization. *Mol. Microbiol.* **23**, 861–867.

Darwin, A. J. & Miller, V. L. 1999 Identification of *Yersinia enterocolitica* genes affecting survival in an animal host using signature-tagged transposon mutagenesis. *Mol. Microbiol.* **32**, 51–62.

Deiwick, J., Niklaus, T., Erdogan, S. & Hensel, M. 1999 Environmental regulation of Salmonella pathogenicity island 2 gene expression. *Mol. Microbiol.* **31**, 1759–1773.

de Lorenzo, V. & Timmis, K. N. 1994 Analysis and construction of stable phenotypes in Gram-negative bacteria with Tn5- and Tn10-derived minitransposons. *Meth. Enzymol.* **235**, 386–405.

Dussurget, O. & Smith, I. 1998 Interdependence of mycobacterial iron regulation, oxidative-stress response and isoniazid resistance. *Trends Microbiol.* **6**, 354–358.

Ebel, F., Podzadel, T., Rohde, M., Kresse, A. U., Kramer, S., Deibel, C., Guzman, C. A. & Chakraborty, T. 1998 Initial binding of Shiga toxin-producing *Escherichia coli* to host cells and subsequent induction of actin rearrangements depend on filamentous EspA-containing surface appendages. *Mol. Microbiol.* **30**, 147–161.

Edelstein, P. H., Edelstein, M. A. C., Higa, F. & Falkow, S. 1999 Discovery of virulence genes of *Legionella pneumophila* by using signature-tagged mutagenesis in a guinea pig pneumonia model. *Proc. Natl Acad. Sci. USA* **96**, 8190–8195.

Elliot, S. J., Wainwright, L. A., McDaniel, T. K., Jarvis, K. G., Deng, Y., Lai, L.-C., McNamara, B. P., Donnenberg, M. S. & Kaper, J. B. 1998 The complete sequence of the locus of enterocyte effacement (LEE) from enteropathogenic *Escherichia coli* E2348/69. *Mol. Microbiol.* **28**, 1–4.

Frankel, G., Phillips, A. D., Rosenshine, I., Dougan, G., Kaper, J. B. & Knutton, S. 1998 Enteropathogenic and enterohaemorrhagic *Escherichia coli*: more subversive elements. *Mol. Microbiol.* **30**, 911–921.

Galan, J. E., Ginocchio, C. & Costeas, P. 1992 Molecular and functional characterisation of the *Salmonella* invasion gene *invA*: homology of InvA to members of a new protein family. *J. Bacteriol.* **174**, 4338–4349.

Garcia Vescovi, E., Soncini, F. C. & Groisman, E. A. 1996 Mg^{2+} as an extracellular signal: environmental regulation of *Salmonella* virulence. *Cell* **84**, 165–174.

Greene, C., McDevitt, D., Francois, P., Vaudaux, P. E., Lew, D. P. & Foster, T. J. 1995 Adhesion properties of mutants of *Staphylococcus* aureus defective in fibronectin-binding proteins and studies on the expression of *fnb* genes. *Mol. Microbiol.* **17**, 1143–1152.

Groisman, E. A. & Ochman, H. 1994 How *Salmonella* became a pathogen. *Trends Microbiol.* **5**, 343–349.

Gulig, P. A. & Doyle, T. J. 1993 The *Salmonella typhimurium* virulence plasmid increases the growth rate of salmonellae in mice. *Infect. Immun.* **61**, 504–511.

Hensel, M., Shea, J. E., Gleeson, C., Jones, M. D., Dalton, E. & Holden, D. W. 1995 Simultaneous identification of bacterial virulence genes by negative selection. *Science* **269**, 400–403.

Hensel, M., Shea, J. E., Bumler, A. J., Gleeson, C., Blattner, F. & Holden, D. W. 1997a Analysis of the boundaries of *Salmonella* pathogenicity island 2 and the corresponding chromosomal region of *Escherichia coli* K-12. *J. Bacteriol.* **179**, 1105–1111.

Hensel, M., Shea, J. E., Raupach, B., Monack, D., Falkow, S., Gleeson, C., Kubo, T. & Holden, D. W. 1997b Functional analysis of *ssaJ* and the *ssaK/U* operon, thirteen genes encoding components of the type III secretion apparatus of *Salmonella* pathogenicity island 2. *Mol. Microbiol.* **24**, 155–167.

Hensel, M., Shea, J. E., Waterman, S. R., Mundy, R., Nikolaus, T., Banks, G., Vasquez-Torres, A., Gleeson, C., Fang, F. C. & Holden, D. W. 1998 Genes encoding putative effector proteins of the type III secretion system of *Salmonella* pathogenicity island 2 are required for bacterial virulence and proliferation in macrophages. *Mol. Microbiol.* **30**, 163–174.

Hensel, M., Hinsley, A. P., Nikolaus, T., Sawers, G. & Berks, B. C. 1999 The genetic basis of tetrathionate respiration in *Salmonella typhimurium*. *Mol. Microbiol.* **32**, 275–287.

Hueck, C. J. 1998 Type III protein secretion systems in bacterial pathogens of animals and plants. *Microbiol. Mol. Biol. Rev.* **62**, 379–433.

Kaniga, K., Bossio, J. C. & Galan, J. E. 1994 The *Salmonella typhimurium* invasion genes *invF* and *invG* encode homologs of the AraC and PulD family of proteins. *Mol. Microbiol.* **13**, 555–568.

Knutton, S., Rosenshine, I., Pallen, M. J., Nisan, I., Neves, B. C., Bain, C., Wolff, C., Dougan, G. & Frankel, G. 1998 A novel EspA-associated surface organelle of enteropathogenic *Escherichia coli* involved in protein translocation into epithelial cells. *EMBO J.* **17**, 2166–2176.

Lowe, A. M., Beattie, D. T. & Deresiewicz, R. L. 1998 Identification of novel staphylococcal virulence genes by *in vivo* expression technology. *Mol. Microbiol.* **27**, 967–976.

Lowy, F. D. 1998 *Staphylococcus* aureus infections. *N. Engl. J. Med.* **339**, 520–532.

Mahan, M. J., Slauch, J. M. & Mekalanos, J. J. 1993 Selection of bacterial virulence genes that are specifically induced in host tissues. *Science* **259**, 686–688.

Mei, J.-M., Nourbakhsh, F., Ford, C. W. & Holden, D. W. 1997 Identification of *Staphylococcus aureus* virulence genes in a murine model of bacteremia using signature-tagged mutagenesis. *Mol. Microbiol.* **26**, 399–407.

Miller, I., Maskell, D., Hormaeche, C., Johnson, K., Pickard, D. & Dougan, G. 1989 Isolation of orally attenuated *Salmonella typhimurium* following Tn*phoA* mutagenesis. *Infect. Immun.* **57**, 2758–2763.

Neyt, C. & Cornelis, G. R. 1999 Insertion of a Yop translocation pore into the macrophage plasma membrane by *Yersinia enterocolitica*: requirement for translocators YopB and YopD, but not LcrG. *Mol. Microbiol.* **33**, 971–981.

Ochman, H. & Groisman, E. A. 1996 Distribution of pathogenicity islands in *Salmonella* spp. *Infect. Immum.* **64**, 5410–5412.

Ochman, H., Soncini, F. C., Solomon, F. & Groisman, E. A. 1996 Identification of a pathogenicity island required for *Salmonella* survival in host cells. *Proc. Natl Acad. Sci. USA* **93**, 7800–7804.

Polissi, A., Pontiggia, A., Feger, G., Altieri, M., Mottl, H., Ferrari, L. & Simon, D. 1998 Large-scale identification of virulence genes from *Streptococcus pneumoniae*. *Infect. Immun.* **66**, 5620–5629.

Rathman, M., Sjaastad, M. D. & Falkow, S. 1996 Acidification of phagosomes containing *Salmonella typhimurium* in murine macrophages. *Infect. Immun.* **64**, 2765–2773.

Rathman, M., Barker, L. P. & Falkow, S. 1997 The unique trafficking pattern of *Salmonella typhimurium*-containing phagosomes in murine macrophages is independent of the mechanism of bacterial entry. *Infect. Immun.* **65**, 1475–1485.

Richter-Dahlfors, A., Buchan, A. M. J. & Finlay, B. B. 1997 Murine salmonellosis studied by confocal microscopy: *Salmonella typhimurium* resides intracellularly inside macrophages and exerts a cytotoxic effect on phagocytes *in vivo*. *J. Exp. Med.* **184**, 569–580.

Schwan, W. R., Coulter, S. N., Ng, E. Y. W., Langhorne, M. H., Ritchie, H. D., Brody, L. L., Westbrock-Wadman, S., Bayer, A. S., Folger, K. R. & Stover, C. K. 1998 Identification and characterization of the PutP proline permease that contributes to *in vivo* survival of *Staphylococcus aureus* in animal models. *Infect. Immun.* **66**, 567–572.

Shea, J. E., Hensel, M., Gleeson, C. & Holden, D. W. 1996 Identification of a virulence locus encoding a second type III secretion system in *Salmonella typhimurium*. *Proc. Natl Acad. Sci. USA* **93**, 2593–2597.

Shea, J. E., Beuzon, C. R., Gleeson, C., Mundy, R. & Holden, D. W. 1999 Influence of the *Salmonella typhimurium* pathogenicity island 2 type III secretion system on bacterial growth in the mouse. *Infect. Immun.* **67**, 213–219.

Smith, J. M., Tang, C. M., Van Noorden, S. & Holden, D. W. 1994 Virulence of *Aspergillus fumigatus* double mutants lacking restrictocin and an alkaline protease in a low-dose model of invasive pulmonary aspergillosis. *Infect. Immun.* **62**, 5247–5254.

Tang, C. & Holden, D. W. 1999 Pathogen virulence genesimplications for vaccines and drug therapy. *Br. Med. Bull.* **55**, 387–400.

Tran Van Nhieu, G., Caron, E., Hall, A. & Sansonetti, P. 1999 IpaC induces actin polymerisation and filopodia formation during *Shigella* entry into epithelial cells. *EMBO J.* **18**, 3249–3262.

Tsolis, R. M., Adams, L. G., Ficht, T. A. & Bumler, A. J. 1999 Contribution of *Salmonella typhimurium* virulence factors to diarrheal disease in calves. *Infect. Immun.* **67**, 4879–4885.

Uchiya, K., Barbieri, M. A., Funato, K., Shah, A. H., Stahl, P. D. & Groisman, E. A. 1999 A *Salmonella* virulence protein that inhibits cellular trafficking. *EMBO J.* **18**, 3924–3933.

Valdivia, R. H. & Falkow, S. 1997 Fluorescence-based isolation of bacterial genes expressed within host cells. *Science* **277**, 2007–2011.

Woestyn, S., Allaoui, A., Wattiau, P. & Cornelis, G. R. 1994 YscN, the putative energiser of the *Yersinia* Yop secretion machinery. *J. Bacteriol.* **176**, 1561–1569.

Zhao, H., Li, X., Johnson, D. E. & Mobley, H. L. 1999 Identification of protease and *rpoN*-associated genes of uropathogenic *Proteus mirabilis* by negative selection in a mouse model of ascending urinary tract infection. *Microbiology* **145**, 185–195.

SALMONELLA INTERACTIONS WITH HOST CELLS: *IN VITRO* TO *IN VIVO*

B. BRETT FINLAY* and JOHN H. BRUMELL

Biotechnology Laboratory, University of British Columbia, Room 237, Wesbrook Building, 6174 University Boulevard, Vancouver, British Columbia, Canada V6T 1Z3

Salmonellosis (diseases caused by *Salmonella* species) have several clinical manifestations, ranging from gastroenteritis (food poisoning) to typhoid (enteric) fever and bacteraemia. *Salmonella* species (especially *Salmonella typhimurium*) also represent organisms that can be readily used to investigate the complex interplay that occurs between a pathogen and its host, both *in vitro* and *in vivo*. The ease with which *S. typhimurium* can be cultivated and genetically manipulated, in combination with the availability of tissue culture models and animal models, has made *S. typhimurium* a desirable organism for such studies. In this review, we focus on *Salmonella* interactions with its host cells, both in tissue culture (*in vitro*) and in relevant animal models (*in vivo*), and compare results obtained using these different models. The recent advent of sophisticated imaging and molecular genetic tools has facilitated studying the events that occur in disease, thereby confirming tissue culture results, yet identifying new questions that need to be addressed in relevant disease settings.

Keywords: pathogenesis; typhoid fever; gastroenteritis; type III secretion

1. *SALMONELLA TYPHIMURIUM*: A MODEL SYSTEM FOR THE STUDY OF INVASIVE PATHOGENS THAT SURVIVE WITHIN A VACUOLAR COMPARTMENT

The pathogenic strategy of *Salmonella* species includes penetration of the mucosal barrier and interaction with cells of the immune system where it functions as an intracellular pathogen. As such, the virulence mechanisms used by these pathogens are necessarily complex, interfacing with diverse host cell types ranging from epitheliod to macrophages, interacting with both their surface and intracellular compartments. Indeed, it is estimated that 4% of the *S. typhimurium* chromosome (about 200 genes) are virulence factors (Bowe *et al.* 1998). These factors include, to date, five pathogenicity islands, numerous smaller pathogenicity 'islets', other virulence factors on

---

*Author for correspondence (bfinlay@interchange.ubc.ca).

the chromosome and at least one virulence plasmid (Groisman & Ochman 1997; Salama & Falkow 1999).

Salmonella species infect a broad range of animals and can cause different diseases in different hosts. For example *Salmonella typhi* causes typhoid fever in humans, an invasive disease that can be fatal. In contrast, *S. typhimurium* usually causes a self-limiting gastroenteritis in humans but induces a systemic disease in mice that is similar to typhoid fever. As such, the study of *S. typhimurium* in mice has provided a valuable animal model for the study of these clinically relevant invasive pathogens that survive within a vacuolar compartment in host cells. However, an often overlooked concept is that this is not a diarrhoea model, but instead a systemic disease model, indicating that different virulence factors may participate in these different diseases. For example, SigD/SopB is not a virulence factor in *S. typhimurium*-infected mice, yet is essential for fluid accumulation in bovine ileal loops infected with *S. dublin* (Galyov *et al.* 1997; Wood *et al.* 1998).

In vitro models of *S. typhimurium* have allowed genetic, cell biological and biochemical analysis of the infection process. By using cultured mammalian cells, investigators have learned much about *Salmonella* interactions with epithelial and macrophage lines. Although these models are consistent, reliable, and relatively easy to use, it is becoming apparent that they also have their limitations and problems. For example, because these cells are studied in isolation from the other cells and factors with which they usually interact (such as cells of the immune system or cytokines), defining the role of the host immune response and other cell responses is not feasible. Cultured cells are usually immortalized, and normal disease processes such as apoptosis in cultured cells are thus affected. Some cell types, such as M cells, which appear to play a major role in *Salmonella* penetration across the intestinal barrier, are difficult if not impossible to grow in tissue culture. Finally, mammalian cells often lose important characteristics upon culturing, such as lack of polarity for most epithelial lines, and lack of an oxidative burst for macrophage lines. However, despite these limitations, these cells have provided enormous amounts of information about how *Salmonella* interacts with them, much of it at the molecular level. A major challenge for the future is to confirm that events studied in tissue culture actually occur during disease. At least for invasion, the events seen with cultured cells look remarkably similar to those described in animal infection models (Jones *et al.* 1994; Takeuchi 1967).

2.   INVASION OF EPITHELIAL CELLS

Studies of *S. typhimurium* infection in mice suggest that M cells are the preferred target of this pathogen, with invasion of these cells occurring at early times (15 min) following inoculation (Jones & Falkow 1996). Binding to M cells is thought to be mediated by the lpf fimbrial operon, though the host receptor for this adhesin is unknown (Baumler *et al.* 1996). However, several other adhesins have been described for *S. typhimurium*, and their functions may be redundant. Interestingly, recent evidence suggests that the host receptor for *S. typhi* may be the cystic fibrosis transmembrane conductance regulator, a chloride channel that is non-functional in patients with cystic fibrosis (Pier *et al.* 1998). However, it does not appear to be a receptor for *S. typhimurium*, despite the striking similarities in invasion loci and mechanism (Mills & Finlay 1994). Invasion of M cells rapidly leads to cellular destruction and the dissemination of *S. typhimurium* to deeper tissues. M cell destruction by *S. typhimurium* may be similar to perforation of the intestine witnessed in typhoid patients infected with *S. typhi* (Jones & Falkow 1996). Although *S. typhimurium* triggers apoptosis in cultured macrophage cells (Chen *et al.* 1996; Monack *et al.* 1996), M cell death appears morphologically to be cytoxic or necrotic in nature.

S. typhimurium and other *Salmonella* species can also invade columnar absorptive cells at their apical surface. Since M cell models remain difficult to study *in vitro*, studies of *Salmonella* invasion have largely been performed with cell lines that resemble absorptive enterocytes. The use of polarized epithelial cell lines such as Caco-2 and T84 have facilitated the study of invasion, although the events appear to be the same in polarized and non-polarized cells. However, polarized cells have also been used to study epithelial monolayer penetration, which cannot be examined in non-polarized cells. Additionally, by adding inflammatory cells such as polymorphonuclear leucocytes (PMNs) to the basolateral surface of polarized cell monolayers, transepithelial PMN migration and other events can be reconstructed *in vitro* (Gewirtz *et al.* 1999; McCormick *et al.* 1993, 1995, 1998). As depicted in Fig. 1, invasion involves denuding of the microvilli and ruffling of the cell surface in both cultured cells and the small bowel. These cell surface rearrangements lead to uptake of the bacterium in large vacuoles that resemble macropinosomes (Fig. 2).

In contrast to other pathogenic bacteria that use adhesinreceptor interactions for their uptake (i.e. a zipper mechanism), *Salmonella* species use a

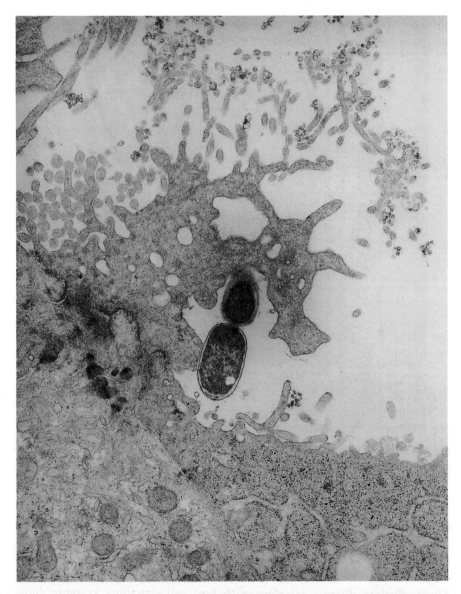

Fig. 1. Invasion of epithelial cells *in vitro*. Transmission electron micrograph illustrating cell surface ruffling of Caco-2 epithelial cells induced by *S. typhimurium*. The bacteria induce dramatic ruffling of the host cell surface at the site of contact with the plasma membrane. These ruffles engulf *S. typhimurium* and lead to their internalization.

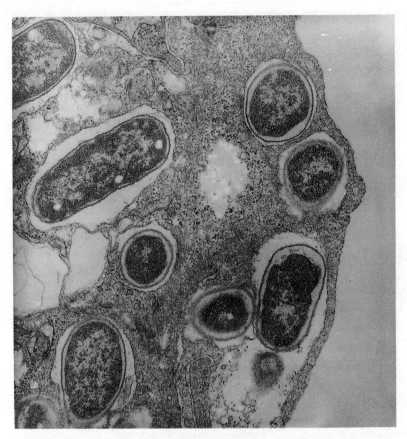

Fig. 2. Invasion of epithelial cells *in vitro*. Shortly following invasion, *S. typhimurium* are present in large membrane-bound vacuoles, morphologically similar to macropinosomes. Depicted are *S. typhimurium* 60 min after invasion of Caco-2 epithelial cells *in vitro*.

type III secretion system to cause host epithelial cell ruffling, which drives their internalization (Collazo & Galan 1997). This secretion system, encoded within *Salmonella* pathogenicity island 1 (SPI-1), delivers a variety of effectors into the host cell that mediate uptake of the pathogen (Fig. 3). These effectors directly interface with host cell signalling systems, independently of receptor–ligand interactions on the host cell surface. As such, invasion by *Salmonella* species could be considered distinct from receptor-mediated phagocytosis (cell eating) and more likened to 'force feeding' of

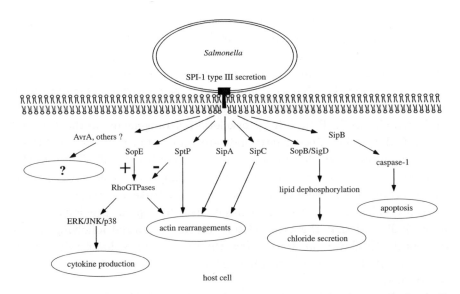

Fig. 3. Invasion of epithelial cells requires delivery of effector molecules into the host cell. Depicted are the translocated effectors of the SPI-1-encoded type III secretion system and the host signalling systems with which they directly interact. These effectors have numerous effects on the host cell, including actin rearrangements (for invasion), production of pro-inflammatory cytokines, chloride secretion, and programmed cell death (apoptosis). Artwork modified from Brumell *et al.* (1999).

the host cell (Brumell *et al.* 1999). *S. typhimurium* mutants lacking SPI-1 activity are unable to invade and destroy M cells and are somewhat attenuated for systemic virulence when delivered orally to mice. Interestingly, injection of these mutants into the peritoneum of these animals (thereby bypassing the need for invasion of the intestine) reveals that these pathogens have normal virulence for causing systemic disease (Galan & Curtiss 1989). Thus, SPI-1 appears to be specialized for gastrointestinal events during the infection process, enabling the pathogen to penetrate epithelial barriers both *in vitro* and *in vivo*.

Recent progress has revealed some of the mechanisms by which the translocated effectors of SPI-1 mediate invasion by *S. typhimurium* in cultured epithelial cells. Crucial to invasion are rearrangements of the actin cytoskeleton, which is accomplished by the actions of at least four effectors. SopE acts as a guanine nucleotide exchange factor for Cdc42 and Rac, Rho family GTPases that regulate the actin cytoskeleton (Hardt *et al.* 1998a).

Interestingly, SopE is not encoded within the SPI-1 yet is translocated by this secretion system into the host cell (Hardt *et al.* 1998*b*). Downstream actions of SopE also include the activation of kinase cascades that lead to the production of pro-inflammatory cytokines. SptP has two functional domains, a tyrosine phosphatase domain and a region with homology to both exoenzyme S from *Pseudomonas aeruginosa* and YopE from *Yersinia* spp. Both domains of SptP initiate cytoskeletal rearrangements in the host cell (Fu & Galan 1998). Recent data indicate that it functions as a GTPase-activating protein, which promotes inactivation of Rho family GTPases (Fu & Galan 1999) . This effect indicates that invading *Salmonella* have the ability to downregulate the signals that mediate cytoskeletal rearrangements once internalized, and provides a mechanism to explain previous results, which indicated that the cytoskeleton returns to normal following bacterial uptake (Finlay *et al.* 1991). How the tyrosine phosphatase domain of SptP contributes to actin rearrangements in the host cell remains to be determined.

In addition to modulating regulators of the host cell actin cytoskeleton, effectors of *S. typhimurium* can also modulate actin dynamics in a direct manner. For example, SipA is an actin-binding protein that lowers its critical concentration and inhibits depolymerization of actin filaments, thereby stabilizing actin polymers at the site of *S. typhimurium* interaction with the host cell (Zhou *et al.* 1999*b*). SipA also forms a complex with T-plastin, resulting in an increase in the actin-bundling activity of this protein (Zhou *et al.* 1999*a*). Recent data suggest that SipC can also bind to actin and mediates bundling of actin filaments directly with its N-terminal domain, in the absence of T-plastin. In addition, the C-terminus of SipC was found to mediate nucleation of actin polymers, a novel function for a bacterial effector protein. Interestingly, SipC was found to associate with lipid bilayers, suggesting that these activities may be directed from the host cell membrane upon delivery by the SPI-1-encoded type III secretion system (Hayward & Koronakis 1999).

Together, these *in vitro* studies suggest that the invasion of *S. typhimurium* involves a very complex and highly calibrated set of interactions between its translocated effectors and the host's actin cytoskeleton. Despite the appeal of these mechanisms, whether these events occur *in vivo* has not been established (not even actin has been shown to accumulate under bacteria in relevant disease models!). However, the similarity of morphological changes that occur in intestines upon infection (ruffling, microvilli denuding, cellular protrusions and macropinocytosis) are quite similar to the tissue culture events, suggesting these molecules function in a similar manner during disease.

SopB from *S. dublin* is a phosphatase, capable of acting on polyphosphoinositides and inositol phosphates (Norris *et al.* 1998). Its phosphatase activity in host cells is thought to activate Ca^{2+}-dependent chloride channels and thereby contribute to fluid secretion induced by *S. dublin* in bovine calf ileal loop models of infection. Though delivered by the SPI-1-encoded type III secretion system, SopB is encoded within a separate pathogenicity island, SPI-5, which appears to be required for enteropathogenicity (Wood *et al.* 1998). The SopB homologue in *S. typhimurium*, SigD, was identified as an invasion gene in SPI-1 mutants (Hong & Miller 1998). This suggests that SopB/SigD may also play a role in invasion. Interestingly, mutations in *sigD* have little if any effect on virulence in the murine typhoid model (Wood *et al.* 1998), yet SopB seems to be required for fluid secretion in the diarrhoea model (Galyov *et al.* 1997). However, *sigD* is induced inside host cells (C. Pfeifer, unpublished data), indicating it may play another, perhaps redundant, role in *S. typhimurium* infections.

Invasion of macrophages by *S. typhimurium* using its SPI-1 secretion system (as opposed to phagocytosis) has been shown to induce rapid apoptosis of the infected cell (Chen *et al.* 1996; Monack *et al.* 1996). Recently, it was shown that SipB, a translocated effector encoded on SPI-1 that also helps to form a pore in the host plasma membrane during secretion, is a key mediator of this effect. SipB binds to and activates caspase-1, a pro-apoptotic protease that converts the precursor form of interleukin-1β to its mature form (Hersh *et al.* 1999). The cytotoxic effect of invasive *S. typhimurium* may be relevant to infection at Peyer's patches, where macrophages are found to underlie target M cells. Apoptosis of infected columnar absorptive cells has also been observed but occurs more slowly and less frequently than that seen in macrophages (Kim *et al.* 1998), prompting the question of why each cell responds differently to invasion by *S. typhimurium*. Interestingly, *S. typhimurium* does not cause apoptosis in cultured epithelial cells. Whether this is a result of using immortalized epithelial cell lines, or because there are fundamental differences between epithelial cells and macrophages that *Salmonella* exploits, remains to be determined.

3.   COLONIZATION OF MACROPHAGES BY
S. TYPHIMURIUM IN A SYSTEMIC MOUSE
MODEL OF INFECTION

Following invasion of host cells *in vitro*, *S. typhimurium* is localized within a membrane compartment known as the *Salmonella*-containing vacuole (SCV). These bacteria are capable of survival and replication within the SCV, eventually killing the host cell and being released into the extracellular medium to infect other cells. The ability of *S. typhimurium* to survive and replicate within macrophages, which possess an arsenal of antimicrobial defences, indicates that these are the host cells colonized during the systemic phase of disease. Indeed, mutants that are incapable of survival in macrophages *in vitro* are avirulent in the mouse model of infection (Fields *et al.* 1986).

To analyse the localization of *S. typhimurium* in a relevant infection model, a low-dose murine infection was used (Richter-Dahlfors *et al.* 1997). This model entails the injection of approximately 100 colony forming units of wild-type *S. typhimurium* into the tail vein of experimental mice and examination of thick liver sections (30 μm) by confocal microscopy at different times following injection. This dose is similar to the lethal dose. A major advantage of this method is that it precludes the need for an artificially high inoculum at late infection times to visualize bacteria in the organs using normal electron microscopic methods, and allows a much larger area of study than conventional electron microscopy. Shortly after appearance in the liver, *S. typhimurium* were localized at inflammatory foci, where infiltrating neutrophils could be seen. At the later stages of infection, three-dimensional confocal projections were used to conclusively establish that the bacteria reside within (as opposed to on) macrophages. Thus, *S. typhimurium* remains a facultative intracellular pathogen during initial infection of epithelial cells in the gut and throughout the duration of infection of solid organs such as the liver.

Infections were found to induce apoptosis of macrophages within the liver, possibly allowing evasion of host immune cell responses in these tissues (Richter-Dahlfors *et al.* 1997). As noted above, *Salmonella* triggers apoptosis in cultured macrophages using the SPI-1 system, with direct contact being required for this event. The apoptosis observed *in vivo* may represent this process, although apoptosis was also observed in liver macrophages that were not infected with bacteria, perhaps suggesting that a secreted factor (cytokine?) may be involved in this process. By triggering host cell apoptosis, the pathogen may be dampening the immune response to the infection (apoptotic cells do not trigger inflammation). Since apoptotic cells are phagocytosed by other macrophages, the pathogen may also be acquiring a fresh host cell to exploit. Alternatively, the host may be depleting or removing the cellular reservoir (macrophages), thereby limiting pathogen spread as a mechanism to control disease. Experimentally distinguishing the role of these events is near to impossible at present, although transgenic mouse lines that are affected in apoptosis (if viable) may help answer this question.

4. INTRACELLULAR TRAFFICKING OF THE SCV

Intracellular trafficking of the SCV differs from that of other internalized particles and may represent a key mechanism for survival and replication within host cells (see Fig. 4). Shortly following invasion, interaction with early endosomes is revealed by delivery of early endosome antigen-1 and the transferrin receptor in cultured epithelial cells (Steele-Mortimer *et al.* 1999). Within 30 min, however, SCVs become uncoupled from the endocytic pathway of the host. As such, the SCV do not fuse with lysosomes and avoid exposure to the harsh degradative enzymes that they contain. While not fusing with lysosomes, *S. typhimurium* SCVs do acquire lysosomal glycoproteins (Lgp). Recent findings suggest that Lgp are acquired through interaction with an unknown compartment and that Rab7 mediates this processing of the SCV in cultured epithelial cells (Méresse *et al.* 1999). Uncoupling from the endocytic pathway is witnessed only with live *S. typhimurium* and is similar in both epithelial (Garcia-del Portillo & Finlay 1995) and macrophage cell lines (Mills & Finlay 1998; Rathman *et al.* 1997), suggesting that these pathogens actively use a common mechanism for altering intracellular trafficking. Confirming the intracellular route taken by SCVs as bacteria penetrate the intestine and live within deeper tissues is extremely difficult. We have been able to visualize SCVs in liver macrophages that stain with Lgp (A. Richtor-Dahlfors, unpublished data), but the resolution and background fluorescence present significant difficulties. Furthermore, synchronizing the infection *in vivo* is difficult. Perhaps the best method to confirm the trafficking of the SCV is to follow it in primary cells such as macrophages *in vitro*.

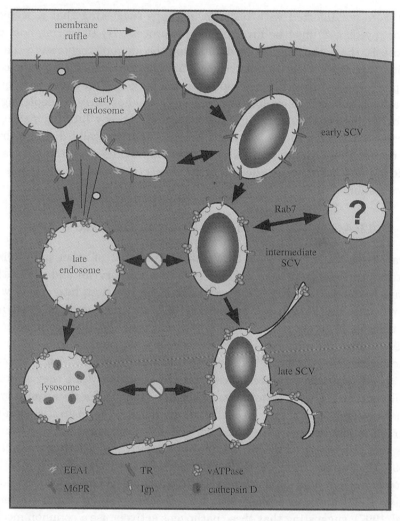

Fig. 4. Intracellular trafficking of *S. typhimurium* SCVs in host cells. Upon entry into host cells, *S. typhimurium* remain in an SCV that interacts transiently with early endosomes. However, these vacuoles do not undergo further interactions with the endosomal system and do not fuse with lysosomes. Instead, the SCV appear to interact with an unknown compartment that mediates delivery of Lgp (such as LAMP-1) but not degradative lysosomal enzymes such as cathepsin D (Steele-Mortimer *et al.* 1999). After several hours, intracellular *S. typhimurium* begin to replicate and long, contiguous tubules (Sifs) are formed. Artwork provided by O. Steele-Mortimer and modified from Steele-Mortimer *et al.* (1999).

In both cultured epithelial cells and macrophages, invasion is followed by a lag phase of 3–4 h, after which the intracellular bacteria begin replication. At this time, *S. typhimurium* are localized within tubular structures known as Sifs (*Salmonella*-induced filaments). These filamentous structures can be visualized in cells by immunostaining for Lgp and appear to require bacterial replication for their formation (Garcia-del Portillo & Finlay 1994). The role of these structures for *S. typhimurium* pathogenesis remains unknown, and current confocal techniques are not of sufficient resolution to see such structures in thick tissue sections, and thus the trafficking and presence of Sifs have not been studied *in vivo*.

5. MECHANISMS FOR SURVIVAL AND REPLICATION OF INTRACELLULAR *S. TYPHIMURIUM*

Many factors have been identified in *S. typhimurium* that are required for survival and replication within host cells and for causing disease (Groisman & Ochman 1997; Salama & Falkow 1999). These include factors which mediate nutrient biosynthesis within the host cell (Hoiseth & Stocker 1981), resistance to reactive intermediates of oxygen and nitrogen in the SCV (Fang *et al.* 1999; Lundberg *et al.* 1999) and DNA repair (Buchmeier *et al.* 1993). Of particular interest for the study of typhoid, is the recent discovery of a type III secretion system within the second pathogenicity island of *S. typhimurium* (SPI-2) (Ochman *et al.* 1996; Shea *et al.* 1996). Many genes encoded within SPI-2 are induced following invasion of host cells and play a critical role in allowing replication of the bacteria within the SCV; type III secretion system apparatus or putative translocated effector mutants are highly attenuated in mouse models of infection (Cirillo *et al.* 1998; Hensel *et al.* 1998). Recent evidence suggests that one putative effector of SPI-2, SpiC, is translocated into the cytosol of infected macrophages and may influence intracellular trafficking of the SCV (Uchiya *et al.* 1999). It is anticipated that the functions of other SPI-2 effectors will be as diverse and complex as those of the SPI-1 effectors. Interestingly, the role of SPI-2 appears limited to systemic phases of disease, since mutants of this type III secretion system had no effect in a diarrhoea model in rabbits (Everest *et al.* 1999).

6. *SALMONELLA* AND THE IMMUNE SYSTEM

The interactions of *Salmonella* species with the immune system are numerous and complex. The use of simple tissue culture cells have yielded a significant amount of (often conflicting) data about cytokine induction, etc. However, by using these cells in isolation, the true nature of bacterial effects on the immune system cannot be easily studied *in vitro*. Some investigators have had significant success by combining a polarized epithelial monolayer with PMNs to study chemotaxis and other events. However, defining the complex events that occur in the immune system following *Salmonella* infection will require more sophisticated technology, such as cytokine arrays, etc. (see §7). It is doubtful that an *in vitro* 'immune response' system can be used to model the complex events of the immune response, and relevant disease models will need to be used to understand further such complex interactions. As discussed below (§7), the choice of the host model is critical for these studies. For example, the standard mouse model for *Salmonella* infections uses a susceptible host, for which the cytokine response differs greatly from that of a resistant host (Lalmanach & Lantier 1999).

7. THE FUTURE

The biggest challenge to this field in the immediate future is to establish which events, identified using tissue culture models, actually occur *in vivo*, and more importantly, their contribution to disease. Continuing to use techniques such as *in vivo* expression technology (IVET), signature-tagged mutagenesis, and differential fluorescence induction to search for genes that are expressed *in vivo* will certainly increase the number of genes that need to be studied. Tissue culture cells will no doubt be extremely useful in delineating how these molecules function at a molecular level. However, clever experiments will be needed to determine how these molecules actually function *in vivo*, and to define their contribution to disease. The use of transgenic animals will certainly provide additional tools to study these events and confirm their functions. Perhaps even expressing individual bacterial virulence factors in a transgenic setting will enhance these studies.

Another major technological advance that will impact significantly on the study of pathogenesis is genomics (pathogenomics). It is anticipated that we will soon have the entire genomic sequence of *S. typhi* and *S. typhimurium*. This will enable arrays to be constructed to follow the expression of the entire bacterial genome during interactions with host cells. However, such a technique will encounter many technical hurdles. For example, isolating bacterial mRNA from a small number of *Salmonella* in the context of infected tissue or in the intestine (where significant normal flora are present) will be a major challenge. Synchronizing the infection will be difficult, and arrays will only give an average expression profile. As was learned using IVET, just because a particular gene is expressed *in vivo*, it does not necessarily follow that it has a critical role in disease. Perhaps the biggest problem will be the vast amount of information these techniques will generate, and deciding which leads to follow. Transforming this information to the study of individual genes will be an immense task. Given that several dozen virulence factors have been identified in *Salmonella* already, yet we only know the molecular function of a handful, perhaps examining the expression of a 'mini-array' of *Salmonella* virulence factors would be more effective (and technically realistic). Within a few years, the entire genome of mice and humans will also be available. This will enable the host response to a pathogen to be examined at a genomic level. Arrays already exist containing several hundred host genes of various gene families. Again, determining the interplay of individual host genes that are induced during infection will require significant effort. However, by using defined *Salmonella* mutants, determining the host response to a particular virulence factor should be possible. Array technology will also miss those factors that are modified post-transcriptionally (such as phosphorylation events) during infection, so alternative techniques will need to be developed to detect these modifications.

As discussed above, the choice of disease model is critical to extending *in vitro* observations to *in vivo* settings. The most popular model of *Salmonella* infection is the murine typhoid model. However, this model is not a diarrhoeal model, which is the predominant *Salmonella* disease in humans in developed countries, especially as typhoid fever decreases due to use of the successful vaccines. Additionally, this model uses a susceptible animal with a defective host resistance gene, NRAMP-1. Ironically, no *Salmonella* virulence factor has ever been identified in a resistant (functional NRAMP-1) murine host! Clearly this is an issue that needs to be addressed.

Salmonella species show varying degrees of host specificity. For example, *S. typhi* is a human-specific pathogen with no known animal infections. In contrast, *S. typhimurium* causes enteritis in humans and systemic disease in susceptible mice. Other *Salmonella* species are animal specific (e.g. poultry pathogens), yet others cause a similar disease in a wide range of animal hosts. Finally, some animals are carriers of *Salmonella* species and cause no disease symptoms in these hosts. Clearly different virulence factors are at play in these different settings, and elucidating these differences between hosts remains a major challenge. It is obvious that infected tissue culture models will not provide the answers for such a diverse range of host infection strategies, emphasizing that we have only begun to appreciate the complexity and diversity of *Salmonella* pathogenesis.

8.   CONCLUSIONS

Salmonella species cause a wide range of effects on different host cells. An immense amount has been learned about how these pathogens interact with isolated cultured host cells. However, the major challenge will be to extend and confirm these findings to relevant infection and disease models, and define the contribution to disease of these events. Only by using this approach will we truly begin to define the important molecular mechanisms of diseases caused by *Salmonella*.

We thank the members of our laboratory who have contributed to this field. Work in the laboratory of B.B.F. is supported by operating grants from the Medical Research Council of Canada, the Canadian Bacterial Disease Centre of Excellence and a Howard Hughes International Research Scholar award. J.H.B. is supported by a Natural Science and Engineering Research Council of Canada fellowship and is an honorary fellow of the Killam Memorial Foundation.

REFERENCES

Baumler, A. J., Tsolis, R. M. & Heffron, F. 1996 The lpf fimbrial operon mediates adhesion of *Salmonella typhimurium* to murine Peyer's patches. *Proc. Natl Acad. Sci. USA* **93**, 279–283.

Bowe, F., Lipps, C. J., Tsolis, R. M., Groisman, E., Heffron, F. & Kusters, J. G. 1998 At least four percent of the *Salmonella typhimurium* genome is required for fatal infection of mice. *Infect. Immun.* **66**, 3372–3377.

Brumell, J. H., Steele-Mortimer, O. & Finlay, B. B. 1999 Bacterial invasion: force feeding by *Salmonella*. *Curr. Biol.* **9**, R277–R280.

Buchmeier, N. A., Lipps, C. J., So, M. Y. & Heffron, F. 1993 Recombination-deficient mutants of *Salmonella typhimurium* are avirulent and sensitive to the oxidative burst of macrophages. *Mol. Microbiol.* **7**, 933–936.

Chen, L. M., Kaniga, K. & Galan, J. E. 1996 *Salmonella* spp. are cytotoxic for cultured macrophages. *Mol. Microbiol.* **21**, 1101–1115.

Cirillo, D. M., Valdivia, R. H., Monack, D. M. & Falkow, S. 1998 Macrophage-dependent induction of the *Salmonella* pathogenicity island 2 type III secretion system and its role in intracellular survival. *Mol. Microbiol.* **30**, 175–188.

Collazo, C. M. & Galan, J. E. 1997 The invasion-associated type-III protein secretion system in *Salmonella* review. *Gene* **192**, 51–59.

Everest, P., Ketley, J., Hardy, S., Douce, G., Khan, S., Shea, J., Holden, D., Maskell, D. & Dougan, G. 1999 Evaluation of *Salmonella typhimurium* mutants in a model of experimental gastroenteritis. *Infect. Immun.* **67**, 2815–2821.

Fang, F. C. (and 10 others) 1999 Virulent *Salmonella typhimurium* has two periplasmic Cu, Zn-superoxide dismutases. *Proc. Natl Acad. Sci. USA* **96**, 7502–7507.

Fields, P. I., Swanson, R. V., Haidaris, C. G. & Heffron, F. 1986 Mutants of *Salmonella typhimurium* that cannot survive within the macrophage are avirulent. *Proc. Natl Acad. Sci. USA* **83**, 5189–5193.

Finlay, B. B., Ruschkowski, S. & Dedhar, S. 1991 Cytoskeletal rearrangements accompanying Salmonella entry into epithelial cells. *J. Cell Sci.* **99**, 283–296.

Fu, Y. & Galan, J. E. 1998 The *Salmonella typhimurium* tyrosine phosphatase SptP is translocated into host cells and disrupts the actin cytoskeleton. *Mol. Microbiol.* **27**, 359–368.

Fu, Y. & Galan, J. E. 1999 A *Salmonella* protein antagonizes Rac-1 and CDC42 to mediate host-cell recovery after bacterial invasion. *Nature* **401**, 293–297.

Galan, J. E. & Curtiss, R. D. 1989 Cloning and molecular characterization of genes whose products allow *Salmonella typhimurium* to penetrate tissue culture cells. *Proc. Natl Acad. Sci. USA* **86**, 6383–6387.

Galyov, E. E., Wood, M. W., Rosqvist, R., Mullan, P. B., Watson, P. R., Hedges, S. & Wallis, T. S. 1997 A secreted effector protein of *Salmonella dublin* is translocated into eukaryotic cells and mediates inflammation and fluid secretion in infected ileal mucosa. *Mol. Microbiol.* **25**, 903–912.

Garcia-del Portillo, F. & Finlay, B. B. 1994 *Salmonella* invasion of nonphagocytic cells induces formation of macropinosomes in the host cell. *Infect. Immun.* **62**, 4641–4645.

Garcia-del Portillo, F. & Finlay, B. B. 1995 Targeting of *Salmonella typhimurium* to vesicles containing lysosomal membrane glycoproteins bypasses compartments with mannose 6-phosphate receptors. *J. Cell Biol.* **129**, 81–97.

Gewirtz, A. T., Siber, A. M., Madara, J. L. & McCormick, B. A. 1999 Orchestration of neutrophil movement by intestinal epithelial cells in response to *Salmonella typhimurium* can be uncoupled from bacterial internalization. *Infect. Immun.* **67**, 608–617.

Groisman, E. A. & Ochman, H. 1997 How *Salmonella* became a pathogen. *Trends Microbiol.* **5**, 343–349.

Hardt, W. D., Chen, L. M., Schuebel, K. E., Bustelo, X. R. & Galan, J. E. 1998a *S. typhimurium* encodes an activator of Rho GTPases that induces membrane ruffling and nuclear responses in host cells. *Cell* **93**, 815–826.

Hardt, W. D., Urlaub, H. & Galan, J. E. 1998b A substrate of the centisome 63 type III protein secretion system of *Salmonella typhimurium* is encoded by a cryptic bacteriophage. *Proc. Natl Acad. Sci. USA* **95**, 2574–2579.

Hayward, R. D. & Koronakis, V. 1999 Direct nucleation & bundling of actin by the SipC protein of invasive Salmonella. *EMBO J.* **18**, 4926–4934.

Hensel, M., Shea, J. E., Waterman, S. R., Mundy, R., Nikolaus, T., Banks, G., Vazquez-Torres, A., Gleeson, C., Fang, F. C. & Holden, D. W. 1998 Genes encoding putative effector proteins of the type III secretion system of *Salmonella* pathogenicity island 2 are required for bacterial virulence and proliferation in macrophages. *Mol. Microbiol.* **30**, 163–174.

Hersh, D., Monack, D. M., Smith, M. R., Ghori, N., Falkow, S. & Zychlinsky, A. 1999 The *Salmonella* invasin SipB induces macrophage apoptosis by binding to caspase-1. *Proc. Natl Acad. Sci. USA* **96**, 2396–2401.

Hoiseth, S. K. & Stocker, B. A. 1981 Aromatic-dependent *Salmonella typhimurium* are non-virulent and effective as live vaccines. *Nature* **291**, 238–239.

Hong, K. H. & Miller, V. L. 1998 Identification of a novel *Salmonella* invasion locus homologous to *Shigella ipgDE*. *J. Bacteriol.* **180**, 1793–1802.

Jones, B. D. & Falkow, S. 1996 Salmonellosis: host immune responses and bacterial virulence determinants. *A. Rev. Immunol.* **14**, 533–561.

Jones, B. D., Ghori, N. & Falkow, S. 1994 *Salmonella typhimurium* initiates murine infection by penetrating and destroying the specialized epithelial M cells of the Peyer's patches. *J. Exp. Med.* **180**, 15–23.

Kim, J. M., Eckmann, L., Savidge, T. C., Lowe, D. C., Witthoft, T. & Kagnoff, M. F. 1998 Apoptosis of human intestinal epithelial cells after bacterial invasion. *J. Clin. Invest.* **102**, 1815–1823.

Lalmanach, A.-C. & Lantier, F. 1999 Host cytokine response and resistance to *Salmonella* infection. *Microb. Infect.* **1**, 719–726.

Lundberg, B. E., Wolf Jr, R. E., Dinauer, M. C., Xu, Y. & Fang, F. C. 1999 Glucose 6-phosphate dehydrogenase is required for *Salmonella typhimurium* virulence and resistance to reactive oxygen and nitrogen intermediates. *Infect. Immun.* **67**, 436–438.

McCormick, B. A., Colgan, S. P., Delp-Archer, C., Miller, S. I. & Madara, J. L. 1993 *Salmonella typhimurium* attachment to human intestinal epithelial monolayers: transcellular signalling to subepithelial neutrophils. *J. Cell Biol.* **123**, 895–907.

McCormick, B. A., Miller, S. I., Carnes, D. & Madara, J. L. 1995 Transepithelial signaling to neutrophils by salmonellae: a novel virulence mechanism for gastroenteritis. *Infect. Immun.* **63**, 2302–2309.

McCormick, B. A., Parkos, C. A., Colgan, S. P., Carnes, D. K. & Madara, J. L. 1998 Apical secretion of a pathogen-elicited epithelial chemoattractant activity in response to surface colonization of intestinal epithelia by *Salmonella typhimurium*. *J. Immunol.* **160**, 455–466.

Méresse, S., Steele-Mortimer, O., Finlay, B. B. & Gorvel, J. P. 1999 The rab7 GTPase controls the maturation of *Salmonella typhimurium*-containing vacuoles in HeLa cells. *EMBO J.* **18**, 4394–4403.

Mills, S. D. & Finlay, B. B. 1994 Comparison of *Salmonella typhi* and *Salmonella typhimurium* invasion, intracellular growth and localization in cultured human epithelial cells. *Microb. Pathogen.* **17**, 409–423.

Mills, S. D. & Finlay, B. B. 1998 Isolation and characterization of *Salmonella typhimurium* and *Yersinia pseudotuberculosis*-containing phagosomes from infected mouse macrophages: *Y. pseudotuberculosis* traffics to terminal lysosomes where they are degraded. *Eur. J. Cell Biol.* **77**, 35–47.

Monack, D. M., Raupach, B., Hromockyj, A. E. & Falkow, S. 1996 *Salmonella typhimurium* invasion induces apoptosis in infected macrophages. *Proc. Natl Acad. Sci. USA* **93**, 9833–9838.

Norris, F. A., Wilson, M. P., Wallis, T. S., Galyov, E. E. & Majerus, P. W. 1998 SopB, a protein required for virulence of *Salmonella dublin*, is an inositol phosphate phosphatase. *Proc. Natl Acad. Sci. USA* **95**, 14057–14059.

Ochman, H., Soncini, F. C., Solomon, F. & Groisman, E. A. 1996 Identification of a pathogenicity island required for *Salmonella* survival in host cells. *Proc. Natl Acad. Sci. USA* **93**, 7800–7804.

Pier, G. B., Grout, M., Zaidi, T., Meluleni, G., Mueschenborn, S. S., Banting, G., Ratcliff, R., Evans, M. J. & Colledge, W. H. 1998 *Salmonella typhi* uses CFTR to enter intestinal epithelial cells. *Nature* **393**, 79–82.

Rathman, M., Barker, L. P. & Falkow, S. 1997 The unique trafficking pattern of *Salmonella typhimurium*-containing phagosomes in murine macrophages is independent of the mechanism of bacterial entry. *Infect. Immun.* **65**, 1475–1485.

Richter-Dahlfors, A., Buchan, A. M. J. & Finlay, B. B. 1997 Murine salmonellosis studied by confocal microscopy: *Salmonella typhimurium* resides intracellularly inside macrophages and exerts a cytotoxic effect on phagocytes *in vivo*. *J. Exp. Med.* **186**, 569–580.

Salama, N. R. & Falkow, S. 1999 Genomic clues for defining bacterial pathogenicity. *Microb. Infect.* **1**, 615–619.

Shea, J. E., Hensel, M., Gleeson, C. & Holden, D. W. 1996 Identification of a virulence locus encoding a second type III secretion system in *Salmonella typhimurium*. *Proc. Natl Acad. Sci. USA* **93**, 2593–2597.

Steele-Mortimer, O., Méresse, S., Gorvel, J.-P., Toh, B.-H. & Finlay, B. B. 1999 Biogenesis of *Salmonella typhimurium*-containing vacuoles in epithelial cells involves interactions with the early endocytic pathway. *Cell. Microbiol.* **1**, 33–51.

Takeuchi, A. 1967 Electron microscope studies of experimental *Salmonella* infection. I. Penetration into the intestinal epithelium by *Salmonella typhimurium*. *Am. J. Pathol.* **50**, 109–136.

Uchiya, K., Barbieri, M. A., Funato, K., Shah, A. H., Stahl, P. D. & Groisman, E. A. 1999 A *salmonella* virulence protein that inhibits cellular trafficking. *EMBO J.* **18**, 3924–3933.

Wood, M. W., Jones, M. A., Watson, P. R., Hedges, S., Wallis, T. S. & Galyov, E. E. 1998 Identification of a pathogenicity island required for *Salmonella* enteropathogenicity. *Mol. Microbiol.* **29**, 883–891.

Zhou, D., Mooseker, M. S. & Galan, J. E. 1999a An invasion-associated *Salmonella* protein modulates the actin-bundling activity of plastin. *Proc. Natl Acad. Sci. USA* **96**, 10176–10181.

Zhou, D., Mooseker, M. S. & Galan, J. E. 1999b Role of the *S. typhimurium* actin-binding protein SipA in bacterial internalization. *Science* **283**, 2092–2095.

DISCUSSION

C. W. Keevil (Centre for Applied Microbiology and Research, Porton Down, Wiltshire, UK). The mouse model of *S. typhimurium* infection has provided clear benefits in elucidating the role of potential virulence determinants at the molecular level. As you have pointed out, however, the model does not describe gastroenteritis. Recent studies show that *S. typhimurium* mutants which are avirulent in the mouse model continue to colonize cattle. Consequently, can you comment on the continued suitability of the mouse model for studying infection in man?

B. B. Finlay. *S. typhimurium* infection in mice is a relatively good model of typhoid fever. However, it is not a good model for gastroenteritis. Despite this, it has led to some significant advances in virulence determinants and there is a surprising amount of overlap in virulence determinants between those needed to cause gastroenteritis and those needed for systemic typhoid disease. Both mouse and cattle are complementary models, and I believe that studies comparing and contrasting results with both models will elucidate further the true virulence determinants of salmonellosis.

Impact of the New Methods

IN VIVO GENE EXPRESSION AND THE ADAPTIVE RESPONSE: FROM PATHOGENESIS TO VACCINES AND ANTIMICROBIALS

DOUGLAS M. HEITHOFF, ROBERT L. SINSHEIMER,
DAVID A. LOW and MICHAEL J. MAHAN*

*Department of Molecular, Cellular and Developmental Biology,
University of California, Santa Barbara 93106, CA, USA*

Microbial pathogens possess a repertoire of virulence determinants that each make unique contributions to fitness during infection. Analysis of these *in vivo*-expressed functions reveals the biology of the infection process, encompassing the bacterial infection strategies and the host ecological and environmental retaliatory strategies designed to combat them (e.g. thermal, osmotic, oxygen, nutrient and acid stress). Many of the bacterial virulence functions that contribute to a successful infection are normally only expressed during infection. A genetic approach was used to isolate mutants that ectopically expressed many of these functions in a laboratory setting. Lack of DNA adenine methylase (Dam) in *Salmonella typhimurium* abolishes the preferential expression of many bacterial virulence genes in host tissues. Dam⁻ *Salmonella* were proficient in colonization of mucosal sites but were defective in colonization of deeper tissue sites. Additionally, Dam⁻ mutants were totally avirulent and effective as live vaccines against murine typhoid fever. Since *dam* is highly conserved in many pathogenic bacteria that cause significant morbidity and mortality worldwide, Dams are potentially excellent targets for both vaccines and antimicrobials.

Keywords: vaccines; immunity; DNA adenine methylase; pathogenesis; *Salmonella*

1. INTRODUCTION

Similar to that of all microbes, a pathogen's primary objective is simply to survive and replicate (Falkow 1997). Key attributes that distinguish microbial pathogens from other micro-organisms include their ability to gain access to, replicate within, and persist at host sites that are forbidden to commensal species (Falkow 1997; Heithoff *et al.* 1997*b*). Such interactions within the host are often associated with pathological lesions; overt symptoms and disease are a consequence of a virulent organism achieving its goals using all evolutionary adaptations at its disposal. Thus, it can be argued that any function that contributes to the fitness of a pathogen within

---

*Author for correspondence (mahan@lifesci.ucsb.edu).

a host, or to its transmission to new hosts, may be viewed as a virulence factor. Since pathogenesis is a multifactorial process that is not restricted to a single pathway from infection to mortality, important virulence functions are often missed in standard virulence assays that require overt host injury (e.g. LD_{50}). Such virulence functions include those that are redundant, depend on the nutritional status of the host, depend on the presence of other host micro-organisms, and/or are involved in the transmission of the microbe to a new host. Thus, although the loss of some virulence functions may not cause a measurable change of host injury, they may make important contributions to the fitness of an organism within its natural host.

2.   THE ADAPTIVE RESPONSE AND COORDINATE CONTROL OF GLOBAL VIRULENCE REGULONS

A more expansive definition of virulence functions includes factors involved in the adaptive response to environmental stresses since pathogens are exposed to alterations in temperature, osmolarity, oxygen tension, pH and nutrient availability during infection (Mahan *et al.* 1996; Mekalanos 1992; Miller *et al.* 1989). Indeed, several classical virulence functions are coordinately expressed with other genes in response to well-characterized environmental stresses duplicated in the laboratory. Consequently, such coordinately regulated genes can be considered members of global virulence regulons even if each individual gene does not make a direct contribution to pathogenesis. Moreover, independent of the actual (often unknown) *in vivo* signals, classification of bacterial genes based on regulatory patterns in a laboratory setting may reflect coordinate expression at a given anatomical site as well as a possible functional relationship within the host (e.g. survival within the small intestine or macrophage phagosome). For example *in vitro*, production of cholera toxin is coordinately regulated by the same environmental and genetic signals (e.g. pH, temperature, osmolarity, transcriptional activator ToxR) as the toxin co-regulated pilus (*tcp*); both of these virulence factors are required for the colonization and/or infection of the small intestine (Mahan *et al.* 1996; Waldor & Mekalanos 1996).

Additionally, environmental conditions can be established in a laboratory setting that appear to mimic micro-environments encountered within the host. Virulence genes that respond to these signals often belong to well-characterized global regulatory networks, and the changes that cells undergo in response to these environmental signals may themselves directly contribute to virulence. For example, low pH and low Mg^{2+} have been associated with *Salmonella*-containing vacuoles (Garcia-del Portillo *et al.* 1992) and have been shown to be relevant signals for bacterial genes presumed to function within the macrophage (Garcia Vescovi *et al.* 1996; Heithoff *et al.* 1999a; Soncini & Groisman 1996; Soncini *et al.* 1996). Thus, classification of virulence genes based on their coordinate response to environmental signals provides a means to understand the pathogenesis of the infecting microbe, the ecology of host–pathogen interactions, and the spatial and functional relationship between bacterial gene products and the cognate anatomical host sites at which they operate during infection. Moreover, investigation of gene expression patterns reveals the inherent versatility of pathogens to modify the expression of a given gene(s) at various host sites or within the context of different animal hosts.

Investigation into global adaptive responses controlling virulence gene expression may reveal novel regulatory mechanisms, virulence functions and mechanisms of transmission that each uniquely contribute to the fitness of a pathogen. For example, the ability to reduce the production of classical virulence functions such as adhesins, invasins and toxins may be as critical to bacterial survival as is the ability to produce them. This downregulation may be important to delay immune recognition, exercise economy and facilitate transmission to a new host. Additionally, downregulation of classical virulence functions in response to host environmental cues that favour bacterial growth (availability of nutrients, exposure to optimal temperature, oxygen tension and osmolarity) may be highly significant for bacterial fitness. Such a strategy may result in a subclinical infection that significantly enhances bacterial load while avoiding overt host injury. In time the disease path, however, would elicit immune recognition and other associated host response mechanisms that inhibit bacterial growth, which may ultimately force bacteria to take up residence at another anatomical site or within another host. Information gleaned from defining bacterial evolutionary adaptations, coupled with probing bacterial genes expressed within the host, may provide a means to understand further these microbial–host interactions.

3.   METHODS TO INVESTIGATE *IN VIVO* GENE EXPRESSION

Analysis of bacterial–host interactions is complicated by the fact that not all virulence factors confer a selective advantage at the same stage of infection or at the same anatomical site within the host. Accordingly, expression of certain virulence factors must be modulated in response to signals determining the transition from external environments to growth within the natural host and to other environmental signals encountered throughout the infection process. Moreover, individual responses may not all exist in the same bacterial cell or, alternatively, the overall response may be the additive result of different responses occurring in different cells. Thus, mimicking these environments in the laboratory is problematic since the micro-environments within host tissues are both complex and dynamic; gene expression is a consequence of all of these contributions.

Many of these limitations have been overcome, in part, by recent *in vivo* expression methods that allow the isolation of bacterial genes that are expressed during infection. There are many fundamentally different methods for identifying *in vivo*-expressed genes (Heithoff *et al.* 1997*b*). A survey of *in vivo* expression methods reveals that the most effective approach is to use them in combination, since they each contribute uniquely to our understanding of pathogenesis, virulence and infection.

(a) In vivo *expression technology*

In vivo expression technology (IVET) is a promoter trap wherein bacterial promoters drive the expression of a gene that is required for virulence; complementation in the animal demands elevated levels of expression compared with growth on a laboratory medium (Mahan *et al.* 1993, 1995). The advantage is that the β-galactosidase reporter system can be used to explore functional, spatial and regulatory relationships between genes normally expressed during infection. The disadvantage is that mutants in desired genes must be constructed individually and subsequently tested for virulence.

(b) *Signature-tagged mutagenesis*

Signature-tagged mutagenesis (STM) is a negative selection strategy in which a pool of tagged insertion mutants is used to inoculate an animal;

mutations represented in the initial inoculum but not recovered from an infected animal are required for infection (Hensel *et al.* 1995). The advantage is that individual sequence tags serve as insertion mutations that can be tested in virulence assays. The disadvantage is that the mutations are genetically intractable for further analysis.

(c) *Differential display*

Differential display (DD) is a subtractive hybridization strategy in which bacterial cDNAs from infected tissue are hybridized against a cDNA library constructed from laboratory-grown bacteria (Abu Kwaik & Pederson 1996; Plum & Clarke-Curtiss 1994); the resulting host-specific cDNAs are used as probes to isolate *in vivo*-expressed bacterial genes. The advantage is that DD is applicable to many important pathogens lacking defined genetic systems. The disadvantage is that mutants in desired genes must be constructed individually and subsequently tested for virulence.

(d) *Known virulence genes*

Table 1 shows that many known virulence genes have been identified using *in vivo* expression methods in a wide variety of host–pathogen systems. Examples include adhesins and colonization factors making up type III secretion systems, which are involved in host-contact-dependent secretion of virulence functions; *tcp*, required for *Vibrio cholerae* colonization of the mouse small intestine; and B2, a *Proteus mirabilis* virulence factor required for bladder colonization in a murine model of urinary tract infection. Other virulence functions are involved in intracellular and/or systemic survival, including *spvB*, a *Salmonella* plasmid-encoded virulence function required for growth at systemic sites of infection; *eml*, a *Legionella pneumophila* virulence function involved in the early stages of macrophage infection; and SAP2, a *Candida albicans* virulence function involved in late-stage disseminated infection.

The most common classes of *in vivo*-expressed genes are those involved in nutrient acquisition and the stress response, suggesting that these genes contribute significantly to the fitness of the micro-organism during infection. Examples include the acquisition of metals such as Fe^{2+}, Mg^{2+}, Cu^{2+}, the acquisition and synthesis of nucleotides and co-factors, and the induction of genes involved in DNA repair and thermo-, osmotic and acid

Table 1. Microbial genes expressed during infection.

(Listed are functions or attributes of *in vivo*-expressed genes, their known or inferred role in pathogenesis, the host tissue or condition in which they were selected, and the methodology and organism from which they were recovered. DFI (differential fluorescence induction) is an IVET-like selection strategy involving the green fluorescent protein reporter.)

adherence, colonization, invasion and secretion

gene	function	role in pathogenesis	tissue	method	species	references
SPI-2	type III secretion system	invasion and systemic survival	mouse spleen, macrophage line	STM, DFI	*Salmonella typhimurium*	Hensel et al. (1995), Shea et al. (1996), Valdivia & Falkow (1997)
yscCLRU, *virG*, *lcrV*	type III secretion system	systemic survival	mouse spleen	STM	*Yersinia enterocolitica*	Darwin & Miller (1999)
bscN	type III secretion system	colonization	rat trachea	DD	*Bordetella bronchiseptica*	Yuk et al. (1998)
rhi-18	type III secretion system	colonization	sugar-beet roots	IVET	*Pseudomonas fluorescens*	Rainey (1999)
tcpAEFT	toxin co-regulated pilus	adhesion and colonization	mouse intestine	STM	*Vibrio cholerae*	Chiang & Mekalanos (1998)
EPA1	adherence	adhesion	epithelial cell line	STM	*Candida glabrata*	Cormack et al. (1999)
B2	protease/collagenase	colonization	mouse bladder	STM	*Proteus mirabilis*	Zhao et al. (1999)
iviVI-A, *iviVI-B*	Tia-like, PfEMP1-like	adhesion and invasion	mouse spleen	IVET	*Salmonella typhimurium*	Heithoff et al. (1997a)

Table 1. (*Continued*)

gene	function	role in pathogenesis	tissue	method	species	references
*ivi*131–19	Hag2-like	adhesion and invasion	rat lung, mouse liver	IVET	*Pseudomonas aeruginosa*	Handfield *et al.* (2000)
cutinase-like gene	cutinase	invasion	pepper roots	DD	*Phytophthora capsici*	Munoz & Bailey (1998)
secE	secretion	protein export	pig lung	IVET	*Actinobacillus pleuropneumoniae*	Fuller *et al.* (1999)
intracellular and systemic survival						
spvRAD, *spvB*	plasmid virulence	systemic survival	mouse spleen	STM, IVET	*Salmonella typhimurium*	Hensel *et al.* (1995), Heithoff *et al.* (1997a)
eml	early macrophage infection	macrophage survival	macrophage line	DD	*Legionella pneumophila*	Abu Kwaik & Pederson (1996)
SAP2	secreted aspartic proteases	disseminated candidiasis	mouse spleen	IVET	*Candida albicans*	Staib *et al.* (1999)
dotB	organelle trafficking	macrophage survival	guinea-pig lung and spleen	STM	*Legionella pneumophila*	Edelstein *et al.* (1999)
icmX	intracellular multiplication	macrophage survival	guinea-pig lung and spleen	STM	*Legionella pneumophila*	Edelstein *et al.* (1999)
mig	unknown	macrophage survival	macrophage line	DD	*Mycobacterium avium*	Plum *et al.* (1997)
Rv0288	ESAT.6	T cell antigen	unknown	DD	*Mycobacterium tuberculosis*	Rindi *et al.* (1999)

Table 1. (*Continued*)

gene	function	role in pathogenesis	tissue	method	species	references
Rv2770c	PPE protein family	antigenic variation and	unknown	DD	*Mycobacterium tuberculosis*	Rindi *et al.* (1999)
MTV041.29	PPE protein family	inhibit antigen processing	unknown	DD	*Mycobacterium tuberculosis*	Rivera-Marrero *et al.* (1998)
Rv1345	polyketide synthase	lipid and/or metabolite synthesis	unknown	DD	*Mycobacterium tuberculosis*	Rindi *et al.* (1999)
nutrient acquisition and synthesis						
metals						
entF	siderophore	Fe^{2+} uptake	macrophage line	IVET	*Salmonella typhimurium*	Heithoff *et al.* (1997a)
irp1	siderophore	Fe^{2+} uptake	mouse spleen	STM	*Yersinia enterocolitica*	Darwin & Miller (1999)
fhuA, cirA	iron transport	Fe^{2+} uptake	mouse intestine, macrophage line	IVET	*Salmonella typhimurium*	Heithoff *et al.* (1997a)
fyuA	iron transport	Fe^{2+} uptake	mouse Peyer's patch	IVET	*Yersinia enterocolitica*	Young & Miller (1997)
fptA	iron transport	Fe^{2+} uptake	mouse liver	IVET	*Pseudomonas aeruginosa*	Wang *et al.* (1996)
pydD	pyoverdin synthesis	Fe^{2+} acquisition	rat lung and mouse liver	IVET	*Pseudomonas aeruginosa*	Handfield *et al.* (2000)

Table 1. (*Continued*)

gene	function	role in pathogenesis	tissue	method	species	references
np20	Fur-like	Fe^{2+} starvation response	mouse liver	IVET	*Pseudomonas aeruginosa*	Wang et al. (1996)
mgtA, *mgtB*	Mg^{2+} transport	Mg^{2+} uptake	mouse spleen, macrophage line	IVET	*Salmonella typhimurium*	Heithoff et al. (1997a)
mgtE	Mg^{2+} transport	Mg^{2+} uptake	mouse intestine	STM	*Vibrio cholerae*	Chiang & Mekalanos (1998)
iviX	Cu^{2+} transport	Cu^{2+} homeostasis	macrophage line	IVET	*Salmonella typhimurium*	Handfield et al.
nucleotides						
carAB, *purDL*	pyrimidine, purine synthesis	*de novo* requirement	mouse spleen	IVET, STM	*Salmonella typhimurium*	Mahan et al. (1993), Hensel et al. (1995)
purCELK	purine synthesis	*de novo* requirement	mouse lung	STM	*Streptococcus pneumoniae*	Polissi et al. (1998)
purDHK	purine synthesis	*de novo* requirement	mouse intestine	STM	*Vibrio cholerae*	Chiang & Mekalanos (1998)
purL	purine synthesis	*de novo* requirement	mouse spleen	STM	*Staphylococcus aureus*	Mei et al. (1997)
vacB, *vacC*	mRNA, tRNA processing	post-transcriptional regulation	mouse spleen, mouse intestine	IVET	*Salmonella typhimurium*	Heithoff et al. (1997a)

Table 1. (Continued)

gene	function	role in pathogenesis	tissue	method	species	references
vacC	tRNA processing	post-transcriptional regulation	mouse Peyer's patch	IVET	Yersinia enterocolitica	Young & Miller (1997)
cofactors						
hemA	haem synthesis	peroxide resistance	mouse spleen	IVET	Salmonella typhimurium	Heithoff et al. (1997a)
hemD	haem synthesis	peroxide resistance	mouse Peyer's patch	IVET	Yersinia enterocolitica	Young & Miller (1997)
cobI	vitamin B$_{12}$ synthesis	carbon source use	mouse liver	IVET	Pseudomonas aeruginosa	Wang et al. (1996)
iviXVII	B$_{12}$ synthesis, propanediol use	carbon source use	mouse spleen	IVET	Salmonella typhimurium	Heithoff et al. (1997a)
bioB	biotin synthetase	biotin synthesis	mouse intestine	STM	Vibrio cholerae	Chiang & Mekalanos (1998)
stress response						
DNA repair						
recD	recombination and repair	macrophage survival	macrophage line	IVET	Salmonella typhimurium	Heithoff et al. (1997a)
recB	recombination and repair	macrophage survival	mouse Peyer's patch	IVET	Yersinia enterocolitica	Young & Miller (1997)
radA	recombination and repair	macrophage survival	mouse lung	STM	Streptococcus pneumoniae	Polissi et al. (1998)

Table 1. (*Continued*)

gene	function	role in pathogenesis	tissue	method	species	references
recA	recombination and repair	macrophage survival	mouse liver and spleen	STM	*Staphylococcus aureus*	Coulter *et al.* (1998)
mutL	DNA repair	intestine survival	mouse Peyer's patch	IVET	*Yersinia enterocolitica*	Young & Miller (1997)
hexA	DNA repair	systemic survival	mouse liver and spleen	STM	*Staphylococcus aureus*	Coulter *et al.* (1998)
environmental						
ompR/envZ	osmoregulation	osmotic protection	mouse spleen	STM	*Salmonella typhimurium*	Hensel *et al.* (1995)
otsA	trehalose synthesis	thermo/osmotic protection	mouse intestine	IVET	*Salmonella typhimurium*	Heithoff *et al.* (1997a)
rhi-12	IsfA-like	oxidative stress	sugar-beet roots	IVET	*Pseudomonas fluorescens*	Rainey *et al.* (1997)
rhiI	glycine–betaine binding	osmotic protection	sugar-beet roots	IVET	*Pseudomonas fluorescens*	Rainey *et al.* (1997)
cadC	cadaverine synthesis	acid tolerance	mouse intestine and spleen	IVET	*Salmonella typhimurium*	Heithoff *et al.* (1997a)
cadA	cadaverine synthesis	acid tolerance	rabbit intestine	IVET	*Vibrio cholerae*	Merrell & Camilli (2000)
dnaJ	protein chaperone	heat shock response	mouse spleen	STM	*Yersinia enterocolitica*	Darwin & Miller (1999)
fadB	fatty acid degradation	clear pro-inflammatory	mouse spleen	IVET	*Salmonella typhimurium*	Mahan *et al.* (1995)

Table 1. (*Continued*)

gene	function	role in pathogenesis	tissue	method	species	references
hylC	secreted lipase	fatty acids	mouse intestine	IVET	*Vibrio cholerae*	Camilli & Mekalanos (1995)
lip	lipid degradation		mouse kidney	IVET	*Staphylococcus aureus*	Lowe et al. (1998)
cfa aas,	membrane modifications	membrane repair	mouse intestine and spleen, macrophage cell	IVET, DFI	*Salmonella typhimurium*	Heithoff et al. (1997a), Valdivia & Falkow (1996)
ivi134–21	membrane modifications	protein targeting	mouse liver and rat lung	IVET	*Pseudomonas aeruginosa*	Handfield et al. (2000)
chemotaxis						
iviIV	signal receptor	chemoreception	mouse intestine	IVET	*Vibrio cholerae*	Camilli & Mekalanos (1995)
np9	response regulator	chemoreception	mouse liver	IVET	*Pseudomonas aeruginosa*	Wang et al. (1996)
regulatory functions						
phoP	virulence regulator	invasion and macrophage survival	mouse spleen	IVET	*Salmonella typhimurium*	Heithoff et al. (1997a)
toxT	transcription activator	colonization regulator	mouse intestine	STM	*Vibrio cholerae*	Chiang & Mekalanos (1998)
pmrB	polymyxin resistance	neutrophil survival	macrophage line	IVET	*Salmonella typhimurium*	Heithoff et al. (1997a)

Table 1. (*Continued*)

gene	function	role in pathogenesis	tissue	method	species	references
pspC	phage shock protein regulation	unknown	mouse spleen	STM	*Yersinia enterocolitica*	Darwin & Miller (1999)
agrA	accessory gene regulation	virulence gene regulation	mouse kidney	IVET	*Staphylococcus aureus*	Lowe et al. (1998)
hrc-7	AcrR-like	efflux pump regulator	mouse Peyer's patch	IVET	*Yersinia enterocolitica*	Young & Miller (1997)
rhi-3	CopRS-like	copper-inducible regulator	sugar-beet roots	IVET	*Pseudomonas fluorescens*	Rainey (1999)
B5	nitrogen response regulation	colonization	mouse bladder	STM	*Proteus mirabilis*	Zhao et al. (1999)
membrane/cell wall						
rfb	lipopolysaccharide synthesis	cell integrity	mouse spleen	IVET, STM	*Salmonella typhimurium*	Mahan et al. (1993), Hensel et al. (1995)
rfb	lipopolysaccharide synthesis	cell integrity	mouse intestine	STM	*Vibrio cholerae*	Chiang & Mekalanos (1998)
galE	lipopolysaccharide synthesis	cell integrity	mouse spleen	STM	*Yersinia enterocolitica*	Darwin & Miller (1999)

Table 1. (*Continued*)

gene	function	role in pathogenesis	tissue	method	species	references
femA, femB	peptidoglycan cross-linking	cell wall assembly	mouse spleen	STM	*Staphylococcus aureus*	Mei *et al.* (1997)
pbp2	peptidoglycan cross-linking	cell wall assembly	mouse kidney	IVET	*Staphylococcus aureus*	Lowe *et al.* (1998)
pbp2	penicillin-binding protein	cell wall	mouse kidney	IVET	*Staphylococcus aureus*	Lowe *et al.* (1998)
2.9-71pB	lipoprotein	unknown	rat peritoneum	DD	*Borrelia burgdorferi*	Akins *et al.* (1998)
nlpD	lipoprotein	unknown	mouse spleen	STM	*Yersinia enterocolitica*	Darwin & Miller (1999)

tolerance. These *in vivo*-expressed genes reveal important insights into host ecology, which plays a dual role in pathogenesis: to induce the activation of bacterial genes that complement nutrient-limiting conditions in the host and to induce the activation of other virulence genes required for immediate survival and spread to subsequent anatomical sites of infection.

Determination of the genetic and environmental factors that control virulence genes and the host site(s) in which they are expressed provides clues to both the intracellular environment and possible functions and/or functional relationships of these genes at these specific anatomical sites. For example, many IVET-selected bacterial strains recovered from infected spleens and/or from cultured macrophages were grouped based on their response to environmental and genetic signals that are believed to be involved in bacterial gene expression in the macrophage phagosome (low pH, low Mg^{2+}) (Mahan *et al.* 1996; Mekalanos 1992; Miller *et al.* 1989). Intracellular expression studies showed that each of the low pH and low Mg^{2+} responsive *in vivo*-induced (*ivi*) fusion strains is induced upon entry into and growth within three distinct mammalian cell lines, including murine macrophages and two cultured human epithelial cell lines (Heithoff *et al.* 1999*a*). This suggests that this class of coordinately expressed *ivi* genes responds to general intracellular signals that are present both in initial and in progressive stages of infection and may reflect their response to similar vacuolar micro-environments in these cell types. Determination of the genetic and environmental factors that regulate gene expression at their cognate host site(s) provides clues to both the intracellular environment and possible gene functions at these anatomical sites. Moreover, the functions of some of these genes may change dependent upon the context of the animal, tissue, cell type or subcellular compartment.

4. PILI EXPRESSION, DNA METHYLATION AND INTEGRATION WITH THE ADAPTIVE RESPONSE

Pili (fimbriae) are appendages that allow bacteria to adhere to host cells and to other bacteria. Environmental conditions controlling pili expression can be established in the laboratory setting that appear to mimic, at least in part, micro-environments encountered within the host (reviewed by Krabbe *et al.* 2000). The expression of many pili types is controlled by complex and overlapping regulatory circuitry involving a DNA methylation system that integrates pili expression with the adaptive response. Such regulation is best understood in uropathogenic *Escherichia coli* wherein DNA adenine methylase (Dam) controls the expression of pyelonephritis-associated pili (Pap), which are essential for colonization of the urogenital tract. Genes encoding Pap are reversibly switched between the unexpressed state and the expressed state, termed phase variation, a process that is exquisitely sensitive to environmental signals such as temperature and carbon source and to a number of DNA-binding proteins. Pap phase variation is principally controlled by Dam, which binds to the sequence GATC in double-stranded DNA and methylates adenine at the *N-6* position (Marinus 1996). Dam controls several important and diverse biological processes including DNA replication, methyl-directed mismatch repair (MDMR), transposition and the expression of pili required for colonization of host tissues (Marinus 1996; Van der Woude *et al.* 1996). The molecular basis of this versatility lies in its ability to chemically modify DNA (by methylation), which imparts additional information to the DNA primary sequence.

Although Dam methylates most of the approximately 18 000 GATC sites in the *E. coli* chromosome, certain GATC sites are protected from methylation by the binding of regulatory proteins at or near these sites, forming specific DNA methylation patterns similar to those observed in eukaryotes (Hale *et al.* 1994; Tavazoie & Church 1998). Studies on the *pap* operon have shown that DNA methylation patterns directly control the ON–OFF switch regulating Pap expression in uropathogenic *E. coli* (Braaten *et al.* 1992, 1994; Van der Woude *et al.* 1992, 1996; Van der Woude & Low 1994). Methylation of one of the GATC sites in the *pap* regulatory region inhibits the binding of the global regulator leucine responsive regulatory protein (Lrp), providing a mechanism by which Dam controls *pap* gene expression (Nou *et al.* 1993; Van der Woude *et al.* 1996). Both genetic analysis and DNA database searching for conserved GATC box motifs present in *pap* indicates that there are at least 16 additional pili operons in various *E. coli* and *Salmonella* strains that are probably under DNA methylation control (Krabbe *et al.* 2000; Van der Woude & Low 1994). In addition to regulating pili gene expression, Dam has also been reported to control the expression of the phase variable Agg43 outer membrane protein in *E. coli*. In this case it appears that methylation of a regulatory DNA GATC site inhibits the binding of the redox-sensing transcriptional activation OxyR, here acting as a repressor (Henderson *et al.* 1997), tying the DNA methylation state to the oxidative stress response.

Examination of pili regulatory mechanisms reveal that a series of DNA–protein and protein–protein interactions make up a regulatory network that integrates pili expression with adaptive response pathways. Various environmental stimuli control the expression of DNA-binding proteins such as Lrp. Lrp binds to specific sequences containing GATC sites, blocking their methylation and forming specific DNA methylation patterns (Braaten *et al.* 1994). These DNA methylation patterns may constitute a type of 'bacterial memory' since they are heritable to subsequent generations. Thus pili expression serves an important paradigm to understand the complex regulatory circuitry that integrates environmental signals with the *in vivo* response of bacterial pathogens.

5. COMBINING THE GLOBAL REGULATORY ADAPTIVE AND METHYLATION RESPONSE SYSTEMS WITH IVET TO REVEAL BACTERIAL VIRULENCE STRATEGIES

Combining global regulatory adaptive and methylation response systems with *in vivo* expression methods yields many powerful experimental tools to reveal bacterial infection strategies and the ensuing host responses. For example, conceptually, the IVET strategy selects for genes that are ON inside the mouse and OFF outside the mouse. Such coordinate expression in response to the transition from host tissues to the laboratory setting suggested a master genetic switch(es) controlling *ivi* gene expression. Accordingly, *lac* fusion technology inherent in the IVET approach was exploited to screen for regulatory mutants that exhibited the phenotype: ON inside the mouse and ON outside the mouse. It was anticipated that this class of mutants should not undergo differential expression in host tissues, and consequently, such mutants should be severely attenuated in their ability to cause disease.

(a) *Dam is a global regulator of bacterial genes induced during infection*

Mutations in Dam answered this genetic selection, repressing the expression of greater than 20% of over 100 *Salmonella ivi* genes (Heithoff *et al.* 1999*b*). Several of these Dam-regulated *ivi* genes have been shown to be involved, or have been implicated, in virulence, and are controlled by environmental signals in the laboratory that are presumed to reflect micro-environments within the host (e.g. low pH, low Mg^{2+} and/or low iron). These results indicate that Dam is a global regulator of *Salmonella* gene expression and that the *dam*-regulated *ivi* genes constitute a *dam* regulon as suggested by Marinus (1996).

(b) *Dam is essential for pathogenesis*

Since Dam is a master switch controlling genes that are specifically induced during infection, its effect on virulence was determined in a murine model of typhoid fever. Dam plays an essential role in *Salmonella* pathogenesis as strains that lack Dam contain a severe virulence defect (greater than 10 000-fold) in that they are unable to cause murine typhoid fever (Heithoff *et al.* 1999*b*). Mutational analysis indicates that the reduction in virulence of Dam⁻ *Salmonella* strains was not due to the loss of methyl-directed mismatch repair (MDMR), or Lrp, since *mutS* and *lrp* mutants are fully virulent. Thus, Dam's global regulatory role in *Salmonella* virulence is clearly distinct from its role in MDMR and its role with Lrp controlling pili expression in *E. coli* (Braaten *et al.* 1994).

(c) *Dam, PhoP and DNA methylation patterns*

Salmonella pathogenesis is known to be controlled by PhoP, a DNA-binding protein that acts as both an inducer and repressor of specific virulence genes (reviewed by Groisman & Heffron 1995). Binding of regulatory proteins to DNA can form DNA methylation patterns by blocking methylation of specific Dam target sites (GATC sequences) (Van der Woude *et al.* 1998). Thus, it was determined whether binding of PhoP (or a PhoP-regulated protein) to specific DNA sites blocks methylation of these sites by Dam, resulting in an alteration in the DNA methylation pattern (following *Mbo*I cleavage and pulsed-field gel analysis). *Salmonella* PhoP$^+$ and PhoP$^-$ DNA methylation patterns showed distinct differences, suggesting that binding of PhoP blocked methylation at specific GATC sites (Heithoff *et al.* 1999*b*). Moreover, *Salmonella* GATC sites protected from methylation are likely to be within gene regulatory regions since almost all GATC sites protected from methylation in *E. coli* are in 5' non-coding DNA regions that are presumably involved in the control of gene expression (Hale *et al.* 1994; Tavazoie & Church 1998). Methylation of specific GATC sites in gene regulatory regions could modulate the binding of regulatory proteins to DNA, which in turn provides a mechanism to control virulence gene expression.

(d) *Dam$^-$ mutants serve as live attenuated vaccines*

Live attenuated vaccines contain living organisms that are benign but can replicate in host tissues and presumably express many natural immunogens similar to those generated during infection. This interaction elicits a protective response just as if the immunized individual had been previously exposed to the disease. Most of the work defining attenuating mutations for the construction of live bacterial vaccines has been done in *Salmonella* spp., since they establish an infection by direct interaction with the gut-associated lymphoid tissue, resulting in a strong humoral immune response. They also invade host cells and thus are capable of eliciting a strong cell-mediated response (Hassan & Curtiss 1997; Hormaeche *et al.* 1996; Miller *et al.* 1990).

Since Dam represses the expression of many *S. typhimurium* genes that are normally only expressed in animal hosts, in the absence of Dam, these genes and their cognate proteins should be ectopically expressed by bacteria when grown on laboratory media (Heithoff *et al.* 1999*b*). Consequently, Dam$^-$ strains were tested for their protective capacity as live attenuated vaccines. Mice immunized with Dam$^-$ mutants were completely protected against a wild-type challenge of 10 000-fold above the LD$_{50}$ of virulent *Salmonella* (Heithoff *et al.* 1999*b*), indicating that Dam$^-$ mutants serve as live vaccines against murine typhoid fever. Additionally, Dam$^-$ *Salmonella* were shown to survive in Peyer's patches of the mouse small intestine at wild-type levels, yet were unable to invade and/or survive in systemic tissues (Heithoff *et al.* 1999*b*). Moreover, it was recently shown that Dam$^-$

bacteria are mildly deficient in the invasion of epithelial cells and the secretion of type III effector molecules and are less cytotoxic to M cells, yet
they retain their protective capacity (Garcia-del Portillo *et al.* 1999). Taken
together, these data indicate how the loss of a single enzyme can completely
block the ability of a pathogen to cause disease yet elicit a fully protective
immune response.

These studies offer two exciting opportunities for further research.
(i) Dams are highly conserved in many pathogenic bacteria that cause significant morbidity and mortality worldwide including *V. cholerae* (cholera),
Shigella spp. (dysentery), *Haemophilus influenzae* (meningitis; ear infection), *Salmonella typhi* (typhoid fever), and pathogenic *E. coli* and *Yersinia
pestis* (plague) (Heithoff *et al.* 1999b). If Dam works as a master switch in
these organisms as it does in *Salmonella*, the construction of Dam-based
live vaccines and antimicrobials that inhibit the Dam enzyme may result
in a new generation of therapeutic agents to combat these infectious diseases. (ii) Since Dam is a global repressor of genes expressed during infection, Dam⁻ mutants of these organisms may inappropriately express
many immunogens, which may elicit a cross-protective response against
related pathogenic strains. Moreover, it may be possible to construct Dam-
based vaccines in strains that are less harmful to humans, yet capable
of conferring full protection against related pathogenic strains. Such vaccines would exploit the benefits of the high levels of protection elicited by
live vaccines, while reducing the risk of infection to immunocompromised
individuals.

6.   CONCLUSIONS AND UNANSWERED QUESTIONS

The reactinogenicity (e.g. symptoms such as diarrhoea) of current live bacterial vaccine candidates and the reality that many individuals within the
human population are immunocompromised clearly warrants the search for
additional vaccines that offer better protection, are longer lasting, and have
less toxicity. Use of *in vivo* expression methods in combination with the
adaptive and/or methylation global regulatory response systems provides a
powerful means to understand pathogenesis, virulence and infection. Such
a multifaceted experimental strategy has broad application towards the design of a new generation of vaccines and antimicrobials to target a wide
variety of infectious agents.

Important unanswered questions for future research include: (i) Are mutants lacking Dam attenuated in other pathogens, and if so, do they elicit a fully protective immune response? (ii) Will Dam⁻ vaccines constructed in less pathogenic strains elicit a cross-protective response to related pathogenic strains? (iii) Can protein fusions to Dam-repressed immunogens be used in Dam⁻ strains to elicit a protective response to heterologous antigens? (iv) Can other master switches of *in vivo* gene expression be identified, and if so, can they be used to construct novel vaccines and antimicrobials?

This work was supported by National Institutes of Health (NIH) grant AI36373, Santa Barbara Cottage Hospital Research Foundation and private donations from Jim and Deanna Dehlsen (M.J.M.), NIH grant AI23348 (D.A.L.) and the Cancer Centre of Santa Barbara (D.M.H.).

REFERENCES

Abu Kwaik, Y. & Pederson, L. L. 1996 The use of differential display-PCR to isolate and characterize a *Legionella pneumophila* locus induced during the intracellular infection of macrophages. *Mol. Microbiol.* **21**, 543–556.

Akins, D. R., Bowrell, K. W., Caimano, M. J., Norgard, M. V. & Radolf, J. D. 1998 A new animal model for studying Lyme disease spirochetes in a mammalian host adapted state. *J. Clin. Invest.* **101**, 2240–2250.

Braaten, B. A., Platko, J. V., Van der Woude, M. W., Simons, B. H., de Graaf, F. K., Calvo, J. M. & Low, D. A. 1992 Leucine-responsive regulatory protein controls the expression of both the pap and fan pili operons in *Escherichia coli. Proc. Natl Acad. Sci. USA* **89**, 4250–4254.

Braaten, B. A., Nou, X., Kaltenbach, L. S. & Low, D. A. 1994 Methylation patterns in *pap* regulatory DNA control pyelonephritis-associated pili phase variation in *E. coli. Cell* **76**, 577–588.

Camilli, A. & Mekalanos, J. J. 1995 Use of recombinase gene fusions to identify *Vibrio cholerae* genes induced during infection. *Mol. Microbiol.* **18**, 671–683.

Chiang, S. L. & Mekalanos, J. J. 1998 Use of signature-tagged transposon mutagenesis to identify *Vibrio cholerae* genes critical for colonization. *Mol. Microbiol.* **27**, 797–805.

Cormack, B. P., Ghori, N. & Falkow, S. 1999 An adhesin of the yeast pathogen *Candida glabrata* mediating adherence to human epithelial cells. *Science* **285**, 578–582.

Coulter, S. N., Schwan, W. R., Ng, E. Y. W., Langhorne, M. H., Ritchie, H. D., Westrock-Wadman, S., Hufnagle, W. O., Folger, K. R., Bayer, A. S. & Stover, C. K. 1998 *Staphylococcus aureus* genetic loci impacting growth and survival in multiple infection environments. *Mol. Microbiol.* **30**, 393–404.

Darwin, A. J. & Miller, V. L. 1999 Identification of *Yersinia enterocolitica* genes affecting survival in an animal host using signature-tagged transposon mutagenesis. *Mol. Microbiol.* **32**, 51–62.

Edelstein, P. H., Edelstein, M. A. C., Higa, F. & Falkow, S. 1999 Discovery of virulence genes of *Legionella pneumophila* by using signature-tagged mutagenesis in a guinea pig pneumonia model. *Proc. Natl Acad. Sci. USA* **96**, 8190–8195.

Falkow, S. 1997 What is a pathogen? *Am. Soc. Microbiol. News* **63**, 356–359.

Fuller, T. E., Shea, R. J., Thacker, B. J. & Mulks, M. H. 1999 Identification of *in vivo* induced genes in *Actinobacillus pleuropneumonia. Microb. Pathogen.* **27**, 311–327.

Garcia-del Portillo, F., Foster, J. W., Maguire, M. E. & Finlay, B. B. 1992 Characterization of the micro-environment of *Salmonella typhimurium*-containing vacuoles within MDCK epithelial cells. *Mol. Microbiol.* **6**, 3289–3297.

Garcia-del Portillo, F., Pucciarelli, M. G. & Casadesus, J. 1999 DNA adenine methylase mutants of *Salmonella typhimurium* show defects in protein secretion, cell invasion, and M cell cytotoxicity. *Proc. Natl Acad. Sci. USA* **96**, 11 578–11 583.

Garcia Vescovi, E., Soncini, F. C. & Groisman, E. A. 1996 Mg^{2+} as an extra-cellular signal: environmental regulation of *Salmonella* virulence. *Cell* **84**, 165–174.

Groisman, E. A. & Heffron, F. 1995 Regulation of *Salmonella* virulence by two-component regulatory systems. In *Two-component signal transduction* (ed. J. A. Hoch & T. J. Silhavy), pp. 319–332. Washington, DC: American Society for Microbiology Press.

Hale, W. B., Van der Woude, M. W. & Low, D. A. 1994 Analysis of nonmethylated GATC sites in the *Escherichia coli* chromosome and identification of sites that are differentially methylated in response to environmental stimuli. *J. Bacteriol.* **176**, 3438–3441.

Handfield, M., Sanschagrin, F. C., Cardinal, G., Mahan, M. J., Woods, D. E. & Levesque, R. C. 2000 *In vivo* induced genes and iron uptake in *Pseudomonas aeruginosa. Infect. Immun.* (In the press.)

Hassan, J. O. & Curtiss III, R. 1997 Efficacy of a live avirulent *Salmonella typhimurium* vaccine in preventing colonization and invasion of laying hens by *Salmonella typhimurium* and *Salmonella enteritidis. Avian Dis.* **41**, 783–791.

Heithoff, D. M., Conner, C. P., Hanna, P. C., Julio, S. M., Hentschel, U. & Mahan, M. J. 1997*a* Bacterial infection as assessed by *in vivo* gene expression. *Proc. Natl Acad. Sci. USA* **94**, 934–939.

Heithoff, D. M., Conner, C. P. & Mahan, M. J. 1997*b* Dissecting the biology of a pathogen. *Trends Microbiol.* **5**, 463–520.

Heithoff, D. M., Conner, C. P., Hentschel, U., Govantes, F., Hanna, P. C. & Mahan, M. J. 1999a Coordinate intracellular expression of *Salmonella* genes induced during infection. *J. Bacteriol.* **181**, 799–807.

Heithoff, D. M., Sinsheimer, R. L., Low, D. A. & Mahan, M. J. 1999b An essential role for DNA adenine methylation in bacterial virulence. *Science* **284**, 967–970.

Henderson, I. R., Meehan, M. & Owen, P. 1997 A novel regulatory mechanism for a novel phase-variable outer membrane protein of *Escherichia coli. Adv. Exp. Med. Biol.* **412**, 349–355.

Hensel, M., Shea, J. E., Gleeson, C., Jones, M. D., Dalton, E. & Holden, D. W. 1995 Simultaneous identification of bacterial virulence genes by negative selection. *Science* **269**, 400–403.

Hormaeche, C. E., Mastroeni, P., Harrison, J. A., Demarco de Hormaeche, R., Svenson, S. & Stocker, B. A. D. 1996 Protection against oral challenge three months after i.v. immunization of BALB/c mice with live Aro *Salmonella typhimurium* and *Salmonella enteritidis* vaccines is serotype (species)-dependent and only partially determined by the main LPS O antigen. *Vaccine* **14**, 251–259.

Krabbe, M., Weyand, N. & Low, D. A. 2000 Environmental control of pili gene expression. In *Bacterial stress responses* (ed. G. Storz & R. Hengge-Aronis). Washington, DC: American Society for Microbiology Press. (In the press.)

Lowe, A. M., Beattie, D. T. & Deresiewicz, R. L. 1998 Identification of novel *Staphylococcal virulence* genes by *in vivo* expression technology. *Mol. Microbiol.* **27**, 967–976.

Mahan, M. J., Slauch, J. M. & Mekalanos, J. J. 1993 Selection of bacterial virulence genes that are specifically induced in host tissues. *Science* **259**, 666–668.

Mahan, M. J., Tobias, J. W., Slauch, J. M., Hanna, P. C., Collier, J. R. & Mekalanos, J. J. 1995 Antibiotic-based selection for bacterial genes that are specifically induced during infection of a host. *Proc. Natl Acad. Sci. USA* **92**, 669–673.

Mahan, M. J., Slauch, J. M. & Mekalanos, J. J. 1996 Environmental regulation of virulence gene expression in *Escherichia, Salmonella* and *Shigella* spp. In Escherichia coli *and* Salmonella. *Cellular and molecular biology*, 2nd edn (ed. F. C. Neidhardt, R. Curtiss III, J. L. Ingraham, E. C. C. Lin, K. B. Low, B. Magasanik, W. S. Reznikoff, M. Riley, M. Schaecter and H. E. Umberger), pp. 2803–2815. Washington, DC: American Society for Microbiology Press.

Marinus, M. G. 1996 Methylation of DNA. In Escherichia coli *and* Salmonella. *Cellular and molecular biology*, 2nd edn (ed. F. C. Neidhardt, R. Curtiss III, J. L. Ingraham, E. C. C. Lin, K. B. Low, B. Magasanik, W. S. Reznikoff, M. Riley, M. Schaecter and H. E. Umberger), pp. 782–791. Washington DC: American Society for Microbiology Press.

Mei, J. M., Nourbakhsh, F., Ford, C. W. & Holden, D. W. 1997 Identification of *Staphylococcus aureus* virulence genes in a murine model of bacteraemia using signature-tagged mutagenesis. *Mol. Microbiol.* **26**, 399–407.

Mekalanos, J. J. 1992 Environmental signals controlling the expression of virulence determinants in bacteria. *J. Bacteriol.* **174**, 1–7.

Merrell, D. S. & Camilli, A. 2000 The *cadA* gene of *Vibrio cholerae* is induced during infection and plays a role in acid tolerance. *Mol. Microbiol.* **34**, 836–849.

Miller, J. F., Mekalanos, J. J. & Falkow, S. 1989 Coordinate regulation and sensory transduction in the control of bacterial virulence genes. *Science* **243**, 916–922.

Miller, S. I., Mekalanos, J. J. & Pulkkinen, W. S. 1990 *Salmonella* vaccines with mutations in the *phoP* virulence regulon. *Res. Microbiol.* **141**, 817–821.

Munoz, C. I. & Bailey, A. M. 1998 A cutinase encoding gene from *Phytophthora capsici* isolated by differential display RT-PCR. *Curr. Genet.* **33**, 225–230.

Nou, X., Skinner, B., Braaten, B., Blyn, L., Hirsh, D. & Low, D. A. 1993 Regulation of pyelonephritis-associated pili phase variation in *Escherichia coli*: binding of the PapI and Lrp regulatory proteins is controlled by DNA methylation. *Mol. Microbiol.* **7**, 545–553.

Plum, G. & Clarke-Curtiss, J. E. 1994 Induction of *Mycobacterium avium* gene expression following phagocytosis by human macrophages. *Infect. Immun.* **62**, 476–483.

Plum, G., Brenden, M., Clarke-Curtiss, J. E. & Pulverer, G. 1997 Cloning, sequencing, and expression of the mig gene of *Mycobacterium avium*, which codes for a secreted macrophage induced protein. *Infect. Immun.* **65**, 4548–4557.

Polissi, A., Pontiggia, A., Feger, G., Altieri, M., Mottle, H., Ferrari, L. & Simon, D. 1998 Large scale identification of virulence genes from *Streptococcus pneumoniae*. *Infect. Immun.* **66**, 5620–5629.

Rainey, P. B. 1999 Adaptation of *Pseudomonas fluorescens* to the plant rhizosphere. *Environ. Microbiol.* **1**, 243–257.

Rainey, P., Heithoff, D. M. & Mahan, M. J. 1997 Single-step conjugative cloning of bacterial fusions involved in microbial–host interactions. *Mol. Gen. Genet.* **256**, 84–87.

Rindi, L., Lari, N. & Garzelli, C. 1999 Search for genes potentially involved in *Mycobacterium tuberculosis* virulence by mRNA differential display. *Biochem. Biophys. Res. Commun.* **258**, 94–101.

Rivera-Marrero, C. A., Burroughs, M. A., Masse, R. A., Vannberg, F. O., Leimbach, D. L., Roman, J. & Murtagh, J. J. 1998 Identification of genes differentially expressed in *Mycobacterium tuberculosis* by differential display PCR. *Microb. Pathogen.* **25**, 307–316.

Shea, J. E., Hensel, M., Gleeson, C. & Holden, D. W. 1996 Identification of a virulence locus encoding a second type III secretion system in *Salmonella typhimurium*. *Proc. Natl Acad. Sci. USA* **93**, 2593–2597.

Soncini, F. C. & Groisman, E. A. 1996 Two-component regulatory systems can interact to process multiple environmental signals. *J. Bacteriol.* **178**, 6796–6801.

Soncini, F. C., Garcia Vescovi, E., Solomon, F. & Groisman, E. A. 1996 Molecular basis of the magnesium deprivation response in *Salmonella typhimurium*: identification of PhoP-regulated genes. *J. Bacteriol.* **178**, 5092–5099.

Staib, P., Kretschmar, M., Nichterlein, T., Kohler, G., Michel, S., Hof, H., Hacker, J. & Morschhauser, J. 1999 Host-induced, stage specific virulence gene activation in *Candida albicans* during infection. *Mol. Microbiol.* **32**, 533–546.

Tavazoie, S. & Church, G. M. 1998 Quantitative whole-genome analysis of DNA–protein interactions by *in vivo* methylase protection in *E. coli*. *Nat. Biotechnol.* **16**, 566–571.

Valdivia, R. H. & Falkow, S. 1996 Bacterial genetics by flow cytometry: rapid isolation of *Salmonella typhimurium* acid-inducible promoters by differential fluorescence induction. *Mol. Microbiol.* **22**, 367–389.

Valdivia, R. H & Falkow, S. 1997 Fluorescence-based isolation of bacterial genes expressed within host cells. *Science* **277**, 2007–2011.

Van der Woude, M. W. & Low, D. A. 1994 Leucine-responsive regulatory protein and deoxyadenosine methylase control the phase and expression of the *sfa* and *daa* pili operons in *Escherichia coli*. *Mol. Microbiol.* **11**, 605–618.

Van der Woude, M. W., Braaten, B. A. & Low, D. A. 1992 Evidence for global regulatory control of pilus expression in *Escherichia coli* by Lrp and DNA methylation: model building based on analysis of *pap*. *Mol. Microbiol.* **6**, 2429–2435.

Van der Woude, M., Braaten, B. & Low, D. A. 1996 Epigenetic phase variation of the *pap* operon in *Escherichia coli*. *Trends Microbiol.* **4**, 5–9.

Van der Woude, M., Hale, W. B. & Low, D. A. 1998 Formation of DNA methylation patterns: nonmethylated GATC sequences in the *gut* and *pap* operons. *J. Bacteriol.* **180**, 5913–5920.

Waldor, M. K. & Mekalanos, J. J. 1996 Lysogenic conversion by a filamentous phage encoding cholera toxin. *Science* **272**, 1910–1914.

Wang, J., Mushegian, A., Lori, S. & Jin, S. 1996 Large scale isolation of candidate virulence genes of *Pseudomonas aeruginosa* by *in vivo* selection. *Proc. Natl Acad. Sci. USA* **93**, 10 434–10 439.

Young, G. & Miller, V. L. 1997 Identification of novel chromosomal loci affecting *Yersinia enterocolitica* pathogenesis. *Mol. Microbiol.* **25**, 319–328.

Yuk, M. H., Harvill, E. T. & Miller, J. F. 1998 The *bvgAS* virulence control system regulates type III secretion in *Bordetella bronchiseptica*. *Mol. Microbiol.* **28**, 945–959.

Zhao, H., Li, X., Johnson, D. E. & Mobley, H. L. 1999 Identification of protease and *rpoN*-associated genes of uropathogenic *Proteus mirabilis* by negative selection in a mouse model of ascending urinary tract infection. *Microbiology* **145**, 185–195.

CHALLENGE OF INVESTIGATING BIOLOGICALLY RELEVANT FUNCTIONS OF VIRULENCE FACTORS IN BACTERIAL PATHOGENS

RICHARD MOXON* and CHRISTOPH TANG

Molecular Infectious Diseases Group,
Oxford University Department of Paediatrics,
Institute of Molecular Medicine, John Radcliffe Hospital,
Oxford OX3 9DS, UK

Recent innovations have increased enormously the opportunities for investigating the molecular basis of bacterial pathogenicity, including the availability of whole-genome sequences, techniques for identifying key virulence genes, and the use of microarrays and proteomics. These methods should provide powerful tools for analysing the patterns of gene expression and function required for investigating hostmicrobe interactions *in vivo*. But, the challenge is exacting. Pathogenicity is a complex phenotype and the reductionist approach does not adequately address the eclectic and variable outcomes of hostmicrobe interactions, including evolutionary dynamics and ecological factors. There are difficulties in distinguishing bacterial 'virulence' factors from the many determinants that are permissive for pathogenicity, for example those promoting general fitness. A further practical problem for some of the major bacterial pathogens is that there are no satisfactory animal models or experimental assays that adequately reflect the infection under investigation. In this review, we give a personal perspective on the challenge of characterizing how bacterial pathogens behave *in vivo* and discuss some of the methods that might be most relevant for understanding the molecular basis of the diseases for which they are responsible. Despite the powerful genomic, molecular, cellular and structural technologies available to us, we are still struggling to come to grips with the question of 'What is a pathogen?'

Keywords: microbial infections; bacterial virulence; pathogenesis; genomics

'... although microorganisms grown *in vitro* are conveniently studied, in this field of pathogenicity they can be incomplete and even misleading.' (Smith 1972)

1. INTRODUCTION

From the moment we are born, humans are exposed to a myriad of microorganisms, some of which become resident on skin, mucous membranes and in the gastrointestinal tract — the process of colonization. In most instances, microbes and man coexist in mutually benign or even symbiotic partnership; commensal organisms may stimulate host immunity and pro-

---

*Author for correspondence (moxon@paediatrics.ox.ac.uk).

vide some essential nutrients and co-factors. Rarely, these resident or newly acquired organisms result in disease in which specific interactions between host and microbe culminate in damage to host tissues and impairment of health. This potential of microbes to injure and even kill their hosts is an essential characteristic of pathogens and represents a constant challenge to human health in every part of the globe.

But, this simple perspective does not do justice to the challenge of defining pathogenicity, a complex phenotype that is dependent on the interactions between a spectrum of different host and microbial molecules. This mutuality of host and microbe, and its implicit co-evolutionary implications, are key concepts. The distinction between commensal and pathogenic behaviour is not a sharp cut-off, but a continuum and must take into account hosts that are compromised (opportunistic infections) and those with competent innate and acquired immune mechanisms. Falkow (1997) defines a pathogen as 'any micro-organism whose survival is dependent upon its capacity to replicate and persist on or within another species by actively breaching or destroying the cellular or humoral host barrier that ordinarily restricts or inhibits other micro-organisms'. Recognizing additional complexity, he emphasizes the importance of (i) ecological factors, for example the capacity to reach a unique host niche free from microbial competition, (ii) variations in host clearance mechanisms, and (iii) evolutionary factors that determine the efficiency with which microbes are transmitted to new, susceptible hosts. But, elusive though it may be, the definition of what is a pathogen is not just an issue of semantics, but a conceptual issue of substance. What we understand by pathogenicity determines and, importantly, confines how we think and therefore has major implications for the theoretical and experimental approaches that we deploy to understand its biological basis (Read *et al.* 1998). (Some use pathogenicity and virulence interchangeably. Others use pathogenicity as a qualitative term, the capacity to damage host tissues or decrease host fitness, and virulence as a quantitative term, for example the challenge dose of organisms that kills a proportion of hosts under defined experimental conditions.)

The survival of pathogenic bacteria depends on transmission from one host to another. This requires factors that facilitate dissemination, translocation and survival between hosts, as well as colonization (Lipsitch & Moxon 1997). The relationship between the microbial factors supporting fitness, commensal and virulence behaviour is complex. Early in the last century, Topley (1919) recognized that selection for enhanced transmissibility might drive pathogens towards heightened pathogenicity under circumstances where the virulence of the pathogen was linked to its rate of replication and transmission.

Ewald (1994) has built upon this general thesis and suggested that water and food are 'cultural vectors' and, by analogy to the transmission of parasites by insects, play an important role in the evolution of bacterial virulence. Our behaviour — living in towns, working in institutions, infringing on rainforests, waging wars, travelling in jet plane — sare major factors affecting microbial transmission and therefore the evolution of pathogenic bacteria. The evolution of pathogenicity is all about trade-offs (Levin & Svanborg-Eden 1990). Natural selection may favour properties that facilitate successful transmission of a microbe, the major driving force being the number of secondary infections that result from a primary infection (Anderson & May 1982). However, singular explanations of the evolution of virulence are too simplistic. The meningococcus, for example, is one of the most virulent microbes of man. But it is not a simple matter to explain how the devastating virulence of this organism benefits its transmission to other humans, because the population of bacteria *in vivo* (for example in blood or spinal fluid) of a fatal case of meningococcal disease is not transmitted to other persons. What drives the evolution of its virulent behaviour?

Over the past three decades, molecular biology has brought about a revolution in the methodology of experiments on pathogenic bacteria. The classical approach to investigate a pathogen, enshrined in Koch's postulates, was transformed by advances in molecular biology (Falkow 1988). Most conspicuously, the availability of molecular genetic techniques encouraged an approach based on the use of genetically defined strains that resulted in an explosion of information on key microbial characteristics contributing to pathogenicity. However, reductionism may obscure key concepts. Too often, pathogenicity is approached as if it could be captured through identifying and characterizing a subset of the genes of a particular microbe. While appealing, this perspective is naive. The challenge of coming to grips with pathogenicity is not so very different from trying to fathom the genetic basis of, say, a champion athlete. Nobody would seriously consider that any fixed set of genes would be common to those who run 100 m in less than 10 s! Our colleagues in human genetics face similar challenges. Consider single gene disorders such as cystic fibrosis or phenylketonuria and the contrast with polygenic diseases such as hypertension or diabetes mellitus. In the field of pathogenicity, there are parallels. For example, tetanus is a disease of a single gene that can be precisely reproduced by injecting an experimental animal with the purified toxin. Contrast tetanus with the pathogenesis of bacterial meningitis in which multiple genes are implicated in a complex series of interdependent, sequential steps in pathogenesis. Crucial to our understanding of these complexities is the need to translate reductionist methods for investigation of pathogens into experiments that enlighten *in vivo* function.

Over many years, few have championed this issue more tenaciously than Harry Smith. In numerous reviews, he has lamented how investigators failed to grasp this nettle. His message is insistent, unambiguous and compelling: stop procrastinating on what might happen *in vivo* and get your teeth into an effort to find out what does happen (a paraphrase of Smith (1996)). The current symposium seems a judiciously timed opportunity to review and renew our efforts to meet the challenge of a better understanding of what a pathogen is. In this presentation, we will offer a personal perspective on what we know about the behaviour of pathogenic bacteria *in vivo* and our thoughts about how we might make progress in the future. As with all biology, the challenge is as exciting as it is daunting. Although we subscribe to the view that our knowledge, where possible, should seek to delineate functions at the molecular level, we caution that the reductionist's dream may be the holist's nightmare!

2. BACTERIAL PATHOGENS *IN VIVO*

Molecules that interact with host factors in the pathogenic process can be classified as follows (Falkow 1988; Moxon 1997):

  (i) tropism for host tissues, or more specifically for particular cells of the host (adhesins);

 (ii) invasion of host tissues by dissemination on, within or through host cells (invasins);

(iii) facilitating microbial survival of the host clearance mechanisms (evasins);

(iv) the strategies for acquiring nutrients and co-factors for *in vivo* growth (pabulins);

(v) damaging, directly or indirectly, host tissues (cytotoxins).

In a typical experimental approach, the role of a candidate virulence gene is studied by comparing the behaviour (in a 'biologically relevant' assay) of the wild-type bacterial pathogen with that of a variant lacking this gene. There is a compelling logic and simplicity to this paradigm, but to what extent does it stand up to scrutiny? We need to think carefully about the assumptions underlying the use of isogenic strains. The interpretation of experiments in which a single gene is expressed or not, ignoring for the moment technical genetic issues such as polar effects, may not be straightforward. The loss of one surface molecule may have significant secondary effects on other determinants. For example, capsule-deficient mutants may have altered membrane proteins (Loeb & Smith 1980), and truncated forms of lipopolysaccharide (LPS) may affect the distribution of membrane phospholipids (Zwahlen *et al.* 1985). Despite the obvious merits of using isogenic strains, these caveats and their implications cannot be ignored.

However, by far the most difficult issue is the selection of an experimental system that allows biologically relevant insights into the function of virulence factors. The best system is to investigate well-characterized bacterial pathogens in spontaneous or experimental infections of their natural hosts. However, for many of the most significant bacterial pathogens, this is not an option. *Haemophilus influenzae* is a good example. This potentially pathogenic bacterium is usually a harmless commensal of the human upper respiratory tract. As far as anyone knows, humans are its only host and therefore appropriate studies on the *in vivo* relevance of its virulence factors are difficult to approach experimentally. Infant rats have been used for many years as an experimental model of *H. influenzae* bacteraemia and meningitis. Table 1 summarizes some of what has been learned about *in vivo* events. However, the limitations of this model must be emphasized. To give a few examples: in almost all of these experiments, the bacterial inoculum was prepared by growing organisms in rich, artificial medium with a surfeit of essential nutrients and co-factors. Yet, it is well known that *H. influenzae* undergoes a phenotypic shift *in vivo*. When organisms were obtained directly from the nasopharynx, or from the blood of bacteraemic animals, they showed relative resistance to complement-mediated killing

Table 1. Usefulness of infant rats as a model for investigating the pathogenesis of *H. influenzae* bacteraemia and meningitis.

bacteria reach the central nervous system by the haematogenous route, not via the cribrifrom plate, following intranasal challenge (Ostrow *et al.* 1979)

the probability of meningitis is directly related to the concentration of bacteria in the blood and to their exceeding a critical threshold of greater than 103 organisms ml^{-1} blood (Moxon & Ostrow 1977)

the population of organisms recovered from the blood or CSF in the acute phase of bacteraemia and meningitis is the progeny of a few founder bacteria, often a single clone (Moxon & Murphy 1978)

opsonophagocytosis, not bactericidal killing, is the major mechanism of *in vivo* clearance (Weller *et al.* 1978)

organisms grown *in vivo* are relatively resistant to antibody-dependent, complement-mediated killing (*in vitro*), when compared with organisms grown *in vitro* (Shaw *et al.* 1976)

expression of phosphorylcholine, a phase-variable component of LPS, facilitates nasopharyngeal colonization, whereas its absence increases intravascular survival (Weiser & Pan 1998)

expression of type b capsule, *per se*, confers heightened virulence compared with other capsular polysaccharides (Zwahlen *et al.* 1989)

mutations of LPS can attenuate systemic infection even when the bacteria are fully encapsulated (Zwahlen *et al.* 1986)

mean generation time of organisms is not less than 50 min in the acute phase of infection (Moxon 1992)

the blood is a major site of replication of organisms in the pathogenesis of bacteraemia (Rubin *et al.* 1985)

prior nasopharyngeal infection, or induction of inflammation, increases susceptibility to bacteraemia following intranasal challenge (Myerowitz 1981)

decreased concentrations of glucose in CSF during acute meningitis do not result from increased consumption by organisms or inflammatory cells (Moxon *et al.* 1979)

seeding the peritoneal cavity of splenectomized rats with fragments of spleen enhances protection against lethal challenge (Moxon & Schwartz 1980)

serum antibodies to type b capsular polysaccharide facilitate clearance of organisms from the nasopharynx (Moxon & Anderson 1979)

when compared with organisms grown *in vitro* (Shaw *et al.* 1976; Rubin & Moxon 1985). The relevant small molecular weight host factors have been identified and, at least some of the phenotypic correlates are associated with phase-variable phenotypic changes in LPS (Roche & Moxon 1995; Weiser & Pan 1998). The frequency distribution of LPS variants in a challenge

inoculum prepared for experimental assays of infection is therefore critical. For example, LPS variants will be a major factor in determining the
efficiency with which *H. influenzae* colonizes the nasopharynx in experimental infections of infant rats or survive intravascular clearance following systemic inoculation. The LPS of *H. influenzae* has evolved digalactoside structures, gal-α-1-4-gal-β, that mimic the glycosylation residues of human cells
(Virji *et al.* 1990), but in the rat, the prevalent glycosylation on host cells is
gal-α-1-3-gal-β. Whereas rats recognize this epitope, humans do not (Weiser
et al. 1997). Thus, *in vivo*, molecular mimicry dependent on species-specific
differences in host glycosylation patterns may be critical. Many of the
adhesins, invasins and evasins of *H. influenzae* that interact with the
host use ligand–receptor interactions that are specific to human tissues,
but not those of rats. For example, IgA1 proteases cleave the relevant
human, but not rat, immunoglobulin (Plaut *et al.* 1977). However, some
cellular features of humans and rats may be similar. Although not a natural commensal or pathogen of rats, the efficiency with which even a small
inoculum of *H. influenzae* can establish colonization in the experimental
setting is impressive (Moxon & Murphy 1978).

Given the difficulties in identifying relevant models of infection, it is
not surprising that much research that attempts to investigate *in vivo*
relevance has used cultured cells to investigate the molecular basis of
the host–microbe interactions that underlie attachment and invasion. But
these studies can be extremely difficult to interpret. For example, the
interactions of *Salmonella typhimurium* with epithelial cells has been
studied in depth, but there is a considerable body of evidence that the
pathogenesis of *Salmonella typhi* (the pathogenic species of most relevance
to humans) involves entry through M cells (reviewed by Falkow 1996). All
too often, this aspect of *in vivo* relevance and interpretation is given short
shrift. Clear distinctions might need to be made between the scientific merit
of using bacteria as probes for investigating cellular functions, such as
endocytosis or trafficking, and the pragmatic issue of their relevance
to pathogenesis. For example, the pathogenicity of enteropathogenic
Escherichia coli is associated with the effacement of microvilli and
cytoskeletal rearrangements. The latter have been confirmed through ultrastructural studies on jejunal epithelial cells observed in clinical specimens
(Rothbaun *et al.* 1982, 1983). *In vitro* investigations have identified the
molecular details of signalling proteins that induce these changes, which

include protein phosphorylation and inositol and calcium fluxes, and the role of environmental factors such as the effects of temperature, growth phase of the bacteria, calcium and osmolarity (Kenny *et al.* 1997). These investigations offer a detailed coalition of *in vivo* and *in vitro* research, but how these events lead to diarrhoea is unknown!

Environmental factors, such as osmolarity, pH and available nutrients, influence bacterial pathogenicity by affecting the production of virulence determinants and by controlling the growth rate (Miller *et al.* 1989). For example, when *S. typhimurium* is present in the intestinal lumen, several environmental and regulatory conditions modulate the expression of factors required for bacterial entry into host cells. An excellent example of the complex and coordinate regulation of virulence genes *in vivo* is expression of six different invasion genes, located on the pathogenicity island SPI-1, which are coordinately regulated by oxygen, osmolarity, and pH (Bajaj *et al.* 1996). An important point is that the activation of these virulence genes is obligatorily dependent on a multiplicity of environmental cues. Specific, combinatorial responsiveness provides a mechanism whereby bacteria behave in a very organized manner, subject to subtle variations in the different extracellular or intracellular environments encountered in the host. Another fascinating issue is that, as a general rule, physical contact between bacteria and host cells is required for the expression of virulence genes in animal hosts. Upon contact with a host cell, *Yersinia pseudotuberculosis* increases the rate of transcription of certain virulence genes. This microbe–host interaction triggers export of LcrQ, a component of a type III secretion system and a negative regulator of the expression of a family of secreted proteins (Yops). A decrease in the intracellular concentration of LcrQ mediates increased expression of Yops. Thus, the type III secretion system plays a key role in the coordinate elaboration of virulence factors after physical contact with the target cell (Pettersson *et al.* 1996). Interestingly, in microbial–plant interactions, substantial molecular signalling occurs under circumstances where the infecting bacterium is spatially distanced from the target tissues.

Expression of virulence factors may depend on the microbial population attaining a critical density to trigger the elaboration of bacterial cell-to-cell signalling molecules (e.g. *N*-acyl-L-homoserine lactones in *Pseudomonas aeruginosa*) (Winson *et al.* 1995). Although relatively little work has been done *in vivo* to confirm the role of quorum sensing systems (but see Williams *et al.*, this volume) the importance of cell-to-cell signalling within bacterial populations is undoubtedly both neglected and important. *Staphylococcus aureus* uses a global regulator, *agr*, which is activated by secreted autoinducing peptides, to control the expression of its major virulence genes. The autoinducing peptides show sequence variation that affects their specificity. These peptides are thiolactones that either activate or inhibit depending on the conformation of the ligand–receptor interaction, the fine structural details of which are now understood in detail (Ji *et al.* 1997; Balaban *et al.* 1998). This has made it possible to use synthetic variants of these peptides for *in vitro* and *in vivo* assays to establish their relevance and to open the door to novel strategies of infection control (Mayville *et al.* 1999).

There has been much emphasis on the important contributions of gene regulation and regulatory networks. But, gene regulation provides only one example of the genetic mechanisms that have evolved to facilitate bacterial acclimatization to their host. Pathogenic bacteria face especially demanding tests of their adaptive potential, a consequence of the diversity of host polymorphisms (innate immunity) and the capacity to generate extensive repertoires of T and B cells (acquired immunity). Typically, bacterial infections occur within a matter of hours in which bacteria *in vivo* encounter host landscapes of such diversity that prescriptive strategies, such as two-component sensorytransducer systems or classical gene regulation, may be inadequate to encompass the plethora of potential variables. The challenge is all the more striking in that many infections stem from the clonal expansion of a single bacterial cell (Meynell & Stocker 1957; Moxon & Murphy 1978). How then do the small numbers of bacteria that make up a pathogen population in an individual host generate the necessary diversity to adapt to host polymorphisms and immune clearance mechanisms? One strategy is through increasing the mutation frequency of those genes that are involved in critical interactions with their hosts. In many pathogenic bacteria, this hypermutability occurs because selected genes have evolved sequences containing runs of repetitive DNA (Stern *et al.* 1986; Weiser *et al.* 1989). Nucleotide repeats (microsatellites) are highly mutable, because of their propensity to undergo slippage (Streisinger *et al.* 1966; Levinson & Gutman 1987). The resulting mutations, involving spontaneous loss or gain of nucleotides, result in reversible, high-frequency on–off switching of genes through altered transcription, altered binding of RNA

Table 2. Why might we not be identifying relevant (*in vivo*) pathogenicity factors?

factors	reasons
adhesins, invasins	dependent on appropriate 'model' or assay
pabulins	difficulties in distinguishing 'permissive' or 'fitness' genes from virulence factors
evasins	dependent on host, immune status, route of inoculation, etc.
toxins	some methods do not detect, e.g. signature-tagged mutagenesis

polymerase to promoters, or translation, frame shifts (Moxon & Wills 1999). Given that there are several such genetic loci in a single pathogen genome and that the mutations occur at random and independently of one another, the combinatorial effect on the phenotypic diversity of the pathogen population can be substantial. These genes have been called contingency loci, to emphasize their potential to enable at least a few bacteria in a given population to adapt to unpredictable and precipitous contingencies [sic] within, and in transmission between, different host environments (Moxon *et al.* 1994). Traits encoded by contingency genes include those governing recognition by the immune system, motility, attachment to and invasion of host cells, and acquisition of nutrients (Table 2).

A specific example of a contingency gene with a proven role in pathogenesis is *lic1a* in *H. influenzae*. The *lic1* locus is required for the synthesis of choline phosphate (ChoP), a component of *H. influenzae* LPS (Weiser *et al.* 1997). The *lic1a* gene contains multiple copies of a tetranucleotide, CAAT, in the 5'-portion of the gene and, through slippage, *lic1a* switches on and off at high frequency; the result is loss or gain of ChoP$^+$ and ChoP$^-$ LPS glycoforms. In an infant rat model, ChoP$^+$ variants colonize the nasopharynx more efficiently than ChoP$^-$ variants. However, ChoP$^+$ variants are more susceptible to C-reactive protein (CRP)-dependent, complement-mediated killing and variants found in blood and cerebrospinal fluid (CSF) are almost exclusively ChoP$^-$. Thus, the switching of this contingency gene confers an adaptive strategy that is selectively advantageous to the survival of *H. influenzae* at different stages in its pathogenesis (Weiser & Pan 1998).

Survival and commensal or pathogenic behaviour *in vivo* is dependent on the availability of nutrients, and these requirements may be the basis for tissue tropism. This was shown many years ago in the context of *Proteus mirabilis* and urea concentrations in the kidney (MacLaren 1970), and of *Brucellae* and the abundance of erythritol in ungulate foetal tissues (Smith *et al.* 1962). Although much attention has been focused on the provision of iron and its essentiality for bacterial growth, the same cannot be said for other key nutrients. How, for example do bacteria scavenge zinc to provide the metalloproteins for DNA and RNA polymerases? It is only very recently, based on the availability of whole-genome sequences, that information on these important mechanisms has been uncovered for *H. influenzae* and *H. ducreyi* (Lewis *et al.* 1999). Nicotinamide adenine dinucleotide (NAD^+) is also essential for growth of *H. influenzae* and *H. ducreyi* yet, despite its essentiality, the molecular mechanisms by which NAD^+ is taken up and transported remain uncharacterized. A periplasmic pyrophosphatase that hydrolyses NAD^+ or flavin adenine nucleotide to the mononucleotide has been described and analogues capable of inhibiting this enzymatic activity impair growth of *H. influenzae* (Anderson *et al.* 1985). However, much more remains to be learned about the contribution of these essential mechanisms to commensal and virulence behaviour.

Compared with replication *in vitro*, *in vivo* multiplication is complicated by the uncertain contribution of host immunity since population numbers are the result of replication and clearance (Brock 1971). Meynell devised an elegant method for estimating bacterial growth rates *in vivo* that gives the true division rate based on the dilution of non-replicating plasmids. This was used to estimate the growth rate of *S. typhimurium* following intravenous inoculation of mice and their recovery from the spleen (Maw & Meynell 1968). The division rates in the organisms in the spleen proved to be only 5–10% of the maximum observed *in vitro*. Interestingly, however, the death rate of the bacteria in the spleen was also extremely small. Thus, the bacteria presumably could persist in the spleen for long periods. These and similar methods could yield information of great value on many of the problems discussed.

Putative virulence factors expressed *in vitro* may not be expressed *in vivo*. Alternatively, molecules made *in vivo* may not be made *in vitro*. *S. aureus* synthesizes poly-N-succinyl β-1-6 glucosamine (PNSG) as a surface polysaccharide during human and animal infection, but these strains do not produce PNSG *in vitro* (McKenney *et al.* 1999). On the other hand, the *ail* gene of *Y. enterocolitica* facilitates serum resistance and invasion of cultured cells *in vitro*, but when examined in mouse models, had no detectable influence on colonization or LD_{50} following oral challenge (Wachtel & Miller 1995).

Taken together, these observations indicate that although we know many of the questions, there is a considerable shortfall, even neglect, of some of the most fundamental issues concerning the functional relevance of virulence factors. Yet, in addition to the huge body of descriptive, clinical information on bacterial infections — experiments of nature — we have available to us an array of extremely powerful resources and methods to facilitate experiments in the laboratory and the field. However, future studies should take into account the need to consider critically their limitations, as well as their strengths. For example, we can predict that many adhesins and invasins would not be identified by signature-tagged mutagenesis (STM) because the tissues of the animal model may not express the appropriate ligands, or the route of inoculation may bypass relevant tissues. STM does not identify cytotoxins, and redundancy or complementation may obscure the identification of some evasins or pabulins. In the case of *in vivo* expression technology (IVET) and differential fluorescence imaging, we must be cautious in interpreting *in vivo*, but not *in vitro*, gene transcription as a strong indicator of a biologically relevant function. This message has been powerfully affirmed by studies on yeast (Winzeler *et al.* 1999). Furthermore, constraints on export, rather than transcription and translation, are limiting factors of relevance for *in vivo* function of molecules produced by type III secretion systems. The approach to pathogens in the future must secure a greater emphasis on collaboration through research consortia. This makes sense in the context of projects driven by whole-genome sequences, where a range of expertise in genetics, cell biology (including immunology), microscopy and structural biology can be assembled. Importantly, microbiologists, pathologists and infectious diseases physicians have a wealth of experience gathered through years of observation, as well as precious collections of strains and clinical specimens. Too often, communications between the clinic and the molecular biology laboratory are inadequate to bring about the necessary convergence of information and methodologies to execute comprehensive biologically relevant *in vitro* and *in vivo* assays of virulence factors in different hosts and at different stages of an infection.

3.   WHOLE-GENOME SEQUENCING HAS REVOLUTIONIZED STUDIES OF PATHOGENIC BACTERIA

The availability of complete bacterial genome sequences has had a major impact on the opportunities for investigating the biological basis of pathogenicity, although enormous challenges still remain (Strauss & Falkow 1997). The impact of genomics is far reaching because it provides the most economical means of acquiring large amounts of information and because it has forged the creation of new technologies to exploit these data. Furthermore, some pathogens are highly specialized microbes that are difficult to grow in the laboratory. As we move from the era of the gene to the era of the genome, it is essential that we move beyond the functional characterization of isolated genes. To make sure genome analysis fulfils its potential with respect to pathogenesis it is essential to make bioinformatic tools, services, and methodologies accessible to the scientific community (Bork *et al.* 1998). This requires combining traditional reductionist approaches with studies that consider genome-level organization and population biology. Gene context, gene organization and gene acquisition must be considered in the experimental approaches to bacterial pathogens. This will require collaboration between computational biologists and those experimentalists who intimately know their respective organisms at the molecular and whole-organism levels. We should also not forget that the ultimate challenge and opportunity for genomics is the completion of the host genome sequence, when we can evaluate pathogens within the genetic context of their hosts.

Whole-genome sequences make a unique contribution, one that differs and contrasts importantly with the tools of the classical methods and their technological refinements. By providing an inventory of every nucleotide of a pathogen, experimental approaches based on whole-genome sequences can command unprecedented rigour; the great advantage of a 'top-down' approach is that it is comprehensive, a feature that distinguishes it from most other methodologies, such as classical genetics. For example, random mutagenesis, offers a powerful approach; but because mutations in many genes are absolutely or conditionally lethal, many potentially important virulence factors remain obscure. Even worse, we cannot know how many we are missing!

Appropriate informatics is essential to unlock and make available the immense fund of information contained in whole-genome sequences (Field *et al.* 1999). Every genome project has reported the identification of multiple putative virulence-related genes scored by homology. Given the large number of genome sequences now available, it is becoming common to apply 'species-filter' approaches to data, mining for information about genes that control virulence phenotype, a process that entails subtracting all homologues of one genome from another to define common, as opposed to unshared, genes. There are 116 genes in *H. influenzae* that have no homologue in *E. coli* but have a known function or have a similarity to a gene of another genome. Out of these, a search was made for proteins potentially involved in the interaction of *H. influenzae* with its host. Almost half of the open reading frames (ORFs) fulfilled the criteria of being found exclusively in pathogens or having sequence similarity to putative adhesins, invasins or cytotoxins (Huynen *et al.* 1997). An elegant example of the application of microarray methods to virulence concerns bacille Calmette–Guérin (BCG) vaccines that are used as live attenuated vaccines against diseases caused by *Mycobacterium tuberculosis*. Comparative hybridizations between the genomes of *M. tuberculosis*, *M. bovis* and the various daughter BCG strains were undertaken. Twelve novel deletions were found in daughter lineages compared with the progenitor strain, findings that may shed light on the varying effectiveness of different lots of BCG vaccine and their association with attenuated virulence (Behr *et al.* 1999).

It is essential to understand how genes are organized within genomes to produce the complex phenotypic traits underlying commensal or virulence behaviour. One example is the development of databases capable of piecing together metabolic pathways. The concept of the '*in silico*' cell is still in its infancy but will surely be refined so as to give us strong clues as to how bacterial cells can function in particular environments. This hypothesis-generating approach can be coupled to the use of microarrays (see §4), a method allowing genome-based surveys of genetic variation and differences in expression profiles under different conditions, including infected tissues. Another important contribution of whole-genome sequences is the identification of virulence-related genes acquired by gene transfer. These gene acquisitions are often extremely obvious through an analysis of the nucleotide composition of the complete genome by the detection of sequences that vary significantly from those typical of the species (Censini *et al.* 1996).

If the information from genomes is not accessible to those who have the expertise to exploit it, then progress will be thwarted. It has taken a few years to orchestrate the liaison between genome information and the various experimental methods for its exploitation. One of the lessons learned is how essential it is that those who have a command of the biology are involved in whole-genome sequencing projects from the earliest stage. The absence or deficiency of these appropriate collaborations is an important reason why, to date, there are relatively few examples of genome-driven translational biology in the literature, even though the first complete genome sequence of a pathogenic bacterium was completed in 1995. However, this message has been assimilated and consortia are now being organized to undertake comprehensive, genome-based analyses of pathogens with an emphasis on biologically relevant functions. A consortium approach was taken from the outset to sequence *Bacillus subtilis*. This has proved to have great advantages in the subsequent coherence of exploiting the genome information, but another clear lesson is that specialized units or institutes can do the sequencing immeasurably more efficiently and economically. Thus, sequencing projects should be carried out in a seamless collaboration that also provides appropriate expertise in bioinformatics, systematic tagging of all genes (protein-coding and stable RNA genes), mutational, transcriptional and translational studies in concert with *in vitro* and *in vivo* assays.

The availability of the 1.83 megabase-pair sequence of the *H. influenzae* strain Rd genome facilitated significant progress in investigating the biology of its LPS, a major virulence determinant of this human pathogen. By searching the genomic database with sequences of known LPS biosynthetic genes from other organisms, 25 candidate LPS genes were identified and cloned. Construction of mutant strains and characterization of the LPS by reactivity with antibodies, polyacrylamide gel electrophoresis (PAGE) fractionation patterns and electrospray mass spectrometry confirmed a potential role in LPS biosynthesis for the majority of these candidate genes. This allowed a series of virulence studies in an infant rat model to estimate the minimal LPS structure required for intravascular dissemination (Hood *et al.* 1996).

Informatics provides a critical 'hypothesis-generating' starting point for biologically relevant experiments. However, its usefulness is dependent on intelligent and expert application. Sequences identify ORFs, i.e. predicted but not proven genes, whose putative functions can be hypothesized but not assumed without critical appraisal of evidence. Database entries vary enormously in their accuracy, appropriateness and the extent to which they have been validated experimentally. An excellent example to illustrate this point is provided by work on calmodulin, a small Ca^{2+}-binding protein that translates Ca^{2+} signals into cellular responses in yeast. An analysis of mutants indicated that deletion of the gene was lethal but reduction or even complete ablation of calcium binding did not interfere with efficient growth. Thus, calmodulin can perform essential functions without the apparent ability to bind Ca^{2+}. Far from diminishing the importance of genetics and biochemistry, genomics has heightened their importance in investigating the assumptions, extrapolations and speculations that habitually surround assignments of function (Geiser *et al.* 1991).

The power of genomics in raising and refining key questions must be emphasized. For example, a striking finding in the completed genome sequences of every bacterium to date has been the number of genes lacking either any homology to existing sequences in the current databases or with homologies to genes of unknown function. Depending on criteria, we are unable to assign functions to about 15–40% of genes, although the genome sequence of *M. tuberculosis* seems to be an exception. It is reasonable to suppose that many of these genes in bacterial pathogens will be important virulence factors *in vivo*, and in a few instances, this has been shown to be so (Heithoff *et al.* 1997; Martindale *et al.* 2000).

Five of the six genomes published in 1998–1999 were human pathogens, all host adapted. Four of these were obligate intracellular pathogens and their study is providing novel insights into host–pathogen interactions and co-evolution. Many have a lifestyle that is characterized by a relative reduction in genome size, the numbers of genes and the complexity of regulatory elements as well as an increased dependence on the host environment for essential nutrients. Several genome projects are very significant because they mark the beginning of a significant trend in the sequencing of closely related genomes (e.g. *Mycobacteria, Neisseria, Mycoplasma, E. coli, Helicobacter*). This is important because the availability of a complete sequence from a single strain is only the starting point in understanding the genetic diversity of the natural population and its implications for commensal and virulence behaviour.

The information from one index genome of a pathogen must be extended to embrace the diversity of the species. And one obvious drawback of genomics is that usually only one strain of a pathogen is sequenced, although there are exceptions to this (*M. tuberculosis, Helicobacter pylori, Neisseria meningitidis, E. coli*). Nonetheless, the substantial diversity that exists in the natural populations of the bacterial species must be emphasized. In particular, an appreciation of the extent of horizontal transfer of DNA from the global gene pool is critical. Thus, as well as allelic diversity through shift and drift, some strains within a species may possess genes (or sequences within genes) that are denied to others. An example of this is Opc, an adhesin and invasin, found in only a proportion of *N. meningitidis* strains (Seiler *et al.* 1996). Some genes may be crucial to virulence, or contribute to heightened virulence. The capsule of *H. influenzae* is clearly implicated in the potential to cause invasive infections, and of the six distinct capsular polysaccharides, that of serotype b (polyribosyl-ribitol phosphate) heightens pathogenicity in a fashion denied to other capsular serotypes (Moxon & Kroll 1990).

We need to provide a clear understanding of the differences underlying pathogenic and commensal behaviour, or the potential for different diseases. *E. coli* is responsible for many different clinical syndromes of gastrointestinal disease, and to a great extent these differences can be attributed to the acquisition of specific virulence genes through horizontal transfer. *E. coli* well illustrates the depths of our ignorance about *in vivo* behaviour. Enterotoxigenic *E. coli* strains elaborate a toxin that is similar to that of cholera toxin. A great deal of genetic, structural and biochemical data have been accrued on the toxins of both these pathogens, but why after an initially clinically similar prodromal illness do classic cholera strains go on to cause a much more severe, life threatening disease? Another seeming paradox is the production of toxin (PT), a major virulence factor in *Bordetella pertussis*. *In vivo* evidence provides strong support for its key role in pathogenesis since a subunit vaccine based on PT protects against disease. However, *B. parapertussis* lacks PT, but clinical cases of comparable severity occur (Hoppe 1999; He *et al.* 1998). So what is the explanation?

Whole-genome sequences are also opening the door to understanding the importance of higher-order processes (expression patterns, gene regulation, kinetic properties, localization and concentration effects, environmental influences, and fitness contributions). When we do observe what happens *in vivo*, we need to consider not merely qualitative aspects of function (e.g. the potential to induce inflammation), but the quantitative correlates (e.g. the extent to which the factor induces inflammation). Correlation of genotype and virulence potential is not black and white, but replete with subtleties. These issues are critical in our efforts to evaluate vaccine or drug targets (for example, where absolute conservation of an epitope is sought) as contrasted with, for example, novel diagnostic possibilities that depend on identifying variations. Furthermore, we have not been in a position to appreciate the variability in metabolic pathways and these may critically impact on *in vivo* behaviour at different steps in pathogenesis.

4. SYSTEMATIC, GENE-BY-GENE, ANALYSIS OF BACTERIAL PATHOGENS

There is now great excitement among the community of molecular microbiologists concerning prospects for the comprehensive analysis of bacterial pathogens. For many of the major pathogens, there is access to complete genome sequences, appropriate techniques for genetic transfer are in place and importantly, well-validated animal models are available. There is a wealth of accumulated observational data on the biology of infections, as well as many years of accumulated data obtained by microbiologists. Also, there are extensive strain collections, and specimens of infected tissues from biopsies and autopsiesnot forgetting the important insights that can be afforded by studies on ancient DNA!

The potential for substantial progress is reflected in recent publications on functional genomics (Pallen 1999; Fields *et al.* 1999; Winzeler *et al.* 1999). The starting point is the identification, through annotation of the pathogen's genome, of the coding sequences for all proteins and stable RNAs. For a host-adapted pathogenic bacterium, this typically amounts to between 1500 and 2000 putative genes. For many of these genes, no functions are known or the sequences lack any homologies to entries in available general databases. Indeed, incisive data on the function of the majority of these genes will be lacking. Precise deletion of all genes, for example using inverse PCR with sequence tagging (bar-coding) and/or fluorescein labelling, can be attempted using a variety of directed strategies, which are likely to realize some 500–700 mutants. These mutants can be investigated using high throughput screening techniques (see §5(b)), starting with growth in rich and minimal media *in vitro* and thence to tissue or organ culture, embryonated chicken eggs, small animals (rats, mice) and eventually natural hosts including, possibly, humans (Cohen *et al.* 1994). For many of these experiments, pools of deletion mutants can be used in the first instance and aliquots sampled at various times during the growth of cultures or stages of infection. Tags can then be amplified and hybridized to microarrays (see §5(b)). Although apparently simple in outline, such ambitious, comprehensive analyses represent a formidable task. Ideally, this programme of investigation should use a consortium approach (Figure 1).

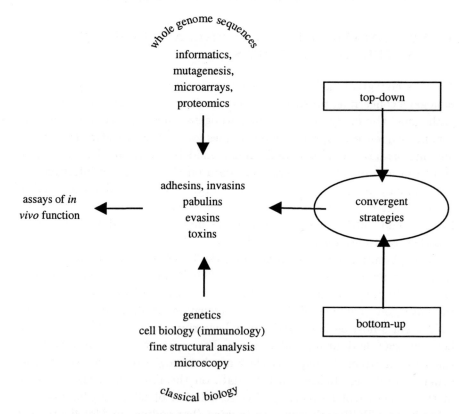

Fig. 1. Convergence of strategies for investigation of bacterial pathogen *in vivo*. The goal is to identify molecules of pathogenic bacteria that are crucial to virulence. Adhesins, invasins, pabulins, evasins and toxins (see §1 for further details of these) can be identified by a combination of genomic and classical genetic approaches. Both genomic and classical genetic approaches have a variety of powerful supporting technologies to facilitate their exploitation. The big challenge is to translate these tools of discovery into an understanding of biological function in pathogenesis.

The pathogenesis of bacterial infections is accompanied by a series of molecular changes in both the bacterium and the host cells. These events result in altered expression of a number of genes. There are now a number of methods for comparing global gene expression profile changes in the bacterium and the host cells. These methods include oligonucleotide arrays (Lockhart *et al.* 1996), serial analysis of gene expression (SAGE) (Velculescu *et al.* 1995), differential cDNA screening (Peitu *et al.* 1996), expressed sequence tag database comparison (Vasmatzis *et al.* 1998), and two-dimensional gel electrophoresis of cellular proteins. Many of these methodologies are labour-intensive and not yet suitable for high-throughput use, for example those dependent on gel-based methods such as SAGE. However, oligonucleotide arrays are a realistic high-throughput methodology, although many technological problems still need to be refined and validated. Nonetheless, there are already many published examples where microarrays have been used to provide identity and expression levels of selected genes simultaneously, providing that the gene sequences are known; i.e. the complete genome sequence of the relevant pathogen is available. The technological advances in the past decade have lead to miniaturization of the appropriate synthesis and attachment chemistry; thousands of DNA or RNA molecules can be arrayed on a few square centimetres of a solid phase, e.g. a glass slide (Cheng *et al.* 1998). This allows monitoring of the expression level of genes.

As a demonstration of the feasibility of this technology, RNA was extracted from *Streptococcus pneumoniae* cultures, with or without exposure to the competence-stimulating peptide (CSP), and labelled with biotin. This allowed investigation of differences in the expression of genes involved in the development of natural competence for DNA transformation by probing a microarray of *S. pneumoniae* genes. Whereas most of the studied genes showed no differences in expression, whether or not they had been exposed to CSP, genes from the competence operon were induced (de Saizieu *et al.* 1998). There are now several demonstrations of the feasibility of this approach to monitor and compare gene expression of organisms in different environmental conditions (Kononen *et al.* 1998; DeRisi *et al.* 1997). However, convincing demonstrations of its application to pathogenic bacteria *in vivo* are yet to come, although we would expect this to be achieved to the point that, within a few years, such an approach is as routine a methodology as automated sequencing.

5.   THE CONVERGENCE OF METHODS

(a) *Gene expression* in vivo

An innovative and powerful method for identifying virulence factors was described and used prior to the availability of whole-genome sequences. IVET was devised to select positively for bacterial genes that are specifically induced when bacteria infect host tissues (Mahan *et al.* 1993). Many of these *in vivo*-induced genes play key roles in the infectious process. Several versions of the IVET principle have been developed, each offering different capabilities including universal (rescue of *purA* auxotrophy (Mahan *et al.* 1993)), intracellular (use of antibiotic selection (Mahan *et al.* 1995)) or stage-specific (use of genetic recombination (Camilli *et al.* 1994)) gene expression in host tissues. A limitation of IVET is that it will not detect many relevant genes, for example, where there are subtle, but biologically significant, *in vitro* versus *in vivo* differences in gene activity. Not surprisingly, many of the genes discovered by IVET are implicated in *in vivo* metabolic or physiological adaptations of pathogenic bacteria (including pabulins), rather than being classical virulence factors (adhesins, invasins, evasins and toxins). The availability of whole genomes is an obvious bonus in that only a few nucleotides of any gene selected by IVET are needed to access its entire coding and flanking sequences. The principles of IVET can also be used in a 'top-down' approach in which the genome sequence can be used to select all or subsets of genes and submit them to microarrays, as one component of a multifaceted high-throughput screening approach.

(b) *STM and complementation of gene libraries*

In STM, transposons carrying unique DNA sequence tags are used to produce a bank of mutants (Hensel *et al.* 1995). The tags from a mixed population of bacterial mutants representing the inoculum (input) and bacteria recovered from infected hosts (output) are detected by amplification, radiolabelling and hybridization analysis. This allows detection of *in vivo* bacterial virulence genes by negative selection; those mutants present in the input pool that cannot be recovered from the output pool are presumably deficient in some function at one or more stages in the infectious process. STM can be adapted to screen site-directed mutants of complete sets, or subsets, of genes based on information from whole-genome sequences. A considerable advantage of STM is that large numbers of mutants can be screened in pools of approximately 100 per animal.

In the absence of well-defined systems for genetic transfer, complementation analysis *in vivo* provides a means for identifying genes associated with virulence. Integrating shuttle cosmid libraries that replicate as multicopy plasmids in *E. coli* and integrate into the mycobacterial chromosome was used to identify key virulence factors using a pair of genetically related *M. tuberculosis* strains, one fully virulent, the other attenuated (Pascopella 1994).

Despite the power of information from genomes and its integration with high-throughput screening of mutants *in vivo*, it is necessary to characterize the role of each of these putative virulence factors using tissue or organ cultures, whole-animal assays, cell biology (including immunology), microscopy and structural biology.

(c) *Immune responses as an indicator of* in vivo *function*

A limiting factor in many functional genomic strategies is the sheer volume of manipulation required to screen for virulence genes and also direct these studies so as to identify genes of *in vivo* relevance. Using host immune responses to indicate genes expressed *in vivo* is both powerful and specific. There are numerous examples of using bacteria obtained from patients or infected animals followed by analysis using sodium dodecyl sulphate PAGE and immunoblotting with convalescent sera (multiple references are cited in Smith (1996)). However, an innovative recent approach is to capitalize upon the surprising finding that amplified PCR fragments can be rendered transcriptionally active when injected into animals, so-called DNA or nucleic-acid immunization (Hoffman *et al.* 1995). For example, given the 4000 genes revealed by the whole-genome sequence of *M. tuberculosis*, each gene could be individually amplified from the genome by PCR. Each PCR product would be annealed to fragments encoding promoter and terminator sequences and pools (approximately 50) of these linear expression elements inoculated into mice (Sykes & Johnston 1999). With minor modifications, we suggest that this approach can be used to carry out a systematic and comprehensive analysis in which the antibodies obtained to *in vivo*-expressed proteins are used to detect the microbial products expressed during infection. Such analysis could use fluorescence for *in situ* work or, more simply, an enzyme-linked immunosorbent assay on *ex vivo* material. In the latter, plates could be coated with extracts of different infected tissues, normal tissues serving as controls, or the same tissues at different stages of infection to obtain a profile of *in vivo* expression during pathogenesis. This approach would not of course be useful for identifying macromolecules such as capsule or LPS and would suffer from the potential loss of epitopes critical to antibody recognition through *in vivo* degradation or changes in conformation.

A different strategy for selecting bacterial genes expressed only *in vivo* which can take advantage of the availability of whole-genome sequence data is the use of expression libraries screened with two types of antibodies. This approach has some advantages for studying pathogens that lack efficient genetic transfer systems. Suk *et al.* (1995) used this approach to identify genes in *Borrelia burgdorferi*. Reactivities of antibodies obtained from animals immunized with killed organisms and from infected hosts were compared. The genomic clones that reacted with antibody from infected animals, but not from immunized animals, indicated genes expressed only in the host.

(d) *Fluorescence-activated cell sorter (FACS) to select bacteria bearing transcriptionally active* gfp *fusions* **in vivo**

Valdivia & Falkow (1996) have devised an elegant strategy that lends itself perfectly to a convergent approach using genomics and high-throughput screening, for example using FACS. The essence of this methodology is that, under particular conditions *in vivo*, transcriptionally active genes fused to *gfp* will fluoresce. FACS can then be used to identify either fluorescent bacteria or host cells containing fluorescent bacteria. The only genetic requirements are that the bacterial pathogen is able to maintain an episomal element and express the functional *gfp*. Although, as initially described, this system used random DNA fragments inserted upstream of a promoterless *gfp* gene, the system could be used for a site-specific strategy using a comprehensive approach based on whole-genome sequences. The fluorescence intensity of individual bacteria grown in tissue culture can be compared with the same bacterial clone after release from infected cells.

This allows selection of bacteria that are transcriptionally active only within the host cell (for example, a macrophage (Valdivia & Falkow 1997)). In the original study, 14 insertion mutants with intracellular-dependent activities were identified. This study was done prior to the availability of the relevant whole-genome sequence of *S. typhimurium*, so the investigators were obliged to characterize these genes the hard way, i.e. by cloning, DNA sequencing and further characterization. However, with the availability of the whole-genome sequences of a pathogen under study, a high-throughput, efficient and comprehensive approach should be feasible since, once the promoter is identified, the entire gene to which it corresponds can be easily isolated. Additionally, it will be possible to identify those promoter fusions that are unstable. An advantage of using fluorescence detection is that there is no dependence upon growth. Many pathogens enter viable, but non-culturable states, e.g. *Vibrio cholerae*. This is therefore a powerful strategy for studying environmentally triggered gene induction in bacteria obtained from infected animals or host cells that accurately reflect conditions of natural infections. As has been stressed, the activity of virulence genes is sensitive to precise combinations of environmental conditions. It is therefore useful to use such strategies to track differences in gene expression at different times and at different stages of pathogenesis.

(e) *Photonic detection of bacterial pathogens in living hosts*

Real-time, non-invasive analyses of pathogenic events can now be performed *in vivo* through marking bacterial pathogens that have been transfected with a plasmid conferring constitutive expression of bacterial luciferase. The transmission of light through opaque tissues *in vivo* has been used for non-invasive optical imaging and spectroscopy in both animals and humans. Further, localization of photons from bioluminescent molecules has been used quantitatively to monitor gene expression in plants (Millar *et al.* 1995) and promoter activity in mammalian cells (Hooper *et al.* 1990). These observations suggested that an optical method for evaluating disease progression is possible. This was elegantly demonstrated by Contag *et al.* (1995) who showed that bioluminescent *Salmonella* could be observed adhering to and entering epithelial cells and macrophages. The detection of photons transmitted through tissues of animals infected with bioluminescent *Salmonella* allowed localization of the bacteria to specific tissues so that progressive infections were distinguished from those that were persistent or abortive. Through this technique, the patterns of bioluminescence suggest that the caecum may play a pivotal role in the pathogenesis of *Salmonella* infection.

(f) *Confocal microscopy*

The traditional tools for examining infectious processes *in vivo* include doseresponses of mortality (LD_{50}), or some appropriate end-point such as bacteraemia, and infection kinetics of organs, such as the liver, lungs or kidneys. These studies, combined with observations from tissue culture experiments do not allow detailed localization of host–pathogen interactions *in vivo*. An excellent example of how more recent techniques can overcome these deficiencies is afforded by studies on the intracellular pathogen *S. typhimurium* (Richter-Dahlfors *et al.* 1997). To pinpoint which particular cell type harbours *S. typhimurium*, confocal laser scanning electron microscopy and computerized image analysis techniques were used to determine bacteria–cell associations at an early stage in the infection of mouse liver cells. This study was able to derive data using relatively small inocula (approximately 100 organisms), a much more convincing scenario than the artificially high challenge doses used in preceding studies. The key to the sensitivity of these experiments was the use of thick sections and the powerful resolution afforded by confocal microscopy and computer-assisted imaging. Using antibodies directed against different host cell types, it was possible to identify the intracellular location of the bacteria and to determine the specific cell type involved.

(g) *Molecular approaches to histopathology*

For studies of bacteria *in vivo*, advances in the application of molecular histopathological techniques to clinical and experimental specimens offer huge potential (Quirke & Mapstone 1999). There are already valuable techniques for the detection of infectious agents in paraffin-embedded tissue, including non-culturable bacteria. Newer thermal cyclers combined with Taqman[TM] fluorescent dyes make amplification and detection of bacterial DNA possible in a few minutes and include quantitative PCR (Belgrader *et al.* 1999). Among the leading-edge technologies, atomic-force microscopy is a type of physical microscopy that allows the shape of a structure under analysis to be visualized down to the atom. It can be used on routine paraffin sections with or without immunochemistry, its power being derived from its ability to image DNA and individual proteins. This opens up the possibility of studying normal and abnormal protein–DNA interactions in real time under physiological conditions. This ability to investigate the surface morphology, at an unprecedented level of resolution, has been used already to study the surface of *E. coli* exposed to antibiotics, but there would seem to be great potential for examining the effects on bacterial morphology of host factors (Braga & Ricci 1998). Through digital storage of information, the image of bacteria can be rotated allowing cross-sections of the bacterium obtained at any point. The initial steps in bacterial adhesion with host cells can be studied to gauge the precise manner in which ligand–receptor, van der Waals and electrostatic interactions are affected by cell surface components (Razatos *et al.* 1998). Interactions between the cantilever tip and confluent monolayers of isogenic bacteria will allow appreciation of subtle effects of cell surface macromolecules, such as LPS and capsular polysaccharide, on host tissues.

6. SUMMARY

The main points are as follows: pathogenicity is a complex entity and its definition is elusive. The strong reductionist culture of today's science, while having great strengths, may obscure the goal of integrating molecular detail into a solid framework of *in vivo* relevance to pathogenesis. For the student of bacterial infections, a critical and convergent approach is needed. Whole-genome sequences of major bacterial pathogens are, or will be, available and can provide an invaluable source of information, and as well as being one of the ways of initiating relevant hypotheses about *in vivo* function. It is crucial that the information from genomes is accessible in 'user friendly' form to provide the basis of a variety of data-mining algorithms and the powerful analyses possible through comparative genomics. A variety of high-throughput screening methods, such as nucleic-acid microarrays and proteomics, are available to make bridges from bacterial genomes to the experimental biology of pathogenic and commensal behaviour. But the crucial studies of *in vivo* function will require not only the relevant biological assays but also a convergence of clinical observations, carefully amassed strain collections and naturally or experimentally infected host tissues. These resources must be integrated with the judicious application of genetics, cell biology, fine structural analyses and high-resolution microscopy. Expertise in physiology, intermediary metabolism and biochemistry has been relatively neglected, but must be resurrected to meet the challenges of the post-genomic era. The task is challenging and may be best met through consortium approaches since many individual research groups are likely to be overwhelmed by the amount of data and resources that emanate from genome projects, as well as the contingent demand for eclectic expertise. Nonetheless, the current era offers unprecedented challenges and excitement to spur our creativity as we try to understand the depths and details of bacterial pathogenicity.

We thank our colleagues in the Molecular Infectious Diseases Group, Department of Paediatrics, for their many useful discussions, Paul Rainey (Department of Plant Sciences, University of Oxford) and Jay Hinton (now at the Institute of Food Research, Norwich) for their insightful comments.

REFERENCES

Anderson, R. M. & May, R. M. (eds) 1982 *Population biology of infectious diseases. Report of the Dahlem workshop on population biology of infectious disease agents.* Berlin: Springer.

Anderson, B. M., Kahn, D. W. & Anderson, C. D. 1985 Studies of the $2' : 3'$-cyclic nucleotide phosphodiesterase of *Haemophilus influenzae. J. Gen. Microbiol.* **131**, 2041–2045.

Bajaj, V., Lucas, R. L., Hwang, C. & Lee, C. A. 1996 Co-ordinate regulation of *Salmonella typhimurium* invasion genes by environmental and regulatory factors is mediated by control of *hilA* expression. *Mol. Microbiol.* **22**, 703–714.

Balaban, N. (and 10 others) 1998 Autoinducer of virulence as a target for vaccine and therapy against *Staphylococcus aureus. Science* **280**, 438–440.

Behr, M. A., Wilson, M. A., Gill, W. P., Salamon, H., Schoolnik, G. K., Rane., S. & Small, P. M. 1999 Comparative genomics of BCG vaccines by whole-genome DNA microarray. *Science* **284**, 1520–1523.

Belgrader, P., Bennett, W., Hadley, D., Richards, J., Stratton, P., Mariella Jr, R. & Milanovich, F. 1999 PCR detection of bacteria in seven minutes. *Science* **284**, 449–450.

Bork, P., Dandekar, T., Diaz-Lazcoz, Y., Eisenhaber, F., Huynen, M. & Yuan, Y. 1998 Predicting function: from genes to genomes and back. *J. Mol. Biol.* **283**, 707–725.

Braga, P. E. & Ricci, D. 1998 Atomic force microscopy: application to investigation of *Escherichia coli* morphology before and after exposure to cefodizime. *Antimicrob. Agents Chemother.* **42**, 18–22.

Brock, T. D. 1971 Microbial growth rates in nature. *Bacteriol. Rev.* **35**, 39–58.

Camilli, A., Beattie, D. T. & Mekalanos, J. J. 1994 Use of genetic recombinations as a reporter of gene expression. *Proc. Natl Acad. Sci. USA* **91**, 2634–2638.

Censini, S., Lange, C., Xiang, Z., Crabtree, J. E., Ghiara, P., Borodovsky, M., Rappuoli, R. & Covacci, A. 1996 *cag*, a pathogenicity island of *Helicobacter pylori*, encodes type I-specific and disease-associated virulence factors. *Proc. Natl Acad. Sci. USA* **93**, 14 648–14 653.

Cheng, J., Sheldon, E. L., Wu, L., Uribe, A., Gerrue, L. O., Carrino, J., Heller, M. J. & O'Connell, J. P. 1998 Preparation and hybridization analysis of DNA/RNA from *E. coli* on microfabricated bioelectronic chips. *Nat. Biotech.* **116**, 541–546.

Cohen, M. S., Cannon, J. G., Jerse, A. E., Charniga, L. M., Isbey, S. F. & Whicker, L. G. 1994 Human experimentation with *Neisseria gonorrhoeae*: rationale, method and implications for the biology of infection and vaccine development. *J. Infect. Dis.* **169**, 532–537.

Contag, C. H., Contag, P. R., Mullins, J. I., Spilman, S. D., Stevenson, D. K. & Benaron, D. A. 1995 Photonic detection of bacterial pathogens in living hosts. *Mol. Microbiol.* **18**, 593–603.

DeRisi, J. L., Iyer, V. R. & Brown, P. O. 1997 Exploring the metabolic and genetic control of gene expression on a genomic scale. *Science* **278**, 680–686.

de Saizieu, A., Certa, U., Warrington, J., Gray, C., Keck, W. & Mous, J. 1998 Bacterial transcript imaging by hybridization of total RNA to oligonucleotide arrays. *Nat. Biotech.* **16**, 45–48.

Ewald, P. 1994 *The evolution of infectious disease*. New York: Oxford University Press.

Falkow, S. 1988 Molecular Koch's postulates applied to microbial pathogenicity. *Rev. Infect. Dis.* **10**, S274–S276.

Falkow, S. 1996 The evolution of pathogenicity in *Escherichia*, *Shigella*, and *Salmonella*. In Escherichia coli *and* Salmonella *cellular and molecular biology*, 2nd edn (ed. F. C. Neidhardt, R. Curtiss III, J. L. Ingraham, E. C. C. Lin, K. B. Low, B. Magasanik, W. S. Reznikoff, M. Riley, M. Schaechter & H. E. Umbarger), pp. 2723–2728. Washington, DC: American Society for Microbiology Press.

Falkow, S. 1997 What is a pathogen? *Am. Soc. Microbiol. News* **63**, 359–365.

Field, D., Hood, D. & Moxon, R. 1999 Contribution of genomics to bacterial pathogenesis. *Curr. Opin. Genet. Dev.* **9**, 700–703.

Fields, S., Kohara, Y. & Lockhart, D. J. 1999 Functional genomics. *Proc. Natl Acad. Sci. USA* **96**, 8825–8826.

Finlay, B. B. & Falkow, S. 1997 Common themes in microbial pathogenicity revisited. *Microbiol. Mol. Biol. Rev.* **61**, 136–169.

Geiser, J. R., Van Tuinen, D., Brockerhoff, S. E., Neff, M. M. & Davis, T. N. 1991 Can calmodulin function without binding calcium? *Cell* **65**, 949–959.

He, Q., Viljanen, M. K., Arvilommi, H., Aittanen, B. & Mertsola, J. 1998 Whooping cough caused by *Bordetella pertussis* and *Bordetella parapertussis* in an immunized population. *J. Am. Med. Assoc.* **280**, 635–637.

Heithoff, D. M., Conner, C. P., Hanna, P. C., Julio, S. M., Hentschel, U. & Mahan, M. J. 1997 Bacterial infection as assessed by *in vivo* gene expression. *Proc. Natl Acad. Sci. USA* **94**, 934–939.

Hensel, M., Shea, J. E., Gleeson, C., Jones, M. D., Dalton, E. & Holden, D. W. 1995 Simultaneous identification of bacterial virulence genes by negative selection. *Science* **269**, 400–403.

Hoffman, S. L., Doolan, D. L., Sedagah, M., Gramzinski, R., Wang, H., Gowda, K., Hobart, P., Margalith, M., Norman, J. & Hedstrom, R. C. 1995 Nucleic acid malaria vaccines. Current status and potential. *Ann. NY Acad. Sci.* **772**, 88–94.

Hood, D. W., Deadman, M. E., Allen, T., Martin, A., Brisson, J. R., Fleischmann, R., Venter, J. C., Richards, J. C. & Moxon, E. R. 1996 Use of the complete genome sequence information of *Haemophilus influenzae* strain Rd to investigate lipopolysaccharide biosynthesis. *Mol. Microbiol.* **22**, 951–965.

Hooper, E. C., Ansorge, R. E., Browne, H. M. & Tomkins, P. 1990 CCD imaging of luciferase gene expression in single mammalian cells. *J. Biolum. Chemilum.* **5**, 123–130.

Hoppe, J. E. 1999 Update on respiratory infection caused by *Bordetella parapertussis*. *Pediatr. Infect. Dis. J.* **18**, 375–381.

Huynen, M. A., Diaz-Lazcoz, Y. & Bork, P. 1997 Differential genome display. *Trends Genet.* **13**, 389–390.

Ji, G., Beavis, R. & Novick, R. P. 1997 Bacterial interference caused by autoinducing peptide variants. *Science* **276**, 2027–2030.

Kenny, B., Abe, A., Stein, M. & Finlay, B. B. 1997 Enteropathogenic *Escherichia coli* protein secretion is induced in response to conditions similar to those in the gastrointestinal tract. *Infect. Immun.* **65**, 2606–2612.

Kononen, J., Bubendorf, L., Kallioniemi, A., Brlund, M., Schraml, P., Leighton, S., Torhorst, J., Mihatsch, M. J., Sauter, G. & Kallioniemi, O.-P. 1998 Tissue microarrays for high-throughput molecular profiling of tumor specimens. *Nat. Med.* **4**, 844–847.

Levin, B. R. & Svanborg-Eden, C. 1990 Selection and evolution of virulence in bacteria: an ecumenical excursion and modest suggestion. *Parasitology* **100**, S103–S115.

Levinson, G. & Gutman, G. A. 1987 Slipped-strand mispairing: a major mechanism for DNA sequence evolution. *Mol. Biol. Evol.* **4**, 203–221.

Lewis, D. A., Klesney-Tait, J., Lumbley, S. R., Ward, C. K., Latimer, J. L., Ison, C. A. & Hansen, E. J. 1999 Identification of the *znuA*-encoded periplasmic zinc transport protein of *Haemophilus ducreyi*. *Infect. Immun.* **67**, 5060–5068.

Lipsitch, M. & Moxon, E. R. 1997 Virulence and transmissibility of pathogens: what is the relationship? *Trends Microbiol.* **31**, 31–37.

Lockhart, D. J. (and 10 others) 1996 Expression monitoring by hybridization to high-density oligonucleotide arrays. *Nat. Biotech.* **14**, 1675–1680.

Loeb, M. R. & Smith, D. H. 1980 Outer membrane protein composition in disease isolates of *Haemophilus influenzae*: pathogenic and epidemiological implications. *Infect. Immun.* **30**, 709–717.

McKenney, D., Pouliot K. L., Wang, Y., Murthy, V., Ulrich, M., Döring, G., Lee, J. C., Goldmann, D. A. & Pier, G. B. 1999 Broadly protective vaccine for *Staphylococcus aureus* based on an *in vivo*-expressed antigen. *Science* **284**, 1523–1527.

MacLaren, D. M. 1970 The influence of urea on the growth of *Proteus mirabilis*. *Guys Hosp. Rep.* **119**, 133–143.

Mahan, M. J., Slauch, J. M., Hanna, P. C., Camilli, A., Tobias, J. W., Waldor, M. K. & Mekalanos, J. J. 1993 Selection for bacterial genes that are specifically induced in host tissues: the hunt for virulence factors. *Infect. Agents Dis.* **2**, 263–268.

Mahan, M. J., Tobias, J. W., Slauch, J. M., Hanna, P. C., Collier, R. J. & Mekalanos, J. J. 1995 Antibiotic-based selection for bacterial genes that are specifically induced during infection of a host. *Proc. Natl Acad. Sci. USA* **92**, 669–673.

Martindale, J., Stroud, D., Moxon E. R. & Tang, C. M. 2000 Genetic analysis of *Escherichia coli* K1 gastrointestinal colonization. *Mol. Microbiol.* (Submitted.)

Maw, J. & Meynell, G. G. 1968 The true division and death rates of *Salmonella typhimurium* in the mouse spleen determined with superinfecting phage P22. *Br. J. Exp. Pathol.* **49**, 597–613.

Mayville, P., Ji, G., Beavis, R., Yang, H., Goger, M., Novick, R. P. & Muir, T. M. 1999 Structure–activity analysis of synthetic autoinducing thiolactone peptides from *Staphylococcus aureus* responsible for virulence. *Proc. Natl Acad. Sci. USA* **96**, 1218–1223.

Meynell, G. G. & Stocker, B. A. D. 1957 Some hypotheses on the aetiology of fatal infections in partially resistant hosts and their application to mice challenged with *Salmonella paratyphi*-B or *Salmonella typhimurium* by intraperitoneal injection. *J. Gen. Microbiol.* **16**, 38–58.

Millar, A. J., Carré, I. A., Straer, C. A., Chua, H.-H. & Kay, S. A. 1995 Circadian clock mutants in *Arabidopsis* identified by luciferase imaging. *Science* **267**, 1161–1163.

Miller, J. F., Mekalanos, J. J. & Falkow, S. 1989 Coordinate regulation and sensory transduction in the control of bacterial virulence. *Science* **243**, 916–922.

Moxon, E. R. 1992 Molecular basis on invasive *Haemophilus influenzae* type b disease. *J. Infect. Dis.* **165**, S77–S81.

Moxon, E. R. 1997 Applications of molecular microbiology to vaccinology. *Lancet* **350**, 1240–1244.

Moxon, E. R. & Anderson, P. 1979 Meningitis caused by *Haemophilus influenzae* in infant rats: protective immunity and antibody priming by gastro-intestinal colonization with *Escherichia coli*. *J. Infect. Dis.* **140**, 471–478.

Moxon, E. R. & Kroll, J. S. 1990 The role of bacterial polysaccharide capsules as virulence factors. In *Bacterial capsules* (ed. K. Jann & B. Jann), pp. 65–85. Berlin: Springer.

Moxon, E. R. & Murphy, P. A. 1978 *Haemophilus influenzae* bacteraemia and meningitis resulting from survival of a single organism. *Proc. Natl Acad. Sci. USA* **75**, 1534–1536.

Moxon, E. R. & Ostrow, P. T. 1977 *Haemophilus influenzae* meningitis in infant rats: role of bacteremia in pathogenesis of age-dependent inflammatory responses in cerebrospinal fluid. *J. Infect. Dis.* **135**, 303–307.

Moxon, E. R. & Schwartz, A. D. 1980 Heterotopic splenic autotransplantation in the prevention of *Haemophilus influenzae* meningitis and fatal sepsis in Sprague–Dawley rats. *Blood* **56**, 842–845.

Moxon, E. R. & Wills, C. 1999 DNA microsatellites: agents of evolution? *Sci. Am.* **280**, 72–77.

Moxon, E. R., Smith, A. L. & Averill, D. 1979 Brain carbohydrate metabolism during experimental *Haemophilus influenzae* meningitis. *Pediatr. Res.* **13**, 52–59.

Moxon, E. R., Rainey, P. B., Nowak, M. A. & Lenski, R. E. 1994 Adaptive evolution of highly mutable loci in pathogenic bacteria. *Curr. Biol.* **4**, 24–33.

Myerowitz, R. L. 1981 Mechanism of potentiation of experimental *Haemophilus influenzae* type B disease in infant rats by influenza A virus. *Lab. Invest.* **44**, 434–441.

Ostrow, P. T., Moxon, E. R., Vernon, N. & Kapko, R. 1979 Pathogenesis of bacterial meningitis: studies on the route of meningeal invasion following *Haemophilus influenzae* inoculation of infant rats. *Lab. Invest.* **40**, 678–685.

Pallen, M. J. 1999 Microbial genomes. *Mol. Microbiol.* **32**, 907–912.

Pascopella, L., Collins, F. M., Martin, J. M., Lee, M. H., Hatfull G. F., Stover, C. K., Bloom, B. R. & Jacobs, W. R. 1994 Use of *in vivo* complementation in *Mycobacterium tuberculosis* to identify a genomic fragment associated with virulence. *Infect. Immun.* **62**, 1313–1319.

Peitu, G., Alibert, O., Guichard, V., Lamy, B., Bios, F., Leroy, E., Mariage-Samxon, R., Houlgatte, R., Soularue, P. & Auffray, C. 1996 Novel gene transcripts preferentially expressed in human muscles revealed by quantitative hybridization of a high density cDNA array. *Genome Res.* **6**, 492–503.

Pettersson, J., Nordfelth, R., Dubinina, E., Bergman, T., Gustafsson, M., Magnusson, K. E. & Wolf-Watz, H. 1996 Modulation of virulence factor expression by pathogen target cell contact. *Science* **273**, 1231–1233.

Plaut, A. G., Gilbert, J. V. & Wistar, R. 1977 Loss of antibody activity in human immunoglobulin A exposed extracellular immunoglobulin A proteases of *Neisseria gonorrhoeae* and *Streptococcus sanguis*. *Infect. Immun.* **17**, 130–135.

Quirke, P. & Mapstone, N. 1999 The new biology: histopathology. *Lancet* **354**, SI26–SI31.

Razatos, A., Ong, Y.-L., Sharma, M. M. & Georgiou, G. 1998 Molecular determinants of bacterial adhesion monitored by atomic force microscopy. *Proc. Natl Acad. Sci. USA* **95**, 11 059–11 064.

Read, A. F. (and 10 others) 1998 What can evolutionary biology contribute to understanding virulence. In *Evolution in health and disease* (ed. S. Stearns), pp. 205–215. Oxford University Press.

Richter-Dahlfors, A., Buchan, A. M. J. & Finlay, B. B. 1997 Murine salmonellosis studied by confocal microscopy: *Salmonella typhimurium* resides intracellularly inside macrophages and exerts a cytotoxic effect on phagocytes *in vivo*. *J. Exp. Med.* **186**, 569–580.

Roche, R. J. & Moxon, E. R. 1995 Phenotypic variation in *Haemophilus influenzae*: the interrelationship of colony opacity, capsule and lipopolysaccharide. *Microb. Pathogen.* **18**, 129–140.

Rothbaun, R., McAdams A., Gianella, R. & Partin, J. 1982 A clinico-pathogenic study of enterocyte-adherent *Escherichia coli*: a cause of protracted diarrhea in infants. *Gastroenterology* **83**, 441–454.

Rothbaun, R., Partin, J., Sallfield, K. & McAdams, A. 1983 An ultrastructural study of enteropathogenic *Escherichia coli* infection in human infants. *Ultrastruct. Pathol.* **4**, 291–304.

Rubin, L. G. & Moxon, E. R. 1985 The effect of serum-factor induced resistance to somatic antibodies on the virulence of *Haemophilus influenzae* type b. *J. Gen. Microbiol.* **131**, 515–520.

Rubin, L. G., Zwahlen, A. & Moxon, E. R. 1985 Role of intravascular replication in the pathogenesis of experimental bacteraemia due to *Haemophilus influenzae* type b. *J. Infect. Dis.* **152**, 307–314.

Seiler, A., Reinhardt, R., Sarkari, J., Caugant, D. A. & Achtman, M. 1996 Allelic polymorphism and site-specific recombination in the *opc* locus of *Neisseria meningitidis*. *Mol. Microbiol.* **19**, 841–856.

Shaw, S., Smith, A., Anderson, P. & Smith, D. H. 1976 The paradox of *Haemophilus influenzae* type b bacteremia in the presence of serum bactericidal activity. *J. Clin. Invest.* **58**, 1019–1029.

Smith, H. 1972 The little-known determinants of microbial pathogenicity. In *Microbial pathogenicity in man and animals. Twenty-second symposium of the Society for General Microbiology, Imperial College, April 1972*, pp. 1–17. Cambridge University Press.

Smith, H. 1996 What happens *in vivo* to bacterial pathogens? *Ann. NY Acad. Sci.* **797**, 77–92.

Smith, H., Williams, A. E., Pearce, J. H., Keppie, J., Harris-Smith, P. W., Fitzgeorge, R. B. & Witt, K. 1962 Foetal erythritol: a cause of the localisation of *Brucella abortus* in bovine contagious abortion. *Nature* **193**, 47–49.

Stern, A., Brown, M., Nickel, P. & Meyer, T. F. 1986 Opacity genes in *Neisseria gonorrhoeae*: control of phase and antigenic variation. *Cell* **47**, 61–71.

Strauss, E. J. & Falkow, S. 1997 Microbial pathogenesis: genomics and beyond. *Science* **276**, 707–712.

Streisinger, G., Okada, Y., Emrich, J., Newton, J., Tsugita, A., Terzaghi, E. & Inouye, M. 1966 Frameshift mutations and the genetic code. *Cold Spring Harb. Symp. Quant. Biol.* **31**, 77–84.

Suk, K., Das, S., Sun, W., Jwang, B., Barthold, S. W., Flavell, R. A. & Fikrig, E. 1995 *Borrelia burgdorferi* genes selectively expressed in the infected host. *Proc. Natl Acad. Sci. USA* **92**, 4269–4273.

Sykes, K. F. & Johnston, S. A. 1999 Linear expression elements: a rapid, *in vivo*, method to screen for gene functions. *Nat. Biotech.* **17**, 355–359.

Topley, W. W. C. 1919 The spread of bacterial infection. *Lancet* (July 5) **ii**, 1–5.

Valdivia, R. H. & Falkow, S. 1996 Bacterial genetics by flow cytometry: rapid isolation of *Salmonella typhimurium* acid-inducible promoters by differential fluorescence induction. *Mol. Microbiol.* **22**, 367–378.

Valdivia, R. H. & Falkow, S. 1997 Fluorescence-based isolation of bacterial genes expressed within host cells. *Science* **277**, 2007–2011.

Vasmatzis, G., Essand, M., Brinkmann, U., Lee, B. & Pastan, I. 1998 Discovery of three genes specifically expressed in human prostate by expressed sequence tag database analysis. *Proc. Natl Acad. Sci. USA* **95**, 300–304.

Velculescu, V. E., Zhang, L., Vogelstein, B. & Kinzler, K. W. 1995 Serial analysis of gene expression. *Science* **270**, 484–487.

Virji, M., Weiser, J. N., Lindberg, A. A. & Moxon, E. R. 1990 Antigenic similarities in lipopolysaccharides of *Haemophilus* and *Neisseria* and expression of digalactoside structure also present on human cells. *Microb. Pathogen.* **89**, 441–450.

Wachtel, M. R. & Miller, V. L. 1995 *In vitro* and *in vivo* characterization of an *ail* mutant of *Yersinia enterocolitica*. *Infect. Immun.* **63**, 2541–2548.

Weiser, J. N. & Pan, N. 1998 Adaptation of *Haemophilus influenzae* to acquired and innate humoral immunity based on phase variation of lipopolysaccharide. *Mol. Microbiol.* **30**, 767–775.

Weiser, J. N., Love, J. M. & Moxon, E. R. 1989 The molecular mechanism of phase-variation of *Haemophilus influenzae* lipopolysaccharide. *Cell* **59**, 657–665.

Weiser, J. N., Shchepetov, M. & Chong, S. T. 1997 Decoration of lipopolysaccharide with phosphorylcholine: a phase-variable characteristic of *Haemophilus influenzae*. *Infect. Immun.* **65**, 943–950.

Weller, P. F., Smith, A. L., Smith, D. H. & Anderson, P. 1978 Role of immunity in the clearance of bacteremia due to *Haemophilus influenzae*. *J. Infect. Dis.* **138**, 427–436.

Wells, D., Sherlock, J. K., Handyside, A. H. & Delhanty, J. D. 1999 Detailed chromosomal and molecular genetic analysis of single cells by whole genome amplification and comparative genomic hybridisation. *Nucl. Acids Res.* **27**, 1214–1218.

Winson, M. K (and 12 others) 1995 Multiple *N*-acyl-L-homoserine lactone signal molecules regulate production of virulence determinants and secondary metabolites in *Pseudomonas aeruginosa*. *Proc. Natl Acad. Sci. USA* **92**, 9427–9431.

Winzeler, E. A. (and 24 others) 1999 Functional characterization of the *S. cerevisiae* genome by gene deletion and parallel analysis. *Science* **285**, 901–906.

Zwahlen, A., Rubin, L. G., Connelly, C. J., Inzana, T. J. & Moxon, E. R. 1985 Alteration of the cell wall of *Haemophilus influenzae* type b by transformation with cloned DNA: association with attenuated virulence. *J. Infect. Dis.* **152**, 485–492.

Zwahlen, A., Rubin, L. G. & Moxon, E. R. 1986 Contribution of lipopolysaccharide to pathogenicity of *Haemophilus influenzae*: comparative virulence of genetically-related strains in rats. *Microb. Pathogen.* **1**, 465–473.

Zwahlen, A., Kroll, J. S., Rubin, L. G. & Moxon, E. R. 1989 The molecular basis of pathogenicity in *Haemophilus influenzae*: comparative virulence of genetically-related capsular transformants and correlation with changes at the capsulation locus *cap*. *Microb. Pathogen.* **7**, 225–235.

VIRULENCE GENE REGULATION INSIDE AND OUTSIDE

VICTOR J. DIRITA[*,‡,§], N. CARY ENGLEBERG[§,¶], ANDREW HEATH[§],
ALITA MILLER[‡], J. ADAM CRAWFORD[†,‡] and ROSA YU[‡]

‡ *Department of Microbiology and Immunology*
§ *Unit for Laboratory Animal Medicine and*
¶ *Division of Infectious Diseases, University of Michigan Medical School,*
Ann Arbor, MI 48103-0620, USA

Much knowledge about microbial gene regulation and virulence is derived from genetic and biochemical studies done outside of hosts. The aim of this review is to correlate observations made *in vitro* and *in vivo* with two different bacterial pathogens in which the nature of regulated gene expression leading to virulence is quite different. The first is *Vibrio cholerae*, in which the concerted action of a complicated regulatory cascade involving several transcription activators leads ultimately to expression of cholera toxin and the toxin-coregulated pilus. The regulatory cascade is active *in vivo* and is also required for maintenance of *V. cholerae* in the intestinal tract during experimental infection. Nevertheless, specific signals predicted to be generated *in vivo*, such as bile and a temperature of 37°C, have a severe down-modulating effect on activation of toxin and pilus expression. Another unusual aspect of gene regulation in this system is the role played by inner membrane proteins that activate transcription. Although the topology of these proteins suggests an appealing model for signal transduction leading to virulence gene expression, experimental evidence suggests that such a model may be simplistic. In *Streptococcus pyogenes*, capsule production is critical for virulence in an animal model of necrotizing skin infection. Yet capsule is apparently produced to high levels only from mutation in a two-component regulatory system, CsrR and CsrS. Thus it seems that in *V. cholerae* a complex regulatory pathway has evolved to control virulence by induction of gene expression *in vivo*, whereas in *S. pyogenes* at least one mode of pathogenicity is potentiated by the absence of regulation.

Keywords: *Vibrio cholerae*; ToxR; *Streptococcus pyogenes*; virulence; regulation

1. INTRODUCTION

Studies on how virulence traits are regulated in bacterial pathogens have generally been guided by either of two intuitions. The first is that genes encoding virulence factors are in fact regulated and that, knowing the virulence gene products, it should be possible to work backwards to the

---

*Author for correspondence (vdirita@umich.edu).
†Present address: Department of Microbiology, University of Maryland School of Medicine, Baltimore, MD 21201, USA.

regulatory factors and thereby to the *in vivo* regulatory parameters. The second guiding intuition has been that correct assumptions may be made about the *in vivo* environment and these will aid in identifying both virulence factors and their regulatory parameters. An example of the first class of intuition was the identification of the regulator for the cholera toxin genes of *Vibrio cholerae*, after it had been demonstrated that high-level expression of cholera toxin is not constitutive (Pearson & Mekalanos 1982). Thus Miller & Mekalanos (1984) cloned the relatively weak promoter for the cholera toxin genes (*ctxAB*) into *Escherichia coli* as an operon fusion to *lacZ*, and subsequently identified a regulatory gene, *toxR*, whose expression in this *E. coli* background resulted in elevated *ctx-lacZ* expression. From this simple experiment sprouted a vast field of observations concerning the regulation of cholera toxin, as well as of several other factors important in the pathogenesis of cholera (Skorupski & Taylor 1997). This system will be explored in more detail in §3(a).

An example of the second intuition underlying regulatory studies in pathogenesis led to classic work on *Shigella* spp., including the observation that cell invasion by these organisms is temperature regulated. Genes encoding temperature-regulated phenotypes were therefore targeted by using a promoterless *lacZ* gene engineered into a transposable element. Screening was for fusions with high activity at 37°C and low activity at 25 or 30°C (Maurelli *et al.* 1984; Maurelli & Curtiss 1984; Maurelli & Sansonetti 1988). This approach led to the identification of several virulence genes encoded on a virulence-associated plasmid, which have been well characterized in the ensuing time-period, and also, eventually, to regulatory elements that control expression of these genes including VirF, VirB and H-NS (Dorman & Porter 1998).

The difference between these examples, of course, is in the relative extent of knowledge available prior to when studies of gene regulation were imposed on each system. Cholera toxin, the signal virulence determinant in *V. cholerae*, had been studied for a long time prior to cloning the genes and subsequent analysis of their regulation. Although the fact of its environmental regulation was well established by the work of Richardson and his colleagues (Evans & Richardson 1968; Callahan & Richardson 1973; Richardson 1969), the study of gene regulation proceeded once ToxR was identified as being required for cholera toxin expression. After that came the assumption that environmental regulation by parameters such as pH, temperature and osmolarity probably proceeds in some fashion through the action of ToxR (Miller *et al.* 1987; Miller & Mekalanos 1988).

Shigella, on the other hand, was less well understood in terms of the factors that contribute to its pathogenicity when the development of gene fusion technology allowed for isolation of temperature-regulated genes. But by focusing on the likely signals that regulate virulence based on the observation that growth temperature influences cell invasiveness, the same outcome as the cholera story was achieved; regulatory elements were identified and the conditions under which they operate to control gene expression were assumed.

Notwithstanding the level of knowledge already in hand upon initiation of regulatory studies, each of the two approaches outlined above is based on a 'virulence-factor' model: that known virulence factors may lead the investigator to the regulatory system controlling them.

What is the value of analysing the animating assumptions that have led to the understanding we currently have of gene regulation in pathogenesis? Because this analysis helps put into perspective the fact that in the post-genomic era the assumptions that underlie regulatory studies in pathogenesis will probably be entirely different ones. To be sure, future work using genomic approaches will take advantage of the large body of work that has been done in gene regulation not only in pathogenic organisms, but also in non-pathogenic organisms. Thus, it is now possible to search complete genomes for homologies to known regulatory elements and knockout each one that is discovered in order to assess its role in virulence. This method, a 'virulence regulator' approach, requires that there be some way for the investigator to analyse the mutant phenotypes, and probably the best way is to use a reliable animal model to ascertain the virulence phenotype of the knockout strains. If introducing a mutation into a homologue of a regulator gene results in altered virulence, then it may be concluded that the gene in question probably regulates another gene that is required for virulence. Once that knowledge is obtained, further molecular or genetic approaches are then necessary to identify those genes and their products.

In either case, the virulence factor model or the virulence regulator model, the *in vivo* relevance of the regulatory system ultimately under consideration is not always straightforward, for even when a regulatory system is shown to be vital for *in vivo* effects leading to virulence, the precise mechanism for why this is so, or the signals that may impinge on the regulatory system, are not always obvious. This is particularly true for regulators of the so-called two-component family, in which a sensor kinase perceives a signal and transmits it through phosphorylation of a response regulator, which is typically a transcription factor. In this family of proteins, the sensor kinase is presumed to be the direct receiver of signals to which the system responds (Parkinson 1993; Parkinson & Kofoid 1992). In many cases of two-component regulators, however, the precise signals that initiate the cascade of events that leads to activation of the response regulator are very often not well understood.

A good example of this is in the well-studied Bvg system of *Bordetella* spp. The activity of the two-component system in these organisms, BvgS (sensor kinase) and BvgA (response regulator) is modulated by signals such as nicotinic acid, Mg^{2+} and temperature, although how these signals are actually identified by BvgS is less clear (Akerley & Miller 1996). Nevertheless, that signal transduction through this system is critical for establishing a normal host–pathogen interaction has been well demonstrated both by isolating mutants of BvgS that signal inappropriately, as if signals were present all the time or not at all (Miller *et al.* 1992), and by converting genes that are normally repressed by the Bvg system into ones that are activated by it. Such ectopic expression, as it has come to be called, profoundly disrupts normal hostmicrobe interaction (Akerley & Miller 1996; Akerley *et al.* 1995).

A more successful example in recent years of attempting to identify the signals that influence regulatory activity during infection has been the characterization of PhoP and PhoQ of *Salmonella* spp. This has been demonstrated to be a sensory (PhoQ)/response (PhoP) system that monitors the levels of available Mg^{2+} (and Ca^{2+}) and activates gene expression when those levels are low. This finding was triggered by the observation that among genes controlled by PhoPQ are a number of genes whose products are involved in magnesium transport (mgt). Subsequently it was demonstrated that magnesium induces a conformational change in PhoQ, the sensor kinase and that the PhoQ periplasmic domain binds magnesium in solution (Garcia *et al.* 1996; Vescovi *et al.* 1997). These observations indicate that magnesium is an important signal for this system and that at low levels, such as those postulated to occur intracellularly, PhoPQ activates virulence gene expression in *Salmonella*.

Two-component gene systems are appealing targets for study along the virulence regulator line because they are widely assumed to be sensory systems triggered by specific signals. Thus, if a two-component system regulates a particular virulence trait, the natural assumption is that the system is responding to some *in vivo* signal to do so. Added to this assumption in the overall appeal of studying two-component systems and their role in virulence is that both the sensor kinase and the response regulator are defined by specific motifs within the primary amino-acid sequence. These are related to domains of the proteins necessary for phosphorylation or for DNA binding and transcription control. It is therefore very straightforward to design primers for the polymerase chain reaction (PCR) that allow for amplification of all of the sensor kinases or response regulators of a given species, which has been done for a number of pathogenic bacteria (Wren *et al.* 1992). The homologies among two-component family members also make it easy to identify these proteins when analysing genome sequence data.

Recent work on a two-component system in *Streptococcus pyogenes*, CsrR and CsrS, suggests a different role for such proteins in governing virulence phenotypes of a pathogen. Rather than functioning as an activator of genes required for pathogenicity, CsrR represses transcription of the genes encoding a hyaluronic acid capsule (*hasAB*) that is strongly associated with abscess formation and subsequent disease in a mouse model of necrotizing fasciitis (Bunce *et al.* 1992; Murley *et al.* 1999). This is an intriguing system as the pathogenesis of skin disease is dependent on acquiring inactivating mutations in *csrR* or the gene encoding its putative sensor kinase gene, *csrS*, thus raising the question of what signals the system may normally respond to and how this signalling may regulate important phenotypes in the interaction between host and pathogen.

After a period in which studies motivated by both virulence factor and virulence regulator models of gene discovery have identified a plethora of regulators required for virulence, demonstrated to be so by relevant animal, and sometimes human, models of infection, the question is less one of whether these are required *in vivo* as much as it is of how these proteins regulate gene expression *in vivo* and what signals may control their activity. Or, more properly, the question may be recast as the following: are signals that stimulate regulatory activity *in vitro* the same ones that stimulate such activity *in vivo*? It seems likely, from the *Salmonella* example, that the answer may ultimately be in the affirmative for all systems under investigation, but only after investigators figure out what the correct signals are. For the remainder of this paper, we will focus on attempts to address this question in two different systems of virulence regulation: the ToxR/ToxT system in *V. cholerae* and the CsrR system in *S. pyogenes*.

2. MATERIAL AND METHODS

(a) *Construction of ToxR deletion derivatives*

PCR products with various amounts of *toxR* were generated using either Taq DNA polymerase (Gibco BRL, Grand Island, NY, USA) or the Expand™ High Fidelity PCR System (Boehringer Mannheim, Indianapolis, IN, USA) using manufacturer's recommended protocols. PCR templates were pVJ21 (Miller *et al.* 1989) or chromsomal DNA from *V. cholerae* strain O395. Synthesized primers were engineered to have added recognition sequences for restriction endonucleases in order to facilitate directional cloning of the products. Products from the PCR were purified by gel electrophoresis followed by extraction with the QIAEX II system (Qiagen, Inc., Valencia, CA, USA). Cloning into expression plasmids was done using standard protocols (Sambrook *et al.* 1989).

(b) *Primer extension of mRNA*

RNA was isolated from bacteria using Trizol Reagent (Gibco BRL). Ten picomoles of primer were end-labelled using 50 μCi [γ-^{32}P]ATP using standard protocols (Sambrook *et al.* 1989). Approximately 2 pmol labelled primer were added to 30 μg RNA and diethylpyrocarbonate-treated water was added to a final volume of 20 μl. Primer extension was carried out using Superscript Reverse Transcriptase (Gibco BRL) as previously described (Higgins & DiRita 1994; Yu & DiRita 1999). Reaction mixtures were resolved in 8% denaturing polyacrylamide gels and visualized by autoradiography following standard protocols (Sambrook *et al.* 1989).

(c) *Mouse infections with group A streptococci*

A dermonecrotic mouse model, described by Barg and co-workers, was used to assess the level of athogenicity of wild-type and *csrRS* mutants of MGAS166, an M1 SpeA2 group A streptococcus strain (Bunce *et al.* 1992; Heath *et al.* 1999). Unless noted, strains were harvested in mid-log growth and concentrated to produce inocula of specific numbers of microbes in 200 μl of suspension. This was injected into the right flank of four-week-old male crl:SKH1 (hairless; hrhr) BR mice (Charles River Laboratories, Wilmington, MA, USA). Mice were weighed before inoculation and every 24 h, and necrotic lesions were measured daily.

3.   RESULTS AND DISCUSSION

(a) *Analysis of the ToxR regulon in* V. cholerae

Identification of ToxR as a major factor in virulence regulation in *V. cholerae* was a prime example of the virulence factor approach. The genes encoding cholera toxin, *ctxAB*, are not well expressed when cloned in the heterologous background of *E. coli*. This observation prompted a genetic screen for factors from *V. cholerae* which, when expressed in an *E. coli* strain having a *ctx-lacZ* operon fusion in it, would result in Lac$^+$ colonies due to activation of the fusion (Miller & Mekalanos 1984). From this screen arose clones that were shown to encode *toxR*, which was later demonstrated to be an unusual regulatory protein in that it resides in the inner membrane and has an amino-terminal, cytoplasmic domain that shares extensive and important homology with the DNA-binding–transcription-activation domains of several response regulator proteins in the two-component family. The carboxy-terminal domain of ToxR is in the periplasmic space and there it probably interacts with another protein called ToxS, also required for *ctxAB* expression (DiRita & Mekalanos 1991; Miller *et al.* 1989); the two proteins are encoded by an operon, *toxRS*. ToxS and the remainder of ToxR beyond the DNA-binding–activation domain do not share homology with other proteins that might provide intuition for how they function. Notably, they lack important conserved residues found in the phosphorylation-dependent two-component regulatory systems.

A search for gene fusions whose pattern of expression was similar to that of cholera toxin ultimately revealed several other genes that were also demonstrated to be regulated by ToxR. These include the toxin-coregulated pilus (TCP), the accessory colonization factor (ACF) and an outer membrane protein (OmpU) (Peterson & Mekalanos 1988; Miller & Mekalanos

1988; Skorupski & Taylor 1997). Another outer membrane protein, OmpT, is regulated oppositely to these other factors and we now know that its gene expression is repressed directly by ToxR (Li *et al.* 2000). Conditions for maximum expression of *ctxAB* in many strains of *V. cholerae* include a temperature of 25–30°C and a relatively acidic pH (6.5 as opposed to 8.5). The temperature optimum in particular is counter-intuitive given that the organism expresses cholera toxin during infection of the small intestine at 37°C. Also counter-intuitive is the fact that bile salts, which, *a priori*, might be considered as a potential signal for *in vivo* stimulation of virulence genes in *V. cholerae*, instead have a strongly repressive effect on expression of the ToxR regulon (Gupta & Chowdhury 1997; Schuhmacher & Klose 1999) (Fig. 1).

ToxR, although required for expression of cholera toxin, TCP, ACF and OmpU, actually controls expression of these genes indirectly through its ability to control expression of another activator, ToxT (DiRita *et al.* 1991).

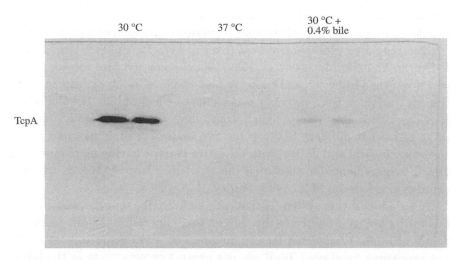

Fig. 1. The effect of temperature and bile salts on expression of the ToxR-regulated TCP. Cultures were grown overnight in Luria–Bertani medium under the indicated conditions and lysed by boiling in sodium dodecyl sulphate polyacrylamide gel electrophoresis buffer. Equivalent amounts of protein were loaded on a 10% polyacrylamide gel and subjected to electrophoresis. The gel was blotted to nitrocellulose and probed with antibodies to the major TCP subunit, TcpA. Equivalent effects on cholera toxin were observed by enzyme-linked immunosorbent assay (not shown).

Regulation of *toxT* transcription is complex and requires another pair of membrane-localized regulatory proteins, TcpP and TcpH, which cooperate with ToxR and ToxS for activation of one of two *toxT* promoters. The other promoter controlling *toxT* expression is several kilobases upstream of the gene in front of the *tcp* operon, which contains the majority of genes for TCP synthesis. This promoter is controlled by ToxT itself. Thus *toxT* expression is controlled by a regulatory loop in which activation by ToxR and TcpP leads to expression from a proximal promoter and subsequent expression is controlled by ToxT from a more distal promoter (Brown & Taylor 1995; Yu & DiRita 1999; Medrano *et al.* 1999). Analysis of *toxT* expression from these different promoters *in vivo* is currently being investigated by Camilli and his co-workers using an *in vivo* reporter system (Merrell & Camilli, this issue).

Signals that control expression of the virulence factors may be overridden by constitutive expression of ToxT, implying that the signalling capacity for the system occurs prior to ToxT expression (DiRita *et al.* 1996). That this is probably so comes from the observation that expression of *tcpPH* is subject to conditional expression by two activators, AphA and AphB, leading to the hypothesis that regulated expression of *tcpPH* in turn controls regulated expression of *toxT* (Skorupski & Taylor 1999; Kovacikova & Skorupski 1999; Murley *et al.* 1999). This is not the case for bile salts, which repress virulence gene expression at the level of ToxT, as constitutive expression of ToxT does not overcome the effect of bile (Schuhmacher & Klose 1999). Regulatory signals controlling the ToxR regulon appear to be somewhat different between the two major epidemic biotypes of *V. cholerae*, classical and El Tor, with the latter biotype having more stringent growth requirements for activating the regulatory loop that leads to *toxT* expression (Medrano *et al.* 1999; Murley *et al.* 1999).

A paradox of this system is that although ToxR was originally identified for its ability to activate the cholera toxin operon in *E. coli* independently of other *V. cholerae* factors, mutants of *V. cholerae* lacking *toxT*, but expressing functional ToxR, do not express cholera toxin in the laboratory (Champion *et al.* 1997). Overexpression of ToxR in the *toxT* mutant background leads to a slight elevation of toxin production, but not to levels seen in wild-type cells. Experiments analysing expression of *ctx* using *in vivo* reporter systems will perhaps shed some light on this *in vitro* paradox.

The localization of several major regulatory factors, ToxR/ToxS and TcpP/TcpH, to the inner membrane leads naturally to the question of what role this placement may have in the regulation of gene expression. Some evidence suggests that the periplasmic domains of both ToxR and TcpP are important for signalling across the inner membrane, and this is indeed an appealing possibility given the receptor-like topology of these proteins. Replacement of the periplasmic domain of ToxR with the periplasmic protein alkaline phosphatase (PhoA) resulted in constitutive expression of cholera toxin, i.e. toxin expression occurred under normally non-permissive conditions (see below). This implies that the periplasmic domain of ToxR is a signal sensing domain that controls, for example, DNA binding or transcription activation by the cytoplasmic domain (Miller *et al.* 1987). Likewise, fusion of the periplasmic protein β-lactamase to the carboxy-terminal domain of TcpP led to constitutive expression of a *toxT–lacZ* gene fusion, again suggesting a role for the periplasmic domain, this time of TcpP, in signalling that leads to virulence gene expression (Häse & Mekalanos 1998).

A slightly different hypothesis is that localization to the membrane is important for the function of these activator proteins, and there is some evidence to support that. Several groups have produced different versions of ToxR that localize their DNA-binding–transcription-activation domain either to the cytoplasm or to the membrane, and these experiments have had different results (DiRita & Mekalanos 1991; Miller *et al.* 1987; Ottemann & Mekalanos 1995; Kolmar *et al.* 1995). These experiments have usually been predicated on the assumption that the multimeric structure of ToxR, which may be conferred by the periplasmic domain, is critical for ToxR function (DiRita & Mekalanos 1991; Miller *et al.* 1987; Ottemann & Mekalanos 1995; Kolmar *et al.* 1995). Therefore such experiments have often tested not solely the role of membrane localization, but also the multimerization status of the protein.

We approached the problem of membrane localization by constructing a series of ToxR derivatives with different degrees of the native protein beyond the DNA-binding–transcription-activation domain but otherwise having no fusion moiety. We specifically addressed the ability of these proteins to control TcpP-directed transcription activation of *toxT*. Based on its similarity to the OmpR response regulator of *E. coli*, this domain is predicted to have the structure of a winged helix, a motif in which an α-helix and two β-strands (which make up the wing) function in DNA recognition. When expressed in *V. cholerae* the winged helix of ToxR exhibited wild-type ToxR activity for regulation of OmpU and OmpT, but was unable to complement a *toxR* mutant for expression of cholera toxin or TcpA, the major subunit of the TCP (data not shown). Primer extension of *toxT* mRNA showed that activation of *toxT* transcription was severely diminished, which is probably the reason for diminished activation of toxin and TcpA. The results of these studies are summarized in Fig. 2, which indicates that membrane localization of the amino-terminal DNA-binding–transcription-activation domain of ToxR is required for activating *toxT* in a TcpP-dependent fashion.

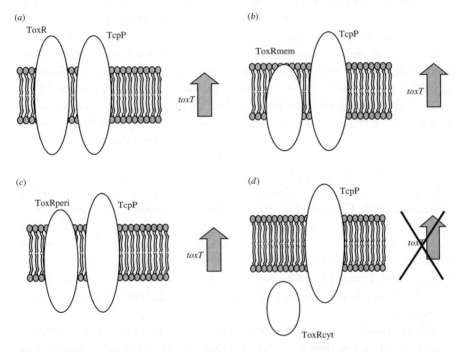

Fig. 2. Membrane localization requirements. Summary of data from primer extension experiments monitoring transcription of *toxT* as described in the text. (*a*)–(*c*) Indicate the ability of membrane-localized forms of ToxR harbouring progressively less of the periplasmic domain to cooperate with membrane-localized TcpP for activation of *toxT* transcription. (*d*) Indicates that the cytoplasmic domain of ToxR (ToxRcyt), with only the DNA-bindingtranscription motif, does not lead to TcpP-dependent activation. The cytoplasmic form does regulate the outer membrane proteins OmpU and OmpT like wild-type, as do all the other forms of ToxR that were tested.

In contrast to the function of the ToxR winged helix domain alone, a construct in which the winged helix was expressed with the remaining cytoplasmic sequences as well as the transmembrane domain, but lacking any periplasmic domain, regulated *omp* gene expression like wild-type and also activated *toxT* transcription in a strictly TcpP-dependent manner. Analysis of other ToxR derivatives demonstrated that there are virtually no requirements for a periplasmic domain on ToxR, as constructs lacking this domain or having other sequences in place of the domain such as alkaline phosphatase, β-lactamase or the yeast leucine zipper domain GCN4 all activated *toxT* transcription and regulate *omp* gene expression like wild-type, similar to what has been observed by others, as described above. Each of these constructs demonstrated dependence on TcpP for their activity, suggesting that co-localization of the DNA-binding domain of ToxR and TcpP to the membrane is necessary and sufficient for activation of *toxT* and subsequent virulence gene regulation.

An appealing hypothesis for the function of this class of regulator is that the periplasmic domain is an important component of function, perhaps because it senses signals from the environment directly and transduces the signal by inducing conformational changes in the domain of the protein required for DNA binding and transcription activation. Evidence for this possibility comes from the observation that cholera toxin expression can be uncoupled from normal regulatory signals *in vitro* when its periplasmic domain was replaced by alkaline phosphatase. In this experiment, Miller *et al.* (1987) showed that toxin levels in cells expressing ToxR-PhoA were unaffected by growth at pH 8.5, a condition in which ToxR-regulated genes are not typically expressed. This result suggests that the periplasmic domain on ToxR may normally play a role in downregulating ToxR function.

Additional support for the periplasmic domain of membrane-localized activators acting as a sensor of environmental signals comes from work with two other such proteins, TcpP and the *E. coli* activator CadC. In the case of TcpP, Häse & Mekalanos (1998) mutagenized *V. cholerae* with the transposon Tn*bla*, which allows for identification of operon fusions to the periplasmic β-lactamase, and screened for isolates that expressed TCP constitutively, i.e. under conditions that typically lead to lack of TCP expression. Among this pool was a strain in which the periplasmic domain of TcpP had been fused to β-lactamase. The reason why this fusion results in an apparently constitutively active protein has yet to be determined, but among the possibilities are that the periplasmic domain regulates the ability of TcpP to activate *toxT* expression in coordination with ToxR under appropriate conditions, and that alteration of the periplasmic domain by fusion to β-lactamase may allow this signal detection process to be bypassed.

Evidence for a role in signal recognition by the periplasmic domain of a membrane-localized transcription activator is more strongly supported by studies with CadC, which regulates gene expression in *E. coli* at pH 5.8 in the presence of lysine. When these conditions are met, CadC activates expression of the *cadBA* operon, which expresses lysine decarboxylase, resulting in production of cadaverine from lysine. Mutant CadC proteins that activate *cadA* expression independently of pH or lysine have lesions in residues within the periplasmic domain of the protein, suggesting a possible direct role in signal recognition by this domain (Dell *et al.* 1994).

Another possibility for how signals may be recognized by this unusual class of proteins is that their membrane location *per se* is a component of their putative ability to sense signals. In this model, the membrane itself may be a co-factor in stimulating transcription activation, such that perturbations to the membrane structure, integrity or bioenergetics during *in vivo* growth, in the case of *V. cholerae*, may stimulate the activity of ToxR and/or TcpP. An intriguing result that supports this hypothesis is that, in their screen of Tn*bla* mutants constitutive for TCP production, Häse & Mekalanos (1999) identified insertions into the gene for a NADH:ubiquinone oxidoreductase (*nqr*). This gene product is probably responsible for the sodium motive force that powers the flagella in some *Vibrio* spp., including *V. cholerae* (Häse & Mekalanos 1999). Subsequent work by these investigators showed that flagellar motility is altered in a *V. cholerae nqr* mutant and that this correlated with upregulation of *toxT* transcription. In addition, upon chemical inhibition of the oxidoreductase, *toxT* transcription was also increased. These results extend previous studies showing that motility and expression of cholera toxin and TCP are oppositely regulated and suggest that TcpP-dependent activation of *toxT* expression is a crucial participant in this process (Häse & Mekalanos 1999).

(b) *Capsule production in group A streptococci as a result of regulatory mutations arising during* in vivo growth

To summarize work on the ToxR/TcpP system described above, it appears that membrane interactions between ToxR and TcpP, as well as membrane-dependent processes linked to flagellar motility are critical for expression of virulence genes in *V. cholerae*. We will turn our attention now to a very different regulatory system controlling invasive behaviour of a Gram-positive pathogen, *S. pyogenes*, or group A streptococci. This pathogen may cause a range of infections from relatively mild ones such as impetigo or pharyngitis to more severe and life-threatening ones such as toxic shock syndrome, necrotizing fasciitis and, through autoimmune sequelae, rheumatic fever. The focus in this review is its role in necrotizing fasciitis, an invasive and potentially rapidly progressing, grave disease that may be fatal if not aggressively treated.

Among a variety of virulence-associated traits that have been characterized in *S. pyogenes*, expression of a hyaluronic-acid capsule is tightly associated with the more severe pathogenicity attributed to the organism. A locus called *csrRS* controls expression of the gene for hyaluronic-acid synthesis, *hasA*. The *csrRS* operon encodes a typical two-component regulatory system including a putative sensor kinase, CsrS and a response regulator CsrR. As opposed to the positive form of regulation by ToxR/TcpP/ToxT in *V. cholerae*, CsrRS regulates capsule production in *S. pyogenes* through repression. Thus, strains carrying mutations in the regulatory locus express higher levels of capsule and are more virulent in a mouse skin abscess model of necrotizing infection (Heath *et al.* 1999). Along with capsule production, other virulence-associated traits are upregulated in *csrRS* mutants of *S. pyogenes*. These are a cysteine protease called pyrogenic exotoxin B, encoded by *speB*, and a gene associated with production of streptolysin S, an oxygen-stable haemolysin, encoded by *sagA*, a streptokinase and a mitogenic factor (Heath *et al.* 1999; Levin & Wessels 1998; Ashbaugh *et al.* 1998).

(i) In vitro *effects of* csrRS *mutation*

CsrRS controls these genes at the level of transcription so that in *csrRS* mutants elevated mRNA for each is detected. In addition, *csrRS* mutants express increased amounts of *csrRS* mRNA, suggesting autoregulation (Fig. 3). Purified CsrR binds to the promoters of *hasA*, *sagA*, and *speB*, but only upon phosphorylation, which may be effected presumably by CsrS *in vivo* but can be done using acyl phosphate *in vitro* (Bernish & Van de Rijn 1999; A. Miller, N. C. Engleberg and V. J. DiRita, unpublished data).

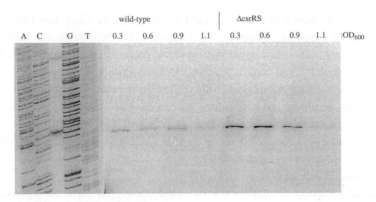

Fig. 3. Primer extension of *csrR*. RNA was isolated from samples taken from cultures of wild-type and *csrRS* mutant *S. pyogenes*, collected at different points in the growth curve (indicated by the OD_{600} in the figure). The RNA was used in primer extension analysis using a primer that corresponds to a sequence within the *csrR* gene upstream of the deletion in the mutant strain.

Despite the autoregulation of *csrRS* observed by primer extension analysis of mRNA isolated from organisms grown *in vitro*, phosphorylated CsrR does not bind to the *csrRS* promoter indicating that regulation of its own operon may involve factors in addition to CsrRS.

Genes regulated by CsrRS are expressed in wild-type cells during late logarithmic growth, similar to regulation of virulence genes in the Gram-positive pathogen *Staphylococcus aureus*. In *csrRS* mutant cells, both the timing and the magnitude of expression is altered with genes being expressed to higher levels beginning at an earlier time in the growth cycle (Heath *et al.* 1999). We hypothesize that phosphorylation of CsrR by CsrS may be regulated during growth of the cells such that during early logarithmic growth a signal leading to CsrS kinase activity is present, thereby resulting in CsrR phosphorylation and repression of virulence genes. As cells progress through the growth curve this signal may be lost and CsrR may become unphosphorylated and derepression of capsule, SpeB and other genes occurs.

(ii) In vivo *behaviour of* csrRS *mutants*

Capsule production by group A streptococci is an important determinant of pathogenicity and strains that express it exhibit the most severe level of virulence in animal models (Ashbaugh *et al.* 1998; Heath *et al.* 1999). Given that CsrRS is a repressor of capsule expression, how does capsule become expressed during *in vivo* growth? One hypothesis is that conditions *in vivo* are such that CsrR is non-phosphorylated and therefore unable to bind DNA and repress *hasA* transcription. That wild-type cultures of group A streptococci grown in static broth prior to inoculation of mice may cause formation of skin abscesses (but not typically more severe, invasive necrotizing disease) suggests that growth conditions may indeed affect the level of virulence and that this may be through control by CsrR, although there is no direct evidence for the latter hypothesis. Another possibility to account for activation of *hasA* and other genes normally repressed by CsrR–P is that mutations in the *csrRS* locus may arise during infection and that these strains lead to elevated levels of tissue invasiveness.

In support of this latter hypothesis is the fact that strains with lesions in *csrRS* are more virulent in a dermo-necrotic mouse model (Heath *et al.* 1999). Significantly more damage, assessed by three different parameters (24 h weight loss, presence of lesions and necrosis within lesions), is noted than when wild-type organisms are injected. Whereas mice injected with 4×10^6 wild-type organisms gained approximately 1 g in weight over 24 h after infection (similar to sham-inoculated controls), mice injected

Table 1. Patterns of mouse skin lesion formation by wild-type and *csrRS* mutants of *S. pyogenes* (Adapted from Heath *et al.* (1999).)

Inoculum	Genotype of strain	24 h weight gain	No. with lesions/ no. inoculated	No. with necrosis/ no. with lesions
2×10^5	uninoculated	0.5 ± 0.6	0/8	0/0
	$csrRS^+$	0.9 ± 0.4	0/6	0/0
	$\Delta csrRS$	-2.1 ± 1.7	3/6	3/3
	$\Delta csrRS,$ $\Delta hasAB$	-0.4 ± 1.3	0/6	0/1
2×10^6	uninoculated	0.6 ± 0.07	0/6	0/0
	$csrRS^+$	-0.4 ± 1.2	0/6	0/0
	$\Delta csrRS$	-4.3 ± 0.6	6/6	6/6
	$\Delta csrRS,$ $\Delta hasAB$	-3.4 ± 0.8	6/6	5/6

with *csrRS* mutants lost over 3 g of weight in the same period. Likewise, all animals infected with *csrRS* mutants developed skin lesions (compared with only one-third of those infected with wild-type); all of these lesions became necrotic by 72 h while none of the lesions caused by wild-type did (Table 1) (Heath *et al.* 1999). These data suggest that derepression of the *csrRS* regulon is an important aspect for an infection to develop into a more life-threatening necrotizing disease. Of interest to us in helping to explain the pathogenesis of lesion formation was the fact that animals infected with double *csrRS/hasAB* mutants, which lack the genes for capsule production, nevertheless produced some lesion formation in infected animals, suggesting that factors other than capsule are involved in the dermonecrotic phenotype.

The development of necrotizing skin infection by the *csrRS* mutant strains is associated with growth of the organisms within the lesion itself. To ask whether the environment within the lesion caused by the mutant was permissive for growth of wild-type organisms, we co-infected mice with both *csrR* mutant and wild-type organisms in numbers at which the wild-type is usually cleared from the animal. Mice developed lesions from this combination of organisms, and bacteria of both genotypes were isolated in numbers significantly higher than the input dose of each (Heath *et al.* 1999). From this experiment we conclude that the presence of the mutant strain influences the survival of the wild-type within the lesion, suggesting that a factor (or factors) other than capsule may contribute to survival within the lesion.

Given that encapsulated strains are hypervirulent and also grow very well within lesions, we asked whether spontaneous *csrRS* mutants could be isolated during infection in the hairless mouse skin model. To do this, wild-type organisms were mixed with Cytodex beads, which enables lesion formation by the wild-type upon subcutaneous inoculation of hairless Br mice (Heath *et al.* 1999). Organisms were isolated from the blood and spleen of these infected animals and screened on plates for capsule production, an easily detectable phenotype on agar media. Six organisms with elevated capsule production were isolated and the nucleotide sequence of the *csrRS* locus was determined in each. All of the isolates had mutations of various classes in either *csrR* or *csrS*. In addition, a hypermucoid isolate from a cutaneous abscess also carried a lesion in the locus as did a spontaneous hypermucoid strain isolated *in vitro*.

These findings suggest that the hypermucoid phenotype of *csrRS* mutants is strongly linked to invasive disease with group A streptococci. Thus, this system represents a rather unusual form of regulation of virulence in a bacterial pathogen, in that a regulator must be inactivated for maximum levels of virulence to occur. There are several questions remaining regarding the CsrRS system of virulence regulation. At present, for example, whether there are important determinants of pathogenicity regulated by CsrRS that we are not currently aware of is not known. Additionally, given that CsrRS is a two-component system we assume that regulatory signals exist and hypothesize that they influence the activity of CsrS for phosphorylation of CsrR. During *in vivo* growth of the wild-type strain, signals very likely keep capsule and other CsrR-regulated genes repressed. Another important question in this system is whether the wild-type organisms play any role in pathogenicity, given that it would appear that infections leading to necrotizing disease are mixed infections of both wild-type and mutant origin. This might be addressed, for example, by determining whether or not mixed infections progress to severe necrotizing disease faster than do clonal infections with only *csrRS* mutants.

REFERENCES

Akerley, B. J. & Miller, J. F. 1996 Understanding signal transduction during bacterial infection. *Trends Microbiol.* **4**, 141–146.

Akerley, B. J., Cotter, P. A. & Miller, J. F. 1995 Ectopic expression of the flagellar regulon alters development of the *Bordetella*–host interaction. *Cell* **80**, 611–620.

Ashbaugh, C. D., Warren, H. B., Carey, V. J. & Wessels, M. R. 1998 Molecular analysis of the role of the group A streptococcal cysteine protease, hyaluronic acid capsule, and M protein in a murine model of human invasive soft-tissue infection. *J. Clin. Invest.* **102**, 550–560.

Bernish, B. & Van de Rijn, I. 1999 Characterization of a two-component system in *Streptococcus pyogenes* which is involved in regulation of hyaluronic acid production. *J. Biol. Chem.* **274**, 4786–4793.

Brown, R. C. & Taylor, R. K. 1995 Organization of the *tcp*, *acf*, and *toxT* genes within a ToxT-dependent operon. *Mol. Microbiol.* **16**, 425–439.

Bunce, C., Wheeler, L., Reed, G., Musser, J. & Barg, N. 1992 A murine model of cutaneous infection with gram-positive cocci. *Infect. Immun.* **60**, 2636–2640.

Callahan, L. T. & Richardson, S. H. 1973 Biochemistry of *Vibrio cholerae* virulence. III. Nutritional requirements for toxin production and the effect of pH on toxin elaboration in chemically defined media. *Infect. Immun.* **7**, 567–574.

Champion, G. A., Neely, M. N., Brennan, M. A. & DiRita, V. J. 1997 A branch in the ToxR regulatory cascade of *Vibrio cholerae* revealed by characterization of *toxT* mutant strains. *Mol. Microbiol.* **23**, 323–331.

Dell, C. L., Neely, M. N. & Olson, E. R. 1994 Altered pH and lysine signalling mutants of *cadC*, a gene encoding a membrane-bound transcriptional activator of the *Escherichia coli cadBA* operon. *Mol. Microbiol.* **14**, 7–16.

DiRita, V. J. & Mekalanos, J. J. 1991 Periplasmic interaction between two membrane regulatory proteins, ToxR and ToxS, results in signal transduction and transcriptional activation. *Cell* **64**, 29–37.

DiRita, V. J., Parsot, C., Jander, G. & Mekalanos, J. J. 1991 Regulatory cascade controls virulence in *Vibrio cholerae. Proc. Natl Acad. Sci. USA* **88**, 5403–5407.

DiRita, V. J., Neely, M., Taylor, R. K. & Bruss, P. M. 1996 Differential expression of the ToxR regulon in classical and El Tor biotypes of *Vibrio cholerae* is due to biotype-specific control over *toxT* expression. *Proc. Natl Acad. Sci. USA* **93**, 7991–7995.

Dorman, C. J. & Porter, M. E. 1998 The *Shigella* virulence gene regulatory cascade: a paradigm of bacterial gene control mechanisms. *Mol. Microbiol.* **29**, 677–684.

Evans, D. J. & Richardson, S. H. 1968 *In vitro* production of choleragen and vascular permeability factor by *Vibrio cholerae. J. Bacteriol.* **96**, 126–130.

Garcia, V. E., Soncini, F. C. & Groisman, E. A. 1996 Mg^{2+} as an extracellular signal: environmental regulation of *Salmonella* virulence. *Cell* **84**, 165–174.

Gupta, S. & Chowdhury, R. 1997 Bile affects production of virulence factors and motility of *Vibrio cholerae. Infect. Immun.* **65**, 1131–1134.

Häse, C. C. & Mekalanos, J. J. 1998 TcpP protein is a positive regulator of virulence gene expression in *Vibrio cholerae. Proc. Natl Acad. Sci. USA* **95**, 730–734.

Häse, C. C. & Mekalanos, J. J. 1999 Effects of changes in membrane sodium flux on virulence gene expression in *Vibrio cholerae. Proc. Natl Acad. Sci. USA* **96**, 3183–3187.

Heath, A., DiRita, V. J., Barg, N. L. & Engleberg, N. C. 1999 A two-component regulatory system, CsrR-CsrS, represses expression of three *Streptococcus pyogenes* virulence factors: hyaluronic acid capsule, streptolysin S, and pyrogenic exotoxin B. *Infect. Immun.* **67**, 5298–5305.

Higgins, D. E. & DiRita, V. J. 1994 Transcriptional control of *toxT*, a regulatory gene in the ToxR regulon of *Vibrio cholerae. Mol. Microbiol.* **14**, 17–29.

Kolmar, H., Hennecke, F., Gotze, K., Janzer, B., Vogt, B., Mayer, F. & Fritz, H. J. 1995 Membrane insertion of the bacterial signal transduction protein ToxR and requirements of transcription activation studied by modular replacement of different protein substructures. *EMBO J.* **14**, 3895–3904.

Kovacikova, G. & Skorupski, K. 1999 A *Vibrio cholerae* LysR homolog, AphB, cooperates with AphA at the *tcpPH* promoter to activate expression of the ToxR virulence cascade. *J. Bacteriol.* **181**, 4250–4256.

Levin, J. C. & Wessels, M. R. 1998 Identification of *csrR/csrS*, a genetic locus that regulates hyaluronic acid capsule synthesis in group A streptococci. *Mol. Microbiol.* **30**, 209–219.

Li, C. C., Crawford, J. A., DiRita, V. J. & Kaper, J. B. 2000 Molecular cloning and transcriptional regulation of *ompT*, a ToxR-repressed gene in *Vibrio cholerae*. *Mol. Microbiol.* **35**, 189–203.

Maurelli, A. T. & Curtiss III, R. 1984 Bacteriophage Mu d1 (Ap^r *lac*) generates *vir-lac* operon fusions in *Shigella flexneri* 2a. *Infect. Immun.* **45**, 642–648.

Maurelli, A. T. & Sansonetti, P. J. 1988 Identification of chromosomal gene controlling temperature regulated expression of *Shigella* virulence. *Proc. Natl Acad. Sci. USA* **85**, 2820–2824.

Maurelli, A. T., Blackmon, B. & Curtiss III, R. 1984 Temperature-dependent expression of virulence genes in *Shigella* species. *Infect. Immun.* **43**, 195–201.

Medrano, A. I., DiRita, V. J., Castillo, G. & Sanchez, J. 1999 Transient transcriptional activation of the *Vibrio cholerae* El Tor virulence regulator toxT in response to culture conditions. *Infect. Immun.* **67**, 2178–2183.

Miller, J. F., Johnson, S. A., Black, W. J., Beattie, D. T., Mekalanos, J. J. & Falkow, S. 1992 Constitutive sensory transduction mutations in the *Bordetella pertussis* bvgS gene. *J. Bacteriol.* **174**, 970–979.

Miller, V. L. & Mekalanos, J. J. 1984 Synthesis of cholera toxin is positively regulated at the transcriptional level by ToxR. *Proc. Natl Acad. Sci. USA* **81**, 3471–3475.

Miller, V. L. & Mekalanos, J. J. 1988 A novel suicide vector and its use in construction of insertion mutations: osmoregulation of outer membrane proteins and virulence determinants in *Vibrio cholerae* requires *toxR*. *J. Bacteriol.* **170**, 2575–2583.

Miller, V. L., Taylor, R. K. & Mekalanos, J. J. 1987 Cholera toxin transcriptional activator ToxR is a transmembrane DNA binding protein. *Cell* **48**, 271–279.

Miller, V. L., DiRita, V. J. & Mekalanos, J. J. 1989 Identification of *toxS*, a regulatory gene whose product enhances ToxR-mediated activation of the cholera toxin promoter. *J. Bacteriol.* **171**, 1288–1293.

Murley, Y. M., Carroll, P. A., Skorupski, K., Taylor, R. K. & Calderwood, S. B. 1999 Differential transcription of the *tcpPH* operon confers biotype-specific control of the *Vibrio cholerae* ToxR regulon. *Infect. Immun.* **67**, 5117–5123.

Ottemann, K. M. & Mekalanos, J. J. 1995 Analysis of *Vibrio cholerae* ToxR function by construction of novel fusion proteins. *Mol. Microbiol.* **15**, 719–731.

Parkinson, J. S. 1993 Signal transduction schemes of bacteria. *Cell* **73**, 857–871.

Parkinson, J. S. & Kofoid, E. C. 1992 Communication modules in bacterial signalling proteins. *A. Rev. Genet.* **26**, 71–112.

Pearson, G. D. & Mekalanos, J. J. 1982 Molecular cloning of *Vibrio cholerae* enterotoxin genes in *Escherichia coli* K-12. *Proc. Natl Acad. Sci. USA* **79**, 2976–2980.

Peterson, K. M. & Mekalanos, J. J. 1988 Characterization of the *Vibrio cholerae* ToxR regulon: identification of novel genes involved in intestinal colonization. *Infect. Immun.* **56**, 2822–2829. (Erratum in *Infect. Immun.* **57**, 660).

Richardson, S. H. 1969 Factors influencing *in vivo* skin permeability factor production by *Vibrio cholerae*. *J. Bacteriol.* **100**, 27–34.

Sambrook, J., Fritsch, E. F. & Maniatis, T. 1989 *Molecular cloning: a laboratory manual*, 2nd edn. Cold Spring Harbor, NY: Cold Spring Harbor Laboratory Press.

Schuhmacher, D. A. & Klose, K. E. 1999 Environmental signals modulate ToxT-dependent virulence factor expression in *Vibrio cholerae*. *J. Bacteriol.* **181**, 1508–1514.

Skorupski, K. & Taylor, R. K. 1997 Control of the ToxR virulence regulon in *Vibrio cholerae* by environmental stimuli. *Mol. Microbiol.* **25**, 1003–1009.

Skorupski, K. & Taylor, R. K. 1999 A new level in the *Vibrio cholerae* virulence cascade: AphA is required for transcriptional activation of the tcpPH operon. *Mol. Microbiol.* **31**, 763–771.

Vescovi, E. G., Ayala, Y. M., DiCera, E. & Groisman, E. A. 1997 Characterization of the bacterial sensor protein phoQ. Evidence for distinct binding sites for Mg^{2+} and Ca^{2+}. *J. Biol. Chem.* **272**, 1440–1443.

Wren, B. W., Colby, S. M., Cubberley, R. R. & Pallen, M. J. 1992 Degenerate PCR primers for the amplification of fragments from genes encoding response regulators from a range of pathogenic bacteria. *FEMS Microbiol. Lett.* **78**, 287–291.

Yu, R. R. & DiRita, V. J. 1999 Analysis of an autoregulatory loop controlling ToxT, cholera toxin, and toxin-coregulated pilus production in *Vibrio cholerae*. *J. Bacteriol.* **181**, 2584–2592.

Evidence for Operation *in Vivo* of Aspects of Pathogenicity Revealed by Recent Work *in Vitro*: Potential Use of New Methods

QUORUM SENSING AND THE POPULATION-DEPENDENT CONTROL OF VIRULENCE

PAUL WILLIAMS[*,†,‡,§], MIGUEL CAMARA[‡], ANDREA HARDMAN[‡],
SIMON SWIFT[*,‡], VICTORIA J. HOPE[‡], KLAUS WINZER[‡],
BARRIE MIDDLETON[‡], DAVID I. PRITCHARD[‡]
and BARRIE W. BYCROFT[‡]

[*] *Institute of Infections & Immunity, Queen's Medical Centre,*
[†] *School of Clinical Laboratory Sciences, Queen's Medical Centre,*
University of Nottingham, Nottingham NG7 2UH, UK
[‡] *School of Pharmaceutical Sciences, University of Nottingham,*
Nottingham NG7 2RD, UK

DEBORAH MILTON

Department of Cell and Molecular Biology, Umeå University,
S901 87 Umeå, Sweden

One crucial feature of almost all bacterial infections is the need for the invading pathogen to reach a critical cell population density sufficient to overcome host defences and establish the infection. Controlling the expression of virulence determinants in concert with cell population density may therefore confer a significant survival advantage on the pathogen such that the host is overwhelmed before a defence response can be fully initiated. Many different bacterial pathogens are now known to regulate diverse physiological processes including virulence in a cell-density-dependent manner through cell–cell communication. This phenomenon, which relies on the interaction of a diffusible signal molecule (e.g. an *N*-acylhomoserine lactone) with a sensor or transcriptional activator to couple gene expression with cell population density, has become known as 'quorum sensing'. Although the size of the 'quorum' is likely to be highly variable and influenced by the diffusibility of the signal molecule within infected tissues, nevertheless quorum-sensing signal molecules can be detected *in vivo* in both experimental animal model and human infections. Furthermore, certain quorum-sensing molecules have been shown to possess pharmacological and immunomodulatory activity such that they may function as virulence determinants *per se*. As a consequence, quorum sensing constitutes a novel therapeutic target for the design of small molecular antagonists capable of attenuating virulence through the blockade of bacterial cell–cell communication.

Keywords: Quorum sensing; *N*-acylhomoserine lactones; virulence; bacteria; infection; cell signalling

---

[§] Author for correspondence (paul.williams@nottingham.ac.uk).

1. INTRODUCTION

A well-prepared army goes into battle with a knowledge of the opposition, significant numbers of troops and excellent lines of communication. In some respects, bacterial pathogens are no different and in recent years it has become clear that bacterial cells are capable of exhibiting much more complex patterns of multicellular behaviour than would perhaps be expected for simple unicellular micro-organisms. The ability of a single bacterial cell to communicate with its neighbours to mount a unified response that is advantageous to its survival in a hostile environment makes considerable sense. Such benefits may include improved access to complex nutrients or environmental niches, collective defence against other competitive microorganisms or eukaryotic host defence mechanisms and optimization of population survival by differentiation into morphological forms better adapted to combating an environmental threat.

For many pathogens, the outcome of the interaction between host and bacterium is strongly affected by bacterial population density. Coupling the production of virulence factors with cell population size ensures the host has insufficient time to mount an effective defence against consolidated attack by the infecting bacterial population. Such a strategy inevitably depends on the capacity of an individual bacterial cell to sense other members of the same species and in response, differentially express specific sets of genes. The generic term 'quorum sensing' is now commonly used to describe the phenomenon whereby the accumulation of a diffusible, low molecular weight signal molecule (sometimes referred to as a 'pheromone' or 'autoinducer') enables individual cells to sense when the minimal population unit or 'quorum' of bacteria has been achieved for a concerted population response to be initiated (Fuqua *et al.* 1994). Quorum sensing is thus an example of multicellular behaviour and modulates a variety of physiological processes, including bioluminescence, swarming, swimming and twitching motility, antibiotic biosynthesis, biofilm differentiation, plasmid conjugal transfer and the production of virulence determinants in animal, fish and plant pathogens (for reviews, see Dunny & Winans 1999; Fuqua *et al.* 1996; Hardman *et al.* 1998; Salmond *et al.* 1995).

Signal transduction through quorum sensing depends on the direct or indirect (via a sensor) activation of a response regulator by a diffusible signal molecule and several chemically distinct families of such molecules have

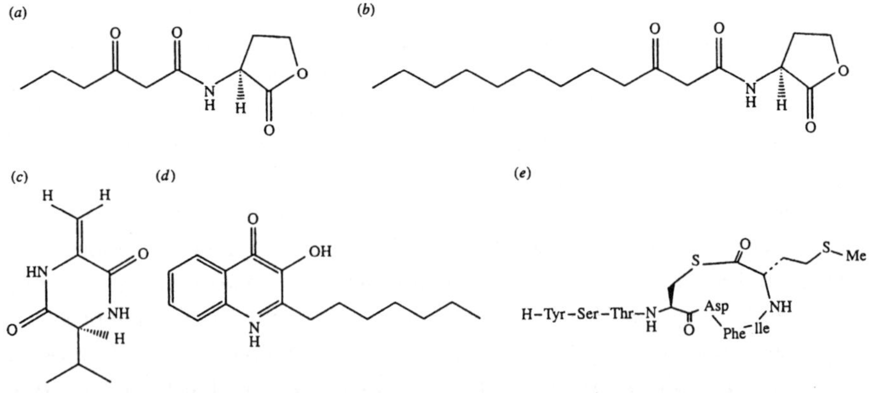

Fig. 1. Structures of some representative quorum-sensing signal molecules. (a) N-(3-oxohexanoyl)-L-homoserine lactone; (b) N-(3-oxododecanoyl)-L-homoserine lactone; (c) cyclo(ΔAla-L-Val); (d) 2-heptyl-3-hydroxy-4-quinolone; and (e) group I *Staphylococcus aureus* cyclic peptide thiolactone.

now been identified. In Gram-negative bacteria, the most intensively investigated family of quorum-sensing signal molecules is the N-acylhomoserine lactones (AHLs; e.g. Figs. 1(a) and (b)), although non-AHL-dependent systems have also been identified. In contrast, Gram-positive bacteria employ quorum-sensing systems in which the signal molecule is often a post-translationally modified peptide (Fig. 1(e); Kleerebezem *et al.* 1997; Mayville *et al.* 1999). With respect predominantly to the AHLs, the focus of this review will be (i) to discuss the role of quorum sensing in the control of virulence gene expression; (ii) to present evidence that quorum sensing occurs *in vivo* during infection; and (iii) to outline the potential of certain quorum-sensing signal molecules to function as virulence determinants *per se*.

2. QUORUM-SENSING SIGNAL MOLECULES IN GRAM-NEGATIVE BACTERIA

Production of the β-lactam antibiotic, 1-carbapen-2-em-3-carboxylic acid (carbapenem) by the terrestrial plant pathogenic bacterium *Erwinia carotovora* was discovered to be regulated by N-(3-oxohexanoyl)-L-homoserine lactone (3-oxo-C6-HSL; Fig. 1(a); Bainton *et al.* 1992a, b). The significance of this finding lay in the fact that 3-oxo-C6-HSL and a related compound, *N*-(3-hydroxybutanoyl) homoserine lactone (3-hydroxy-C4-HSL), had until this time been exclusively known as 'autoinducers' of bioluminescence in the marine bacteria *Vibrio fischeri* and *Vibrio harveyi* (Eberhard *et al.* 1981; Cao & Meighen 1989). In *V. fischeri*, a symbiont found in the light organs of the sepiolid squid *Euprymna scolopes*, the structural and regulatory genes necessary for light production and synthesis of 3-oxo-C6-HSL (the *lux* regulon) are all located on a 9 kb DNA fragment (Engebrecht & Silverman 1984). The *lux* regulon is organized into two divergently transcribed units, separated by an intergenic regulatory region (Engebrecht & Silverman 1984; Stevens & Greenberg 1999). The leftward transcriptional unit consists of the *luxR* gene, encoding LuxR, a positive transcriptional regulator protein. The rightward operon, is comprised of the *luxI* gene, encoding the 3-oxo-C6-HSL synthase, followed by the *luxCDABE* structural genes, which encode the α- and β-subunits of the luciferase (*luxA* and *luxB*) and a fatty-acid reductase complex (*luxCDE*) necessary for light generation. The LuxI and LuxR proteins and a proposed LuxR binding site, a 20 bp inverted repeat called the *lux* box (Stevens *et al.* 1994; Stevens & Greenberg 1999), which is situated within the intergenic region, are required for the primary cell-density-dependent regulation of *lux* gene expression and are essential for the autoinduction response. LuxR is activated by binding 3-oxo-C6-HSL, which in turn induces transcription of *luxICDABE* (Stevens & Greenberg 1999). Homologues of LuxI and LuxR termed CarI (ExpI) and CarR were subsequently identified in *E. carotovora* (Swift *et al.* 1993; Pirhonen *et al.* 1993; McGowan *et al.* 1995). In contrast to the *lux* operon, *carI* (*expI*) is not linked to the *car* structural genes but is located elsewhere on the chromosome and is adjacent to a second LuxR homologue, ExpR (McGowan *et al.* 1995, 1996; Pirhonen *et al.* 1993).

The discovery that 3-oxo-C6-HSL was produced by both marine and terrestrial Gram-negative bacteria suggested that AHL-dependent quorum-sensing systems may be common throughout the bacterial kingdom. To explore this hypothesis, plasmid-based AHL biosensors have been constructed in which the accumulation of an AHL results in the expression of a specific phenotype. Such biosensors usually contain LuxR or a LuxR homologue together with an AHL-activated promoter fused to a reporter gene such as *lacZ* or *luxCDABE*, but lack an AHL synthase (Bainton *et al.* 1992*b*; Pesci *et al.* 1997; Piper *et al.* 1993; Swift *et al.* 1993; Winson *et al.* 1998). Since such strains do not produce any AHLs, expression of the reporter occurs only in the presence of exogenously added AHLs. Using a plasmid-based AHL biosensor carrying *luxRI′::luxAB*, spent culture supernatants were screened for the presence of AHLs. Positive results were obtained with supernatants from *Pseudomonas aeruginosa*, *Serratia marcescens*, *Erwinia herbicola*, *Citrobacter freundii*, *Enterobacter agglomerans* and *Proteus mirabilis* (Bainton *et al.* 1992a; Swift *et al.* 1993). For *P. aeruginosa*, *S. marcescens*, *E. herbicola* and *E. agglomerans*, the inducing compound was isolated and chemically confirmed as 3-oxo-C6-HSL (Bainton *et al.* 1992a). Elucidation of the structure, stereochemistry and synthesis of 3-oxo-C6-HSL produced by *E. carotovora* is fully described in Bainton *et al.* (1992b). These results confirmed, and extended, some earlier indications that the ability of bacteria to produce AHLs might extend beyond the bioluminescent vibrios (Greenberg *et al.* 1979). Since this work, other bacteria now known to produce AHLs include *Aeromonas hydrophila*, *Aeromonas salmonicida*, *Agrobacterium tumefaciens*, *Burkholderia cepacia*, *Chromobacterium violaceum*, *Nitrosomonas europoea*, *Obesumbacterium proteus*, *Pseudomonas aureofaciens*, *Pseudomonas fluorescens*, *Pseudomonas syringae*, *Pseudomonas putida*, *Rhanella aquatilis*, *Hafnia alvei*, *Ralstonia solanacearum*, *Rhizobium etli*, *Rhizobium leguminosarum*, *Rhodobacter spaeroides*, *Vibrio anguillarum*, *Vibrio logei*, *Vibrio metschnikovii*, *Xenorhabdus nematophila*, *Yersinia enterocolitica*, *Yersinia pestis* and *Yersinia pseudotuberculosis* (for reviews, see Dunny & Winans 1999; Hardman *et al.* 1998; Swift *et al.* 1999a, b). In these Gram-negative bacteria, AHLs have been identified with *N*-acyl side chains of 4, 6, 8, 10, 12 and 14 carbons with either an oxo- or hydroxy- or no substituent at the C3 position of the *N*-linked acyl chain. To date, only two compounds with acyl chains containing double bonds have been identified. These are 7,8-cis-*N*-(3-hydroxytetradecenoyl)homoserine lactone and 7,8-cis-*N*-(tetradecenoyl)homoserine lactone produced by *Rhizobium* leguminosarum (Schripsema *et al.* 1996; Gray *et al.* 1996) and *Rhodobacter sphaeroides* (Puskas *et al.* 1997), respectively. Figure 2 presents a schematic diagram illustrating AHL-dependent quorum sensing.

Interestingly, no Gram-positive bacteria have so far been shown to produce AHLs, and pathogenic Gram-negative bacteria which so far have been reported as AHL-negative include *Actinobacillus pleuropneumoniae*, *Escherichia coli*, *Haemophilus influenzae*, *Klebsiella pneumoniae*, *Neiserria*

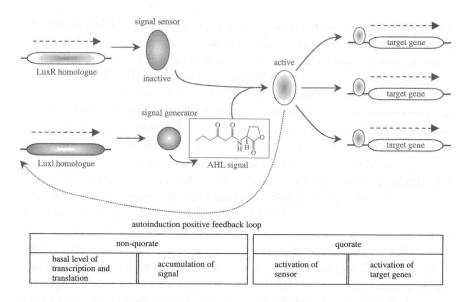

Fig. 2. Schematic representation of AHL-dependent quorum sensing in Gram-negative bacteria.

meningitidis, Salmonella spp., *Vibrio cholerae, Campylobacter jejuni* and *Helicobacter pylori* (Swift *et al.* 1999c). While it is highly likely that these organisms are indeed AHL-negative, the lack of a response could also re- flect the inability of the biosensor employed to respond to the AHL analogue produced. However, for some of these bacteria (e.g. *E. coli, V. cholerae* and *K. pneumoniae*) attempts to use chemical approaches such as electron im- pact mass spectrometry to detect characteristic fragment ions (e.g. the 102 molecular ion fragment corresponding to the homoserine lactone moeity; Camara *et al.* 1998) have proved negative (authors' unpublished data). In addition, perusal of the genome sequence databases for *H. influenzae, H. pylori* and *C. jejuni* reveals no obvious LuxI or LuxR homologues. Intrigu- ingly, *E. coli* and *Salmonella* both possess a LuxR homologue called SdiA (suppressor of division inhibition), which is more than 60% identical to RhlR (VsmR) from *P. aeruginosa* (Wang *et al.* 1991; Ahmer *et al.* 1998; Garcia-Lara *et al.* 1996; Sitnikov *et al.* 1996; Latifi *et al.* 1995). In *E. coli,* SdiA has been proposed to regulate expression of the *ftsQAZ* gene clus- ter, which is required for cell division (Garcia-Lara *et al.* 1996; Sitnikov *et al.* 1996), while the *Salmonella* homologue has been implicated in the expression of several putative virulence determinants (Ahmer *et al.* 1998). Attempts to demonstrate that SdiA functionality depends on activation by an AHL have proved somewhat inconclusive. 3-oxo-C6-HSL, 3-hydroxy-C4-HSL and *N*-decanoyl-DL-homoserine lactone all weakly stimulate the SdiA-dependent P_2 promoter from the *ftsQAZ* gene cluster, although no obvious structure–function relationship is apparent (Sitnikov *et al.* 1996). However, spent *E. coli* culture supernatant is more effective, implying the presence of a diffusible signalling molecule. Furthermore, Garcia-Lara *et al.* (1996) report that *sdiA* itself is regulated by an unidentified, extracellular factor present in spent growth medium.

Further evidence of a role for small, diffusible molecules in the control of cell division in *E. coli* has emerged from the work of Withers & Nordström (1999). They discovered that chromosomal replication could be inhibited by an extracellular factor present in the late logarithmic and early stationary phases of growth. This factor appears to exert its inhibitory activity during initiation of DNA replication directly at each replication fork and does not require transcription or translation. As a consequence, its mode of action must be fundamentally different from all other quorum-sensing systems so far described. This factor is also made by *Salmonella typhimurium* and in contrast to the AHLs, it does not partition into organic solvents, suggesting that it is a highly polar molecule. Coupling the initiation of DNA replication to quorum sensing may offer a bacterium an opportunity to sense the number of cells occupying the same niche such that it is able to slow down replication well in advance of stationary phase. This may be important since it takes time for ongoing replication to proceed to completion. Furthermore, cell survival in the stationary phase and the ability to respond rapidly to environmental changes may be compromised by the presence of stalled replication forks arising as a consequence of insufficient nutrients to complete ongoing replication once stationary phase has been reached (Withers & Nordström 1999).

The existence of an uncharged polar quorum-sensing signal molecule in *E. coli*, which may be related to that described by Withers & Nordström (1999), has been described by Surette & Bassler (1998). This molecule is produced in LB medium only when supplemented with glucose, maximum activity is found in the mid-logarithmic phase of growth and the molecule is rapidly degraded either as the glucose concentration becomes limiting or by the onset of stationary phase. Although no *E. coli* phenotype has been associated with this molecule, it was initially detected using a biosensor based on a quorum-sensing system involved in the regulation of bioluminescence in *Vibrio harveyi*. This free-living marine bacterium can be found in seawater and in the intestines of fish. Although the organization of the *lux* structural genes in *V. harveyi* is identical to that of *V. fischeri*, the mechanism of regulation is quite different (Freeman & Bassler 1999). Bioluminescence in *V. harveyi* is controlled by a sophisticated system involving two integrated quorum-sensing circuits that function via a pair of two-component sensor kinase-response regulator systems. The first of these systems responds to an AHL (3-hydroxy-C4-HSL), the synthesis of which depends not on a LuxI homologue but on LuxLM, an alternative AHL synthase (Bassler 1999; Freeman & Bassler 1999). Signal transduction through 3OH-C4-HSL occurs via LuxN, a membrane-located sensor kinase. The second regulatory system, which senses an, as yet unidentified, signal molecule (termed AI-2), is also sensed via a two-component regulatory pair (LuxP and LuxQ). Sensory information provided via either pathway is relayed to LuxO, a negative regulator of the *lux* structural genes. LuxR, a positive transcriptional activator is also required for *lux* gene expression in *V. harveyi*, but it should be noted that this LuxR protein shares no homology with the *V. fischeri* LuxR. The 3-hydroxy-C4-HSL or AI-2 pathways alone are sufficient for the induction *lux* gene expression in *V. harveyi*, although the kinetics and intensity of response to the two autoinducers are different (Freeman & Bassler 1999). By using *V. harveyi* mutants defective in their ability to respond to either 3-hydroxy-C4-HSL or AI-2, Bassler *et al.* (1997) concluded that the 3-hydroxy-C4-HSL system was highly species-specific, while AI-2 was non-species-specific since they were able to show that many species of bacteria, including laboratory and pathogenic *E. coli* (e.g. O157:H7 strains), *Vibrio cholerae* and *S. typhimurium* produce molecules capable of mimicking the activity of the *V. harveyi* AI-2 autoinducer. Since the chemical identity of these AI-2, *E. coli* or *Salmonella* molecules is not known, it is not clear whether they are identical or if they constitute a new family of quorum-sensing signal molecules. In this context, it is perhaps worth sounding a cautionary note since bacterially derived molecules (cyclic dipeptides) unrelated to AHLs have been identified in culture supernatants as a consequence of their ability to activate AHL biosensors (Holden *et al.* 1999). Interestingly, the *E. coli* laboratory strain DH5α does not produce AI-2-like activity and Surette *et al.* (1999) exploited this finding to clone, by complementation, a gene (termed *luxS*) responsible for AI-2 production common to *V. harveyi, S. typhimurium* and *E. coli*. LuxS is therefore presumed to be an AI-2 synthase and *luxS* homologues are present in a number of Gram-negative and Gram-positive bacteria for which genome databases are available (Surette *et al.* 1999). Whether AI-2 is involved in regulating or activating the *sdiA* gene or gene product, respectively, remains to be established, as does the nature of the signal-transduction system and target structural genes regulated via AI-2 in bacteria other than *V. harveyi.*

3. *N*-ACYL HOMOSERINE LACTONES AND THE REGULATION OF VIRULENCE GENES IN GRAM-NEGATIVE PATHOGENS

Among the AHL producers described above, are several organisms capable of infecting humans and animals. During infection, survival and multiplication in a hostile environment are clearly the priorities of a pathogen which must modulate expression of those genes necessary to establish the organism in a new niche. Parameters such as temperature, pH, osmolarity and nutrient availability are all known to function as environmental signals controlling the expression of coordinately regulated virulence determinants in bacteria (Williams 1988; Mekalanos 1992). A common feature of many bacterial infections is the need for the infecting pathogen to reach a critical cell population density sufficient to overwhelm host immune defences. It is therefore possible that AHLs may be involved in coordinating the control of virulence. The ability of a population of bacteria to coordinate their attack on the host may therefore be a crucial component in the development of infection, particularly by opportunistic pathogens. Interestingly, most of the pathogenic AHL producers so far identified tend to be opportunists which can exist in multiple environments. In contrast, pathogens such as *H. influenzae, N. meningitidis* and *H. pylori*, which uniquely colonize and cause infections in humans, do not appear to employ AHL-mediated quorum sensing (Swift *et al.* 1999c). Among Gram-negative bacteria capable of causing infections in humans and animals, AHL-dependent quorum-sensing circuits have been described in *P. aeruginosa* (Latifi *et al.* 1996; Pesci *et al.* 1997), *B. cepacia* (Lewenza *et al.* 1999), *C. violaceum* (McClean *et al.* 1997), *Yersinia* spp. (Atkinson *et al.* 1999; Swift *et al.* 1999a; Throup *et al.* 1995), *Aeromonas* spp. (Swift *et al.* 1997) and *V. anguillarum* (Milton *et al.* 1997).

The role of AHLs in controlling virulence gene expression is particularly well exemplified in *P. aeruginosa*.

(a) *Pseudomonas aeruginosa*

The pseudomonads exhibit considerable nutritional and metabolic versatility and are found in a wide variety of environmental niches ranging from soil and water to plants and animals (Passador & Iglewski 1995; Williams *et al.* 1996). Among the pathogenic members of this genus, *P. aeruginosa* is the most important human pathogen and is commonly responsible for respiratory tract infections in cystic fibrosis patients as well as blood, skin, eye and genitourinary tract infections in patients immunocompromised by surgery, cytotoxic drugs or burn wounds (Passador & Iglewski, 1995; Williams *et al.* 1996). *P. aeruginosa* produces a wide variety of exoproducts many of which contribute to the virulence of this opportunistic pathogen. These include elastase, alkaline protease, exotoxin A, cytotoxic lectins and phospholipase (Passador & Iglewski 1995) in addition to the phenazine pigment pyocyanin (Hassett *et al.* 1992) and the siderophores, pyoverdin and pyochelin (Meyer *et al.* 1996). The synthesis of these exoproducts is not constitutive but is regulated by the prevailing growth environment with both nutrient deprivation (especially iron-deprivation) and growth rate influencing expression (Brown & Williams 1985). At the genetic level it is now apparent that regulation of the genes for many *P. aeruginosa* virulence determinants including elastase (*lasB*), the LasA protease (*lasA*), alkaline protease (*aprA*), rhamnolipids and exotoxin A (*toxA*), as well as type IV pilus-mediated twitching motility and synthesis of the siderophore pyoverdin, are mediated to varying degrees through quorum sensing (Glessner *et al.* 1999; Pesci & Iglewski 1999; Stintzi *et al.* 1998; Williams *et al.* 1996).

The discovery that *P. aeruginosa* possesses a LuxR homologue (LasR) which regulates *lasB* expression (Gambello & Iglewski 1991) and the isolation of 3-oxo-C6-HSL from *P. aeruginosa* cell-free supernatants by Bainton *et al.* (1992a) provided preliminary evidence of a role for quorum sensing in the regulation of virulence gene expression. Subsequent work from a number of laboratories established that *P. aeruginosa* has two pairs of LuxRI homologues, i.e. LasRI and RhlRI (also termed VsmRI) (Gambello & Iglewski 1991; Passador *et al.* 1993; Ochsner & Reiser 1995; Latifi *et al.* 1995), and four AHLs have been chemically characterized, namely

3-oxo-C6-HSL (Bainton *et al.* 1992a), *N*-3-oxododecanoyl-L-homoserine lactone (3-oxo-C12-HSL; Fig. 1(*b*); Pearson *et al.* 1994), *N*-butanoyl-L-homoserine lactone (C4-HSL; Pearson *et al.* 1995; Winson *et al.* 1995) and *N*-hexanoyl-L-homoserine lactone (C6-HSL; Winson *et al.* 1995). The major signal molecules produced via LasI and RhlI respectively are 3-oxo-C12-HSL and C4-HSL, although in addition, LasI directs 3-oxo-C6-HSL synthesis and RhlI directs C6-HSL synthesis (Pearson *et al.* 1995; Winson *et al.* 1995). Furthermore LasR/3-oxo-C12-HSL is required for the expression of the *rhlRI* locus and *P. aeruginosa* employs a multilayered hierarchical quorum-sensing cascade linking LasR/3-oxo-C12-HSL, RhlR/C4-HSL and the alternative sigma factor RpoS to integrate the regulation of virulence determinants with survival in the stationary phase (Latifi *et al.* 1996). Intriguingly, expression of virulence genes such as that coding for the cytotoxic galactophilic lectin PA-I is directly dependent on both RhlR/C4-HSL and RpoS; mutations in *rhlR*, *rhlI* or *rpoS* all lead to a loss of lectin production, while *rpoS* mutants also exhibit reduced expression of elastase (K. Winzer, S. P. Diggle, C. Falconer, M. Camara and P. Williams, unpublished data). As would be predicted from the hierarchy, mutations in LasR should also lead to the loss of PA-I production. However, in a *lasR* mutant, lectin production was not abolished but substantially delayed when compared with the parent strain, suggesting that the *rhlRI* system is functional in the late stationary phase in the absence of *lasR* (Winzer, K., Diggle, S. P., Camara, M. and Williams, P. unpublished data). Another mechanism for bypassing *lasR* was described by Van Delden *et al.* (1998). They found that when elastase-negative *P. aeruginosa lasR* mutants are grown in a medium in which the sole carbon and nitrogen source provided was casein, spontaneous mutants are generated which suppress the *lasR* mutation such that *rhlRI* functionality and hence synthesis of elastase and other exoproducts is restored.

In *V. fischeri*, the *lux* box, a 20-nucleotide inverted repeat, situated within the intergenic region betweeen *luxR* and *luxI*, is required for the primary regulation of *lux* gene expression and is essential for the autoinduction response (Stevens *et al.* 1994). Similar motifs have been located upstream of the transcriptional start of a number of quorum-sensing-dependent genes in several different organisms (Salmond *et al.* 1995; Fuqua *et al.* 1996; Latifi *et al.* 1995; Pesci & Iglewski 1999) In *Agrobacterium tumefaciens, in vitro* experiments have unequivocally demonstrated that the LuxR homologue

TraR binds to its target *lux* box only when activated by its cognate AHL, 3-oxo-C8-HSL (Zhu & Winans 1999). In *P. aeruginosa*, *lux* box-like sequences have been identified upstream of a number of genes including *lasB*, *rhlI*, *rhlA*, *lasI*, *lasA*, *rhlR*, *lasR* (Latifi *et al.* 1995; Pesci & Iglewski 1999) and *lecA* (K. Winzer, C. Falconer and P. Williams, unpublished data). Thus, these *lux* box sequences appear to be conserved regulatory elements and are likely to be the targets for either or both LasR, RhlR and/or other *P. aeruginosa* LuxR homologues present in the genome database (www.pseudomonas.com) but which have yet to be characterized experimentally. Since none of the *lux* box sequences are identical, *in vitro* binding studies with each purified *P. aeruginosa* LuxR homologue will be required to gain insights in DNA target specificity.

The LasRI and RhlRI systems have also been implicated in the regulation of exoenzyme secretion via the Xcp general secretory pathway and in the control of twitching motility. Chapon-Hervé *et al.* (1997) demonstrated that transcription of the *xcp* genes was reduced in both a *lasR* mutant and in an *rhlRI*-negative background. Control of secretion by quorum sensing presumably enables the organism to deal with the enormous induction of exoproduct (e.g. elastase, LasA protease, exotoxin A) synthesis as the bacterial population reaches a high cell density. Both the *las* and *rhl* quorum-sensing systems are required for type IV pilus-dependent twitching motility and infection by the pilus-specific phage D3112cts (Glessner *et al.* 1999). C4-HSL in particular appears to influence both the export and surface assembly of surface type IV pili, while 3-oxo-C12-HSL plays a role in maintaining cell–cell spacing and associations required for effective twitching motility (Glessner *et al.* 1999). Type IV pili have also been shown to be necessary for promoting the surface attachment of *P. aeruginosa* (O'Toole & Kolter 1998), a necessary pre-requisite to biofilm formation. Indeed, *P. aeruginosa* also employs quorum sensing to control biofilm formation; Davies *et al.* (1998) have shown that production of 3-oxo-C12-HSL via LasI is necessary for the formation of a normal biofilm.

In addition to the AHLs, two additional, chemically distinct, classes of putative quorum-sensing signal molecules have recently been described in *P. aeruginosa*. Holden *et al.* (1999) identified two cyclic dipeptides, cyclo(ΔAla-L-Val) (Fig. 1(*c*)) and cyclo(L-Pro-L-Tyr) present in cell-free culture supernatants via their capacity to weakly activate LuxR-dependent AHL biosensors and to inhibit AHL-mediated swarming in

Serratia liquefaciens. Whether these cyclic dipeptides modulate AHL-dependent quorum sensing in the producer organism or in other organisms occupying the same ecological niche, or indeed whether they function as diffusible signal molecules *per se*, remains to be established. By studying the expression of *lasB* in a *lasR*-negative mutant, Pesci *et al.* (1999) discovered another *P. aeruginosa* quorum-sensing signal molecule unrelated to either the AHLs or cyclic dipeptides. This signal molecule, the synthesis and bioactivity of which is dependent on both the LasRI and RhlRI quorum-sensing systems, was chemically characterized as 2-heptyl-3-hydroxy-4-quinolone (Fig. 1(*d*)), a compound related to the well known class of 4-quinolone antibiotics (Pesci *et al.* 1999). Intriguingly, 4-quinolones such as nalidixic acid and ofloxacin have the capacity to induce high-level shiga toxin expression in enterohaemorrhagic *E. coli* by an as yet unidentified mechanism (Kimmit *et al.* 1999). It is therefore tempting to invoke the existence of a 4-quinolone signalling pathway in *E. coli* to account for these observations.

(b) *The pathogenic Yersiniae*

The genus *Yersinia* which belongs to the *Enterobacteriaceae* contains three species capable of causing disease in both humans and rodents (Cornelis & Wolf-Watz 1997; Swift *et al.* 1999a). *Yersinia pestis* is the causative agent of plague and is transmitted via the bite of an infected flea, while *Yersinia enterocolitica* and *Yersinia pseudotuberculosis* are both food-borne pathogens capable of causing adenitis, septicaemia and gastrointestinal syndromes. Although plague and yersinosis have very different clinical manifestations, the three pathogenic species share many virulence determinants and a tropism for lymphoid tissues (Cornelis & Wolf-Watz 1997; Swift *et al.* 1999a). *Yersinia* virulence determinants include flagellins, lipopolysaccharide and siderophores, as well as gene products important for the attachment and penetration of the intestinal barrier, and a type III secretory pathway which allows the bacterium to inject Yop proteins directly into a target cell (e.g. macrophages and polymorphonuclear leucocytes) cytoplasm and thus evade the cellular immune response (Cornelis & Wolf-Watz 1997; Swift *et al.* 1999a). The regulation of virulence gene expression in the pathogenic *Yersiniae* is complex, with temperature playing a key role; *yop* gene expression, for example, depends on the temperature-dependent interplay between the transcriptional activator VirF, which itself is modulated by a histone-like protein, YmoA (Cornelis & Wolf-Watz 1997). All three pathogenic *Yersinia* species produce AHLs and each possesses at least two LuxRI pairs which, in contrast to the divergent genetic organization in *V. fischeri* and tandem arrangement in *P. aeruginosa*, are transcribed convergently (Atkinson *et al.* 1999; Throup *et al.* 1995; Swift *et al.* 1999a). YenI, YpsI and YpeI from *Y. enterocolitica*, *Y. pseudotuberculosis* and *Y. pestis*, respectively, all direct the synthesis of 3-oxo-C6-HSL and C6-HSL in an approximately equimolar ratio (Throup *et al.* 1995; Atkinson *et al.* 1999; Swift *et al.* 1999a). Inactivation of *yenI* does not influence *yop* gene expression (Throup *et al.* 1995) and a role for quorum sensing in the regulation of virulence in *Y. enterocolitica* has remained elusive. In *Y. pseudotuberculosis* there are two quorum-sensing systems, YpsRI and YtbRI (Atkinson *et al.* 1999). Two phenotypes regulated by YpsRI quorum sensing have been observed; both are repressed in the parent strain and activated in a *ypsR* mutant. At 37°C, but not at 22°C, the *ypsR* mutant, but not the parent strain or the *ypsI* mutant, clump. This suggests that YpsR represses the production of a clumping factor in coordination with temperature changes. Additionally, the major flagellin subunit is overexpressed in the *ypsR* strain and, at 22°C, this is reflected in (i) the hypermotility of the *ypsR* strain on swarm plates, and (ii) the induction of motility in the *ypsR* strain during the logarithmic phase of growth in broth (Atkinson *et al.* 1999). In comparison, the parent strain is not motile at 22°C or 37°C and the *ypsR* mutant is not motile at 37°C. In contrast to the clumping factor, however, the effects on motility are mirrored in a *ypsI* strain. Thus, YpsR appears to act as a repressor of (i) a clumping factor, which is repressed at lower temperatures (i.e. 22°C) and does not require acyl-HSL production for derepression. This may favour the aggregation of cells in a host where increasing numbers could improve (i) colonization and (ii) motility, which is repressed at higher temperatures (i.e. 37°C), and requires acyl-HSL production for derepression to occur in the presence of YpsR. Hence this may facilitate the dispersal and dissemination of bacteria from high-density colonies when removed from the host. Since *Y. pestis* is characteristically non-motile, this poses an interesting question with respect to the function of the homologous quorum-sensing loci. For *Y. pestis*, mutation of the *ypeR* gene (equivalent to the *Y. pseudotuberculosis ypsR* gene) does not appear to influence the expression of major virulence factors including the V antigen, pH6 antigen, Pla (which confers fibrinolytic and coagulase activities) or lipopolysaccharide (Swift *et al.* 1999a).

(c) *A. hydrophila, A. salmonicida* and *V. anguillarum*

While both *A. salmonicida* and *A. hydrophila* are pathogenic for salmonid fish (Fryer & Bartholomew 1996), only *A. hydrophila* causes disease in humans (gastroenteritis and septicaemia; Thornley *et al.* 1997). The virulence of Aeromonas is multifactorial, involving the combined activities of multiple surface-associated macromolecules and secreted exoproteins, including a number of serine and metalloproteases (Swift *et al.* 1999*b*). Many of these exoenzymes are only secreted at high cell densities in the stationary phase, invoking the possibility that they may be regulated via quorum sensing (Swift *et al.* 1999*b*). This has indeed proved to be the case and Swift *et al.* (1997) cloned a divergent pair of LuxRI homologues termed AhyRI and AsaRI in A. hydrophila and *A. salmonicida* respectively. In common with RhlI from *P. aeruginosa*, both AhyI and AsaI direct the synthesis predominantly of C4-HSL along with small amounts of C6-HSL (Swift *et al.* 1997). Mutagenesis of either *ahyI* or *ahyR* results in the loss of C4-HSL production in *A. hydrophila* and a substantial downregulation of both serine and metalloprotease activities, which can be restored in the *ahyI*-negative mutant by supplying exogenous C4-HSL (Swift *et al.* 1999*b*).

Apart from *V. fischeri*, and *V. harveyi*, there is little information on AHL-dependent quorum sensing in other non-luminous Vibrio species. When cell-free supernatants prepared from pathogenic vibrios including *V. cholerae*, *Vibrio parahaemolyticus*, *Vibrio vulnificus* and *V. anguillarum* were screened with AHL *lux*-based biosensors, only *V. anguillarum* gave a positive result (Milton *et al.* 1997; Swift *et al.* 1999*c*). This bacterium is responsible for a terminal haemorrhagic septicaemia known as vibriosis, which is of economic significance in the fish-farming industry. Although the first description of the disease was almost a century ago, not much is known of the pathogenesis of vibriosis and apart from a siderophore transport system, metalloprotease and chemotactic motility, few virulence determinants have been characterized (Crosa 1989; Milton *et al.* 1992, 1996). Milton *et al.* (1997) identified the major *V. anguillarum* AHL as *N*-(3-oxodecanoyl)-L-homoserine lactone (3-oxo-C10-HSL) and described a pair of LuxRI homologues termed VanRI. Mutagenesis experiments and the expression of *vanI* in *E. coli* confirmed that VanI was responsible for 3-oxo-C10-HSL synthesis and that *vanI* expression depended on VanR/3-oxo-C10-HSL. Interestingly, *V. anguillarum* produces an extracellular metalloprotease (EmpA) with 47% identity at the amino-acid level to the elastase of *P. aeruginosa*. Although the *empA* upstream sequence contains a potential *lux* box-like sequence, its expression does not appear to depend on the VanR/3-oxo-C10-HSL since *vanR*, *vanI* and *vanRI* mutants did not show any reduction in protease production (Milton *et al.* 1997). However, the EmpA protease of *V. anguillarum* is induced by gastrointestinal mucus to a level some ninefold greater than is observed in conventional broth (Denkin & Nelson 1999). It is therefore possible that the *vanIR* mutants may respond differently when exposed to mucus. Furthermore, the *vanI* mutant remained capable of weakly activating AHL biosensors suggesting the existence of additional layers of AHL-mediated regulatory hierarchy (Milton *et al.* 1997).

4.   DOES QUORUM SENSING OCCUR *IN VIVO* DURING INFECTION?

Two main approaches can be taken to determine whether AHL-dependent quorum sensing occurs *in vivo* during infection. First, the virulence of mutants with defects in *luxI* or *luxR* homologues can be assessed in appropriate animal models. For mutants with defects in *luxI* homologues, it should be possible to restore virulence by the exogenous provision of the cognate AHL. Second, tissues from infected animals or humans can be extracted and analysed for the presence of AHLs and/or bacterial cells expressing quorum-sensing genes.

In plant pathogens such as *E. carotovora*, there is a direct relationship between virulence and AHL production in that mutants with defects in the quorum-sensing machinery are avirulent (Jones *et al.* 1993; Pirhonen *et al.* 1993) and virulence can be restored by the co-inoculation of 3-oxo-C6-HSL or by expression of a *luxI* homologue in planta (Fray *et al.* 1999). In *P. aeruginosa*, *lasR* mutants have been evaluated in a number of model infections. Using a neonatal mouse model of acute pneumonia, Tang *et al.* (1996) reported that a *lasR* mutant was much less virulent than the parent strain PAO1. In contrast, a *lasR*-negative mutant was as virulent as the wild-type in a murine model of corneal infection (Preston *et al.* 1997). More recently, the nematode *Caenorhabditis elegans* has been exploited to model mammalian bacterial pathogenesis and has been used to identify *P. aeruginosa* virulence factors (Mahajan Miklos *et al.* 1999). By screening *P. aeruginosa* Tn*phoA* mutants using this model, Tan *et al.* (1999) identified a *lasR* mutant which exhibited significantly reduced virulence in the *C. elegans* model, in an Arabidopsis leaf infiltration model and in a mouse full-thickness burn wound infection model. This mutant also failed to accumulate in the gut of *C. elegans*, suggesting that the establishment and/or proliferation of *P. aeruginosa* within the host may also be dependent on quorum sensing. For *P. aeruginosa*, the requirement for a functional LasR protein for virulence in plants, nematodes and mice suggests that at least for this pathogen, quorum sensing is a general feature of pathogenesis in all hosts. Although no data have been published on the virulence of *P. aeruginosa* strains with defects in other components of the quorum-sensing machinery (e.g. *lasI*, *rhlR* and *rhlI*), they are likely to exhibit reduced virulence given the nature of the regulatory hierarchy (Latifi *et al.* 1996; Pesci *et al.* 1997). Wang *et al.* (1996) developed an *in vivo* expression technology (IVET) system for *P. aeruginosa* which was applied to a neutropenic mouse infection model and recovered a number of novel genetic loci that were specifically induced *in vivo*. However, no known quorum-sensing regulatory elements were identified in this study, although only 22 loci were analysed and many more will clearly be involved in sustaining bacterial growth *in vivo*.

In addition to *P. aeruginosa*, two other Gram-negative pathogens with mutations in LuxR and LuxR homologues have been evaluated in experimental animal models of infection. The LD_{50}s for rainbow trout infected either by immersion in seawater or by direct intraperitoneal injection with *V. anguillarum vanI* or *vanR* or *vanIR* mutants were no different to that of the wild-type (Milton *et al.* 1997). This implies that quorum sensing may not play a direct role in vibriosis, although there is a second quorum-sensing system in *V. anguillarum* analogous to that of *V. harveyi* (D. Milton, V. J. Hope, A. Hardman, M. Camara and P. Williams, unpublished data) such that it is possible that the *vanIR* mutations may be compensated for by this alternative system. However, when *V. anguillarum* transformed with a plasmid containing either a *vanR::lacZ* or a *vanI::lacZ* fusion are recovered from the kidneys of infected fish, it is clear from the β-galactosidase levels observed that both genes are being expressed during growth *in vivo* (D. Milton, V. J. Hope, A. Hardman, M. Camara and P. Williams, unpublished data).

Similarly, when Balb/c mice were challenged subcutaneously with either the wild-type *Y. pestis* or the isogenic *ypeR* mutant, the minimum lethal dose was the same for both strains (K. E. Isherwood, P. C. F. Oyston, S. Atkinson, P. Williams, G. S. A. B. Stewart and R. W. Titball, unpublished observations). There was, however, a small increase in time to death implying that the *ypeR* gene makes a contribution to *Y. pestis* pathogenicity, although it is again possible that the second quorum-sensing system compensates for mutations in the first (K. E. Isherwood, P. C. F. Oyston, S. Atkinson, P. Williams, G. S. A. B. Stewart and R. W. Titball, unpublished observations). Alternatively, it is possible that quorum sensing may be more important for the growth and survival of *Y. pestis* in the flea transmission vector than the host.

5.   EVIDENCE FOR AHL PRODUCTION *IN VIVO*

The direct demonstration of quorum sensing *in vivo* could be achieved by the detection of AHLs in the tissues or body fluids of infected animals or humans using AHL biosensors. A number of *lux*-based plasmid AHL biosensors have been constructed (Winson *et al.* 1998). These are based on LuxR, LasR and RhlR, and their target promoters (P_{luxI}, P_{lasI} and P_{rhlI}, respectively) transcriptionally fused to the *luxCDABE* cassette from Photorhabdus luminescens. These sensors emit light in the presence of an activating AHL with sensitivities in the picomolar to nanomolar range. Although they are most sensitive to their natural ligands, they are capable of responding to a range of AHL analogues. Since AHLs are soluble in ethyl acetate and dichloromethane, these organic solvents can be used to extract and concentrate AHLs produced in tissues during growth *in vivo*.

During infection of rainbow trout (*Oncorhynchus mykiss*) with *A. salmonicida* or *A. hydrophila* or *V. anguillarum*, the bacteria become concentrated in the kidney, which can conveniently be removed, homogenized and extracted with dichloromethane. Using this approach, it has been possible to use AHL biosensors to demonstrate the presence of AHLs in infected but not in uninfected tissues (L. Fish, D. Milton, S. Swift and P. Williams, unpublished data). For *A. salmonicida*, the concentration of C4-HSL in the kidney detected using an *rhlRI'::lux* biosensor, was estimated to be approximately 20 nM although this may be an underestimate given that the efficiency of solvent extraction is unlikely to be 100%. In addition, the identity of the AHL is presumed since LuxR homologues will respond to a range of AHL analogues (Winson *et al.* 1998) and therefore definitive proof for the presence of a specific AHL depends on further chemical characterization. Although it was not possible to fully chemically characterize the AHL(s) produced *in vivo* by *A. salmonicida*, extraction of the kidney homogenates from *V. anguillarum* infected rainbow trout provided sufficient material for liquid-chromatography-mass spectrometry. This technique couples the resolving power of HPLC with mass spectrometry such that the mass of the molecular ion and its major component fragments can be determined for a compound with a given retention time. Using this approach, it was possible to unequivocally confirm the presence of 3-oxo-C10-HSL in infected kidneys, a finding which correlated with the activation of the *lasRlasI'::lux*-based AHL biosensor (V. Hope, D.Milton, A. Hardman, M. Camara and P. Williams, unpublished data).

In cystic fibrosis (CF) patients, chronic respiratory tract infection with *P. aeruginosa* or *B. cepacia* leads to destructive lung disease requiring aggressive antibiotic chemotherapy (Govan & Deretic 1996). Several studies have shown that *P. aeruginosa* virulence determinants such as elastase and alkaline protease are produced in the lungs of CF patients (Storey *et al.* 1998). The *lasR* transcript has also been detected in sputum samples (Storey *et al.* 1998), suggesting that AHL-mediated quorum sensing is occurring in the lungs of CF patients and that the sputum from such patients is likely to contain AHLs. By screening solvent-extracted sputum from CF patients infected with *P. aeruginosa*, with a range of AHL biosensors, we have shown that the sputum of patients infected with *P. aeruginosa* contain both short-chain AHLs (probably C4-HSL and C6-HSL) and a molecule which activates the *lasRI'::lux*-based AHL biosensor and co-migrates on thin layer chromatograms with 3-oxo-C12-HSL (B. Middleton, A. Hardman, H. Rogers, A. Knox, M. Camara and P. Williams, unpublished data). Interestingly, we have only been able to obtain mass spectrometry data to confirm the presence of C6-HSL, which is perhaps surprising given that it is produced *in vitro* via RhlI in much smaller quantities than C4-HSL (Winson *et al.* 1995) or 3-oxo-C12-HSL (Pearson *et al.* 1994). In contrast, sputum extracts from CF patients with *Staphylococcus aureus* infections did not activate any of the biosensors consistent with the lack of AHL production in these pathogens (B. Middleton, A. Hardman, H. Rogers, A. Knox, M. Camara and P. Williams, unpublished data). However, extracts from CF patients infected with *B. cepacia*, which produces C8-HSL (Lewenza *et al.* 1999), activated the *luxRI'::lux* biosensor much more strongly than the *lasRI'::lux* biosensor (B. Middleton, A. Hardman, H. Rogers, A. Knox, M. Camara and P. Williams, unpublished data), a result which would be predicted from structure–activity relationships using these biosensors with a range of AHL analogues (Winson *et al.* 1998). Additional preliminary evidence for the production of AHLs *in vivo* was presented by Stickler *et al.* (1998), who tested sections of indwelling urethral catheters that had become colonized with bacterial biofilms. These were removed from patients undergoing long-term indwelling bladder catheterization and tested with an AHL biosensor-based on the *tra* system of *Agrobacterium tumefaciens* (Piper *et al.* 1993). Four out of nine catheter sections gave positive reactions, while unused catheters were negative, suggesting that AHLs were being produced *in situ* in the urinary tract (Stickler *et al.* 1998). However, neither the nature of the biofilm organisms present nor their *in vitro* AHL profiles were presented. Given that the *Agrobacterium* AHL biosensor also responds to molecules such as the cyclic dipeptides described by Holden *et al.* (1999), further work will be required to unequivocally demonstrate AHL production in catheter-associated biofilms.

6.   DO AHLS FUNCTION AS VIRULENCE DETERMINANTS *PER SE*?

Although AHLs have so far largely been considered as effectors of prokaryotic gene expression, they are capable of influencing eukaryotic cell behaviour and potentially modulating disease processes. Cystic fibrosis is a genetic disease characterized by mucus hypersecretion and by chronic bacterial (particularly *P. aeruginosa*) infection and airway inflammation (Govan & Deretic 1996). CF patients carry mutations in a membrane protein termed the CF transmembrane conductance regulator (CFTR), which possesses cyclic AMP-dependent chloride channel activity and is defective in CF. Airway inflammation in CF patients is characterized by the influx and activation of large numbers of neutrophils, which play a major role in the pathology of CF lung disease (Govan & Deretic 1996). Interleukin-8 (IL-8) is a neutrophil-selective stimulus for chemoattraction, adhesion and elastase release and is considered to play an important role in CF (Palfreyman *et al.* 1997). Thus factors which induce IL-8 contribute to the inflammatory process and progressive lung deterioration. Interestingly, the *P. aeruginosa* quorum-sensing signal molecule 3-oxo-C12-HSL has been reported by some

(Di Mango *et al.* 1995), but not others (Palfreyman *et al.* 1997), to stimulate the dose-dependent production of IL-8 by respiratory epithelial cells, albeit at higher levels than are normally produced by *P. aeruginosa* growing in laboratory media (approximately 5 μM; Pearson *et al.* 1995).

Submucosal tracheal gland serous (HTGS) cells are believed to play a major role in the pathophysiology of CF and secrete a number of antibacterial and anti-proteolytic proteins (including lactoferrin, lysozyme and secretory leucocyte proteinase inhibitor (SLPI)). Kammouni *et al.* (1997) have developed a CF HTGS cell line that has retained the CTFR defect. In this cell line, as in normal airway epithelial cells, it is possible to bypass the CTFR defect since calcium-dependent chloride channels remain functional and chloride secretion can be stimulated by nucleotides. ATP, UTP and their analogues have therefore been tested clinically and shown to improve mucociliary clearance in CF. Using both normal and CF HTGS cells, Saleh *et al.* (1999) sought to determine whether AHLs produced by *P. aeruginosa* influenced SLPI secretion. Although they had no effect on the secretion of SLPI in either normal or CF HTGS cells, 3-oxo-C6-HSL and 3-oxo-C12-HSL at picomolar concentrations inhibited the ATP- and UTP-dependent stimulation of SLPI secretion. Similar results were also obtained with C4-HSL and C6-HSL but at much higher concentrations (micromolar range) suggesting that efficient inhibition depends on the presence of a 3-keto substituent and may be independent of acyl chain length. Moreover, it is clear that AHLs such as 3-oxo-C12-HSL are not functioning as receptor antagonists but are able to potently downregulate expression of the nucleotide receptors P2Y2 and P2Y4. The signal transduction pathway involved is not known but may involve leucotrienes, since ibuprofen — an inhibitor of leucotriene production — blocked the action of 3-oxo-C12-HSL (Saleh *et al.* 1999). Furthermore, the high sensitivity of CF HTGS cells compared with normal HTGS cells to 3-oxo-C12-HSL and 3-oxo-C6-HSL suggests a CF-specific defect linked with arachidonate metabolism which, in CF, has long been known to be defective (Saleh *et al.* 1999). Alternatively, the ability of ibuprofen to inhibit the activation and translocation of NF-κB into the nucleus may lead to the downregulation of nucleotide receptor expression (Stuhlmeier *et al.* 1999) and account for the activities of the AHLs. Thus *P. aeruginosa* AHLs can potently modify the behaviour of a cell type (HTGS) believed to make a significant contribution to the pathophysiology of CF.

Apart from the lungs of CF patients, *P. aeruginosa* can cause infections in almost any other body site and Telford *et al.* (1998) sought to determine whether the immune system is capable of responding to AHLs. In murine and human leucocyte immunoassays *in vitro*, 3-oxo-C12-HSL but not 3-oxo-C6-HSL inhibited lymphocyte proliferation and tumor necrosis factor alpha production by lipopolysaccharide stimulated macrophages (Fig. 3(a)). Similarly, neither C4-HSL nor C6-HSL exhibited any inhibitory activity (D. Hooi, D. Pritchard, M. Camara, P. Williams and B. Bycroft, unpublished data). Furthermore, 3-oxo-C12-HSL potently downregulated the

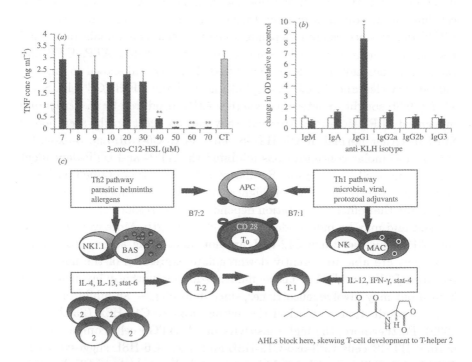

Fig. 3. 3-oxo-C12-HSL suppresses mitogen-induced cell proliferation (Telford *et al.* 1998) and the ability of LPS-stimulated murine macrophages to secrete TNFα (a). Immune suppression may be selective, as murine B cells were stimulated to produce significantly more IgG1, an isotype characteristic of Th-2 activity, following stimulation with keyhole limpet haemocyanin (KLH); (b). Filled bars, + 3-oxo-C12-HSL; open bars, − 3-oxo-C12-HSL. These data suggest that 3-oxo-C12-HSL steers immune responses away from a possibly *Pseudomonas* protective Th-1 phenotype, as shown in (c). APC, antigen-presenting cell; T_0, undifferentiated T cell; NK, natural killer cell; BAS, basophil; MAC, macrophage; CT, control; stat, signal for transduction and activation of transcription; IL, interleukin; IFN, interferon. ** $p < 0.01$; * $p < 0.05$.

production of IL-12, a T helper cell 1 (Th-1) supportive cytokine (Telford *et al.* 1998). At high concentrations (> 70 μM), 3-oxo-C12-HSL inhibited antibody production by keyhole limpet haemocyanin-stimulated spleen cells, but at lower concentrations antibody production was stimulated apparently by increasing the proportion of the immunoglobulin G1 (IgG1) isotype (Fig. 3(*b*)). 3-oxo-C12-HSL also promoted IgE production by interleukin-4-stimulated human peripheral blood mononuclear cells (Telford *et al.* 1998). Since T-cell responses constitute an important component of the host defence against *P. aeruginosa* (Stevenson *et al.* 1996) these data suggest that by modulating T-cell and macrophage functions, 3-oxo-C12-HSL is functioning as a virulence determinant *per se*. More specifically, by switching the T-helper-cell response from the antibacterial Th-1 response (characterized by the secretion of IL-12 and γ-interferon) to a Th-2 response, *P. aeruginosa* may modulate the host inflammatory response to promote its own survival and growth (Fig. 3(*c*)). Such immunomodulatory activity suggests that 3-oxo-C12-HSL and related compounds could find a therapeutic application in the modulation of autoimmune diseases and prevention of septic shock. The molecular basis for the immunomodulatory activity of 3-oxo-C12-HSL is not known. However, it does not appear to be via phosphotyrosine kinases such as p59[fyn] or p56[lck], which are key enzymes in the initial activation of T lymphocytes, since neither enzyme is inhibited by 3-oxo-C12-HSL (Telford *et al.* 1998).

The modulation of T-cell, macrophage and CF HTGS-cell functions by 3-oxo-C12-HSL demonstrates that eukaryotic cells are influenced by AHLs. In addition, Lawrence *et al.* (1999) examined the effect of 3-oxo-C12-HSL and related compounds on the constrictor tone of porcine blood vessels. Over the range of 1–30 μM, 3-oxo-C12-HSL caused a concentration-dependent relaxation of U46619 (a thromboxane mimetic)-induced contractions of the coronary artery, but was markedly less effective in the pulmonary artery. Neither C4-HSL nor homoserine lactone exerted any activity, although the thiolactone analogue *N*-(3-oxododecanoyl)-L-homocysteine thiolactone was as quantitatively effective as 3-oxo-C12-HSL. *N*-3-oxododecanamide, which lacks the lactone ring, possesses only about 30% of the vasorelaxant activity of 3-oxo-C12-HSL (Lawrence *et al.* 1999). 3-oxo-C12-HSL appears to exert its inhibitory effect principally at the level of the smooth muscle, although its mechanism of action remains to be elucidated. This may involve activation of Na^+/K^+-ATPase and the subsequent hyperpolarization of vascular smooth muscle, a mode of action proposed to account for the vasorelaxant properties of the C18 polyunsaturated fatty acid, linoleic acid, on porcine coronary arteries (Pomposiello *et al.* 1998). Given that 3-oxo-C12-HSL is present in the lungs of CF patients chronically colonized with *P. aeruginosa*, the finding that it is a less potent vasorelaxant of pulmonary arterial smooth muscle is perhaps surprising but raises the possibility that other vascular beds may exhibit differential sensitivity to AHLs. Indeed a better indicator of the significance of this vasodilator activity may be obtained by investigating the effects of 3-oxo-C12-HSL on human resistance arteries and small veins rather than the much larger conduit arteries.

The qualitatively similar immunomodulatory and vasodilator activities of the AHLs appear to share a similar dependence on the presence of the C12 acyl chain given the inactivity of the shorter-chain compounds. This implies that their activities in this context differ from the AHL-inhibited nucleotide-dependent stimulation of SLPI secretion in CF HTGS cells, where 3-oxo-C6-HSL was reported to be as active as 3-oxo-C12-HSL. When these observations are taken together, they suggest that 3-oxo-C12-HSL plays a role not only in regulating *P. aeruginosa* virulence gene expression but also the orchestration of eukaryotic cells to maximize the provision of nutrients via the bloodstream while downregulating host defence mechanisms.

7. BLOCKADE OF QUORUM SENSING AS A MEANS OF CONTROLLING INFECTION

As widespread resistance to conventional antibacterial agents continues to pose a major threat, the demand for novel therapeutic approaches to the treatment of infection is increasing. The ability to downregulate virulence gene expression may therefore offer a novel strategy for the treatment or prevention of infection. Our growing understanding that quorum sensing is a generic phenomenon and that a number of pathogens employ diffusible signal molecules to coordinate the control of multiple virulence and survival genes offers a novel chemotherapeutic target (Finch *et al.* 1998). Interference with transmission of the molecular message by a small molecule antagonist which competes for the autoinducer binding site of sensor or transcriptional activator proteins, thereby switching off virulence gene expression and attenuating the pathogen, is an attractive strategy. In this context, the ability of various AHL analogues to inhibit the action of the cognate AHL has been demonstrated (Eberhard *et al.* 1986; McClean *et al.* 1997; Passador *et al.* 1996; Schaefer *et al.* 1996*b*; Swift *et al.* 1997, 1999*b*). For example, C4-HSL-dependent exoprotease production in *A. hydrophila* can virtually be abolished by AHLs with acyl chains of 10, 12 or 14 carbons. Furthermore, Givskov *et al.* (1996) provided evidence that furanone compounds produced by the Australian macroalgae *Delisea pulchra* inhibit AHL-regulated processes including swarming in *S. liquefaciens* and bioluminescence in *V. fischeri*. As the inhibitory effect could be overcome by the addition of an excess of the cognate AHL, the furanones appear to be acting as competitive antagonists. The significant recent advances in defining the enzymatic activity and substrate requirements of LuxI homologues (Moré *et al.* 1996; Schaefer *et al.* 1996*a*; Jiang *et al.* 1998; Parsek *et al.* 1999) also emphasizes the potential of the AHL synthase as an antimicrobial target.

Quorum-sensing blocking agents developed for the prevention or treatment of infections due to AHL-producing Gram-negative bacteria are, however, likely to have a very narrow spectrum of activity since *P. aeruginosa* is the only major human pathogen known to employ AHLs for controlling virulence gene expression. Although many more Gram-negative and some Gram-positive pathogens possess the AI-2/LuxS-based system described by Surette *et al.* (1999), apart from bioluminescence, no phenotype has yet been associated with this quorum-sensing system. Consequently, there is no evidence to suggest that it may constitute a novel antibacterial target. This is not, however, the case for the extracellular factor described by Withers & Nordström (1999) since this molecule is capable of directly inhibiting DNA replication without any requirement for transcription or translation. It is therefore likely that elucidation of the structure of this class of quorum-sensing signal molecules may offer opportunities to design potent antibacterial analogues that are refractory to the relatively rapid turnover observed for the natural molecule. Although no AHL quorum-sensing blocking agents have yet been evaluated *in vivo* in experimental animal models of infection, blockade of the cyclic thiolactone peptide-dependent quorum-sensing system, which controls virulence gene expression in *Staphylococcus aureus*, has been shown to attenuate infection in a mouse skin abscess model (Mayville *et al.* 1999). Given the importance of methicillin-resistant *S. aureus* (MRSA), which are resistant to virtually all clinically available antibiotics including vancomycin (Speller *et al.* 1997), this offers a real possibility for the development of novel anti-staphylococcal chemotherapy that will not be susceptible to conventional antibiotic-resistance mechanisms. However, while such agents would have obvious value in prophylaxis, their therapeutic potential for the treatment of established infections, especially in immunocompromised patients, is less apparent, since an intact host defence may well constitute a necessary prerequisite for clearing the infection.

We dedicate this article to the memory of the late Professor Gordon S. A. B. Stewart, who made many major contributions to the study of quorum sensing. Research in the authors' laboratories has been supported by grants and studentships from the Biotechnology and Biological Sciences Research Council, the Medical Research Council, the Wellcome Trust and the European Union, and these are gratefully acknowledged.

REFERENCES

Ahmer, B. M., Van Reeuwijk, J., Timmers, C. D., Valentine, P. J. & Heffron, F. 1998 *Salmonella typhimurium* encodes an SdiA homolog, a putative quorum sensor of the LuxR family that regulates genes on the virulence plasmid. *J. Bacteriol.* **180**, 1185–1193.

Atkinson, S., Throup, J. P., Stewart, G. S. A. B. & Williams, P. 1999 A hierarchical quorum-sensing system in *Yersinia pseudotuberculosis* is involved in the regulation of motility and clumping. *Mol. Microbiol.* **33**, 1267–1277.

Bainton, N. J. (and 10 others) 1992*a* A general role for the *lux* autoinducer in bacterial cell signalling: control of antibiotic synthesis in *Erwinia*. *Gene* **116**, 87–91.

Bainton, N. J., Stead, P., Chhabra, S. R., Bycroft, B. W., Salmond, G. P. C., Stewart, G. S. A. B. & Williams, P. 1992*b* *N*-(3-oxohexanoyl)-L-homoserine lactone regulates carbapenem antibiotic production in *Erwinia carotovora*. *Biochem. J.* **288**, 997–1004.

Bassler, B. L. 1999 A multichannel two-component signaling relay controls quorum sensing in *Vibrio harveyi*. In *Cell–cell signaling in bacteria* (ed. G. M. Dunny & S. C. Winans), pp. 259–273. Washington, DC: American Society of Microbiology Press.

Bassler, B. L., Greenberg, E. P. & Stevens, A. M. 1997 Cross-species induction of luminescence in the quorum-sensing bacterium *Vibrio harveyi*. *J. Bacteriol.* **179**, 4043–4045.

Brown, M. R. W. & Williams, P. 1985 Influence of environment on envelope properties affecting survival of bacteria in infections. *A. Rev. Microbiol.* **39**, 527–556.

Camara, M., Daykin, M. & Chhabra, S. R. 1998 Detection, purification and synythesis of *N*-acylhomoserine lactone quorum sensing signal molecules. In *Methods in microbiology: bacterial pathogenesis*, vol. 27 (ed. P. H. Williams, G. P. C. Salmond and J. Ketley), pp. 319–330. London: Academic Press.

Cao, J. G. & Meighen, E. A. 1989 Purification and structural identification of an autoinducer for the luminescence system of *Vibrio harveyi*. *J. Biol. Chem.* **264**, 21 670–21 676.

Chapon-Hervé, V., Akrim, M., Latifi, A., Williams, P., Lazdunski, A. & Bally, M. 1997 Regulation of the *xcp* secretion pathway by multiple quorum-sensing modulons in *Pseudomonas aeruginosa*. *Mol. Microbiol.* **24**, 1169–1178.

Cornelis, G. R. & Wolf-Watz, H. 1997 The *Yersinia* Yop virulon: a bacterial system for subverting eukaryotic cells. *Mol. Microbiol.* **23**, 861–867.

Crosa, J. H. 1989 Genetics and molecular biology of siderophore-mediated iron transport in bacteria. *Microbiol. Rev.* **53**, 517–530.

Davies, D. G., Parsek, M. R., Pearson, J. P., Iglewski, B. H., Costerton, J. W. & Greenberg, E. P. 1998 The involvement of cell-to-cell signals in the development of a bacterial biofilm. *Science* **280**, 295–298.

Denkin, S. M. & Nelson, D. R. 1999 The induction of protease activity in *Vibrio anguillarum* by gastrointestinal mucus. *Appl. Environ. Microbiol.* **65**, 3555–3560.

Di Mango, E., Zur, H. J., Bryan, R. & Prince, A. 1995 Diverse *Pseudomonas aeruginosa* gene products stimulate respiratory epithelial cells to produce interleukin-8. *J. Clin. Invest.* **96**, 2204–2210.

Dunny, G. M. & Winans, S. C. 1999 *Cell–cell signaling in bacteria*. Washington, DC: American Society of Microbiology Press.

Eberhard, A., Burlingame, A. L., Kenyon, G. L., Nealson, K. H. & Oppenheimer, N. J. 1981 Structural identification of autoinducer of *Photobacterium fischeri* luciferase. *Biochemistry* **20**, 2444–2449.

Eberhard, A., Widrig, C. A., McBath, P. & Schineller, J. B. 1986 Analogs of the autoinducer of bioluminescence in *Vibrio fischeri*. *Arch. Microbiol.* **146**, 35–40.

Engebrecht, J. & Silverman, M. 1984 Identification of genes and gene products necessary for bacterial bioluminescence. *Proc. Natl Acad. Sci. USA* **81**, 4154–4158.

Finch, R. G., Pritchard, D. I., Bycroft, B. W., Williams, P. & Stewart, G. S. A. B. 1998 Quorum sensing — a novel target for anti-infective therapy. *J. Antimicrob. Chemother.* **42**, 569–571.

Fray, R. G., Throup, J. P., Daykin, M., Wallace, A., Williams, P., Stewart, G. S. A. B. & Grierson, D. 1999 Plants genetically modified to produce *N*-acylhomoserine lactones communicate with bacteria. *Nat. Biotech.* **171**, 1017–1020.

Freeman, J. A. & Bassler, B. L. 1999 A genetic analysis of the function of LuxO, a two-component response regulator involved in quorum sensing in *Vibrio harveyi*. *Mol. Microbiol.* **31**, 665–677.

Fryer, J. L. & Bartholomew, J. L. 1996 Established and emerging infectious diseases of fishas fish move, infections move with them. *Am. Soc. Microbiol. News* **62**, 592–594.

Fuqua, W. C., Winans, S. C. & Greenberg, E. P. 1994 Quorum sensing in bacteria: the LuxR–LuxI family of cell density responsive transcriptional regulators. *J. Bacteriol.* **176**, 269–275.

Fuqua, W. C., Winans, S. C. & Greenberg, E. P. 1996 Census and consensus in bacterial ecosystems: the LuxR–LuxI family of quorum sensing transcriptional regulators. *A. Rev. Microbiol.* **50**, 727–751.

Gambello, M. J. & Iglewski, B. H. 1991 Cloning and characterisation of the *Pseudomonas aeruginosa lasR* gene, a transcriptional activator of elastase expression. *J. Bacteriol.* **173**, 3000–3009.

Garcia-Lara, J., Shang, L. H. & Rothfield, L. I. 1996 An extracellular factor regulates expression of *sdiA*, a transcriptional activator of cell division genes in *Escherichia coli*. *J. Bacteriol.* **178**, 2742–2748.

Givskov, M., DeNys, R., Manefield, M., Gram, L., Maximilien, R., Eberl, L., Molin, S., Steinberg, P. & Kjelleberg, S. 1996 Eukaryotic interference with homoserine lactone-mediated prokaryotic signalling. *J. Bacteriol.* **178**, 6618–6622.

Glessner, A., Smith, R. S., Iglewski, B. H. & Robinson, J. B. 1999 Roles of the *Pseudomonas aeruginosa las* and *rhl* quorum-sensing systems in control of twitching motility. *J. Bacteriol.* **181**, 1623–1629.

Govan, J. R. W. & Deretic, V. 1996 Microbial pathogenesis in cystic fibrosis: mucoid *Pseudomonas aeruginosa* and *Burkholderia cepacia*. *Microbiol. Rev.* **60**, 539–576.

Gray, K. M., Pearson, J. P., Downie, J. A., Boboye, B. B. & Greenberg, E. P. 1996 Cell-to-cell signaling in the symbiotic nitrogen-fixing bacterium *Rhizobium leguminosarum*: autoinduction of stationary phase and rhizosphere-expressed genes. *J. Bacteriol.* **178**, 372–376.

Greenberg, E. P., Hastings, J. W. & Ulitzer, S. 1979 Induction of luciferase synthesis in *Beneckea harveyi* by other marine bacteria. *Arch. Microbiol.* **120**, 87–91.

Hardman, A. M., Stewart, G. S. A. B. & Williams, P. 1998 Quorum sensing and the cell–cell communication dependent regulation of gene expression in pathogenic and non-pathogenic bacteria. *Antony van Leeuwenhoek J. Microbiol.* **74**, 199–210.

Hassett, D. J. Charniga, L., Bean, K., Ohman, D. E. & Cohen, M. S. 1992 Response of *Pseudomonas aeruginosa* to pyocyanin: mechanism of resistance, antioxidant defences and demonstration of a manganese-cofactored superoxide dismutase. *Infect. Immun.* **60**, 328–336.

Holden, M. T. G. (and 16 others) 1999 Quorum sensing cross-talk: isolation and chemical characterization of cyclic dipeptides from *Pseudomonas aeruginosa* and other Gram-negative bacteria. *Mol. Microbiol.* **33**, 1254–1266.

Jiang, Y., Camara, M., Chhabra, S. R., Hardie, K. R., Bycroft, B. W., Lazdunski, A., Salmond, G. P. C., Stewart, G. S. A. B. & Williams, P. 1998 *In vitro* biosynthesis of the *Pseudomonas aeruginosa* quorum sensing signal molecule, *N*-butanoyl-L-homoserine lactone. *Mol. Microbiol.* **28**, 193–203.

Jones, S. (and 13 others) 1993 The *lux* autoinducer regulates the production of exoenzyme virulence determinants in *Erwinia carotovora* and *Pseudomonas aeruginosa*. *EMBO J.* **12**, 2477–2482.

Kammouni, W., Moreau, B., Becq, F., Saleh, A., Pavirani, A., Figarello, C. & Merten, M. D. 1997 A cystic fibrosis tracheal gland cell line, CF-KM4. Correction by adenovirus-mediated CTFR gene transfer. *Am. J. Respir. Cell Mol. Biol.* **20**, 684–691.

Kimmit, P. T., Harwood, C. R. & Barer, M. R. 1999 Induction of type 2 Shiga toxin synthesis in *Escherichia coli* 0157 by 4-quinolones. *Lancet* **353**, 1588–1589.

Kleerebezem, M., Quadri, L. E. N., Kuipers, O. P. & de Vos, W. M. 1997 Quorum sensing by peptide pheromones and two component signal transduction systems in Gram-positive bacteria. *Mol. Microbiol.* **24**, 895–904.

Latifi, A., Winson, M. K., Foglino, M., Bycroft, B. W., Stewart, G. S. A. B., Lazdunski, A. & Williams, P. 1995 Multiple homologues of LuxR and LuxI control expression of virulence determinants and secondary metabolites through quorum sensing in *Pseudomonas aeruginosa* PAO1. *Mol. Microbiol.* **17**, 333–343.

Latifi, A., Foglino, M., Tanaka, K., Williams P. & Lazdunski, A. 1996 A hierarchical quorum sensing cascade in *Pseudomonas aeruginosa* links the transcriptional activators LasR and RhlR (VsmR) to expression of the stationary-phase sigma factor RpoS. *Mol. Microbiol.* **21**, 1137–1146.

Lawrence, R. N., Dunn, W. R., Bycroft, B. W., Camara, M., Chhabra, S. R., Williams, P. & Wilson, V. G. 1999 The *Pseudomonas aeruginosa* quorum sensing signal molecule, *N*-(3-oxododecanoyl)-L-homoserine lactone inhibits poircine arterial smooth muscle contraction. *Br. J. Pharmacol.* **128**, 845–848.

Lewenza, S., Conway, B., Greenberg, E. P. & Sokol, P. A. 1999 Quorum sensing in *Burkholderia cepacia*: identification of the LuxRI homologs CepRI. *J. Bacteriol.* **181**, 748–756.

McClean, K. H. (and 11 others) 1997 Quorum sensing in *Chromobacterium violaceum*: exploitation of violacein production and inhibition for the detection of *N*-acylhomoserine lactones. *Microbiology* **143**, 3703–3711.

McGowan, S., Sebaihia, M., Jones, S., Yu, B., Bainton, N., Chan, P. F., Bycroft, B., Stewart, G. S. A. B., Williams, P. & Salmond, G. P. C. 1995 Carbapenem antibiotic production in *Erwinia carotovora* is regulated by CarR, a homologue of the LuxR transcriptional activator. *Microbiology* **141**, 541–550.

McGowan, S. J., Sebaihia, M., Porter, L. E., Stewart, G. S. A. B., Williams, P., Bycroft, B. W. & Salmond, G. P. C. 1996 Analysis of bacterial carbapenem antibiotic production genes reveals a novel β-lactam biosynthesis pathway. *Mol. Microbiol.* **22**, 415–426.

Mahajan-Miklos, S., Tan, M.-W., Rahme, L. G. & Ausubel, F. M. 1999 Molecular mechanisms of bacterial virulence elucidated using a *Pseudomonas aeruginosa Caenorhabditis elegans* pathogenesis model. *Cell* **96**, 47–56.

Mayville, P., Ji, G., Beavis, R., Yang, H., Goger, M., Novick, R. P. *and* Weir, T. W. 1999 Structureactivity analysis of synthetic autoinducing thiolactone peptides from *Staphylococcus aureus* responsible for virulence. *Proc. Natl Acad. Sci. USA* **96**, 1218–1223.

Mekalanos, J. J. 1992 Environmental signals controlling expression of virulence determinants in bacteria. *J. Bacteriol.* **174**, 1–7.

Meyer, J.-M., Neely, A., Stintzi, A., Georges, C. *and* Holder, I. A. 1996 Pyoverdin is essential for virulence of *Pseudomonas aeruginosa. Infect. Immun.* **64**, 518–523.

Milton, D. L., Nrqvist, A. & Wolf-Watz, H. 1992 Cloning of a metalloprotease gene involved in the virulence mechanisms of *Vibrio anguillarum. J. Bacteriol.* **174**, 7235–7244.

Milton, D. L., O'Toole, R., Hörstedt, P. & Wolf-Watz, H. 1996 Flagellin A is essential for the virulence of *Vibrio anguillarum. J. Bacteriol.* **178**, 1310–1319.

Milton, D. L., Hardman, A., Camara, M., Chhabra, S. R., Bycroft, B. W., Stewart, G. S. A. B. & Williams, P. 1997 *Vibrio anguillarum* produces multiple N-acylhomoserine lactone signal molecules. *J. Bacteriol.* **179**, 3004–3012.

Mor, M. I., Finger, L. D., Stryker, J. L., Fuqua, C., Eberhard, A. & Winans, S. C. 1996 Enzymatic synthesis of a quorum-sensing autoinducer through the use of defined substrates. *Science* **272**, 1655–1658.

Ochsner, U. A. & Reiser, J. 1995 Autoinducer-mediated regulation of rhamnolipid biosurfactant synthesis in *Pseudomonas aeruginosa. Proc. Natl Acad. Sci. USA* **92**, 6424–6428.

O'Toole, G. A. & Kolter, R. 1998 Flagellar and twitching motility are necessary for *Pseudomonas aeruginosa* biofilm development. *J. Bacteriol.* **181**, 1623–1629.

Palfreyman, R. W., Watson, M. L., Eden, C. & Smith, A. W. 1997 Induction of biologically active interleukin-8 from lung epithelila cells by *Burkholderia (Pseudomonas) cepacia* products. *Infect. Immun.* **65**, 617–622.

Parsek, M. R., Val, D. L., Hanzelka, B. L., Cronan, J. E. & Greenberg, E. P. 1999 Acylhomoserine-lactone quorum sensing signal generation. *Proc. Natl Acad. Sci. USA* **96**, 4360–4365.

Passador, L. & Iglewski, B. H. 1995 Quorum sensing and virulence gene regulation in *Pseudomonas aeruginosa. In Virulence mechanisms of bacterial pathogens*, 2nd edn (ed. J. A. Roth, C. A. Bolin, K. A. Brogden, F. C.

Minion and M. J. Wannemuehler), pp. 65–78. Washington, DC: American Society of Microbiology Press.

Passador, L., Cook, J. M., Gambello, M. J., Rust, L. & Iglewski, B. H. 1993 Expression of *Pseudomonas aeruginosa* virulence genes requires cell-to-cell communication. *Science* **260**, 1127–1130.

Passador, L., Tucker, K. D., Guertin, K. R., Journet, M. P., Kende, A. S. & Iglewski, B. H. 1996 Functional analysis of the *Pseudomonas aeruginosa* autoinducer PAI. *J. Bacteriol.* **178**, 5995–6000.

Pearson, J. P., Gray, K. M., Passador, L., Tucker, K. D., Eberhard, A., Iglewski, B. H. & Greenberg, E. P. 1994 Structure of the autoinducer required for expression of *Pseudomonas aeruginosa* virulence genes. *Proc. Natl Acad. Sci. USA* **91**, 197–201.

Pearson, J. P., Passador, L., Iglewski, B. H. & Greenberg, E. P. 1995 A second *N*-acylhomoserine lactone signal produced by *Pseudomonas aeruginosa*. *Proc. Natl Acad. Sci. USA* **92**, 1490–1494.

Pesci, E. C. & Iglewski, B. H. 1999 Quorum sensing in *Pseudomonas aeruginosa*. In *Cell–cell signaling in bacteria* (ed. G. M. Dunny and S. C. Winans), pp. 259–273. Washington, DC: American Society of Microbiology Press.

Pesci, E. C., Pearson, J. P., Seed, P. C. & Iglewski, B. H. 1997 Regulation of *las* and *rhl* quorum sensing in *Pseudomonas aeruginosa*. *J. Bacteriol.* **179**, 3127–3132.

Pesci, E. C., Milbank, J. B., Pearson, J. P., McKnight, S., Kende, A. S., Greenberg, E. P. & Iglewski, B. H. 1999 Quinolone signalling in the cell–cell communication system of *Pseudomonas aeruginosa*. *Proc. Natl Acad. Sci. USA* **96**, 11229–11234.

Piper, K. R., Beck von Bodman, S. & Farrand, S. K. 1993 Conjugation factor of *Agrobacterium tumefaciens* regulates Ti plasmid transfer by autoinduction. *Nature* **362**, 448–450.

Pirhonen, M., Flego, D., Heikinheimo, D. & Palva, E. 1993 A small diffusible signal molecule is responsible for the global control of virulence and exoenzyme production in *Erwinia carotovora*. *EMBO J.* **12**, 2467–2476.

Pomposiello, S. I., Alva, M., Wilde, D. W. & Carretero, O. A. 1998 Linoleic acid induces relaxation and hyperpolarization of the pig coronary artery. *Hypertension* **31**, 615–620.

Preston, M. J., Seed, P. C., Toder, D. S., Iglewski, B. H., Ohman, D. E., Gustin, J. K., Goldberg, J. A. & Pier, G. B. 1997 Contribution of proteases and LasR to the virulence of *Pseudomonas aeruginosa* during corneal infections. *Infect. Immun.* **65**, 3086–3090.

Puskas, A., Greenberg, E. P., Kaplan, S. & Schaefer, A. L. 1997 A quorum sensing system in the free-living photosynthetic bacterium *Rhodobacter sphaeroides*. *J. Bacteriol.* **179**, 7530–7537.

Saleh, A., Figarella, C., Kammouni, W., Marchand-Pinatel, Lazdunski, A., Tubul, A., Brun, P. & Merten, M. D. 1999 *Pseudomonas aeruginosa* quorum sensing signal molecule *N*-(3-oxododecanoyl)-L-homoserine lactone inhibits expression of P2Y receptors in cystic fibrosis tracheal gland cells. *Infect. Immun.* **67**, 5076–5082.

Salmond, G. P. C., Bycroft, B. W., Stewart, G. S. A. B. & Williams, P. 1995 The bacterial enigma: cracking the code of cell–cell communication. *Mol. Microbiol.* **16**, 615–624.

Schaefer, A. L., Val, D. L., Hanzelka, B. L., Cronan Jr, J. E. & Greenberg, E. P. 1996*a* Generation of cell-to-cell signals in quorum sensing: acylhomoserine lactone synthase activity of a purified *V. fischeri* LuxI protein. *Proc. Natl Acad. Sci. USA* **93**, 9505–9509.

Schaefer, A. L., Hanzelka, B. L., Eberhard, A. & Greenberg, E. P. 1996*b* Quorum sensing in *Vibrio fischeri*: probing autoinducer–LuxR interactions with autoinducer analogs. *J. Bacteriol.* **178**, 2897–2901.

Schripsema, J., De Rudder, K. E. E., Van Vliet, T. B., Lankhorst, P. P., de Vroom, E., Kijne, J. W. & Van Brussel, A. A. 1996 Bacteriocin *small of Rhizobium leguminosarum* belongs to a class of *N*-acylhomoserine lactone molecules known as autoinducers and as quorum sensing co-transcriptional factors. *J. Bacteriol.* **178**, 366–371.

Sitnikov, D. M., Schineller, J. B. & Baldwin, T. O. 1996 Control of cell division in *Escherichia coli*: regulation of transcription of *ftsQA* involves both *rpoS* and SdiA-mediated autoinduction. *Proc. Natl Acad. Sci. USA* **93**, 336–341.

Speller, D. C. E., Johnson, A. P., James, D., Marples, R. R., Charlett, A. & George, R. C. 1997 Resistance to methicillin and other antibiotics in isolates of *Staphylococcus aureus* from blood and cerebrospinal fluid in England and Wales 1989–1995. *Lancet* **350**, 323–325.

Stevens, A. & Greenberg, E. P. 1999 Transcriptional activation by LuxR. In *Cell–cell signaling in bacteria* (ed. G. M. Dunny and S. C. Winans), pp. 231–242. Washington, DC: American Society of Microbiology Press.

Stevens, A. M., Doplan, K. M. & Greenberg, E. P. 1994 Synergistic binding of the *Vibrio fischeri* LuxR transcriptional activator domain and RNA polymerase to the *lux* promoter region. *Proc. Natl Acad. Sci. USA* **91**, 12619–12623.

Stevenson, M. M., Kondratieva, T. K., Apt, A. S., Tam, M. F. & Skameme, E. 1996 *In vitro* and *in vivo* T cell responses in mice during broncho-pulmonary infection with mucoid *Pseudomonas aeruginosa*. *Clin. Exp. Immunol.* **99**, 98–105.

Stickler, D. J., Morris, N. S., McLean, R. J. & Fuqua, C. 1998 Biofilms on indwelling urethral catheters produce quorum sensing signal molecules *in situ* and *in vitro*. *Appl. Environ. Microbiol.* **64**, 3468–3490.

Stintzi, A., Evans, K., Meyer, J. M. & Poole, K. 1998 Quorum sensing and siderophore biosynthesis in *Pseudomonas aeruginosa*: *lasR/lasI* mutants exhibit reduced pyoverdine biosynthesis. *FEMS Microbiol. Lett.* **166**, 341–345.

Storey, D. G., Ujack, A. E., Rabin, H. R. & Mitchell, I. 1998 *Pseudomonas aeruginosa lasR* transcription correlates with the transcription of *lasA*, *lasB* and *toxA* in chronic lung infections associated with cystic fibrosis. *Infect. Immun.* **66**, 2521–2528.

Stuhlmeier, K. M., Li, H. & Kao, J. J. 1999 Ibuprofen: a new explanation for an old phenomenon. *Biochem. Pharmacol.* **57**, 313–330.

Surette, M. G. & Bassler, B. L. 1998 Quorum sensing in *Escherichia coli* and *Salmonella typhimurium*. *Proc. Natl Acad. Sci. USA* **95**, 7046–7050.

Surette, M. G., Miller, M. B. & Bassler, B. L. 1999 Quorum sensing in *Escherichia coli*, *Salmonella typhimurium*, and *Vibrio harveyi*: a new family of genes responsible for autoinducer production. *Proc. Natl Acad. Sci. USA* **96**, 1639–1644.

Swift, S. (and 13 others) 1993 A novel strategy for the isolation of *luxI* homologues: evidence for the widespread distribution of a LuxR:LuxI superfamily in enteric bacteria. *Mol. Microbiol.* **10**, 511–520.

Swift, S., Karlyshev, A. V., Durant, E. L., Winson, M. K., Williams, P., Macintyre, S. & Stewart, G. S. A. B. 1997 Quorum sensing in *Aeromonas hydrophila* and *Aeromonas salmonicida*: identification of the LuxRI homologues AhyRI and AsaRI and their cognate signal molecules. *J. Bacteriol.* **179**, 5271–5281.

Swift, S., Isherwood, K. E., Atkinson, S., Oyston, P. C. F. & Stewart, G. S. A. B. 1999a Quorum sensing in *Aeromonas* and *Yersinia*. In *Microbial signalling and communication* (ed. R. England, G. Hobbs, N. Bainton and D. McL. Roberts), pp. 85–104. Cambridge University Press.

Swift, S., Lynch, M. J., Fish, L., Kirke, D. F., Tomas, J. M., Stewart, G. S. A. B. & Williams, P. 1999b Quorum sensing-dependent regulation and blockade of exoprotease production in *Aeromonas hydrophila*. *Infect. Immun.* **67**, 5192–5199.

Swift, S., Williams, P. & Stewart, G. S. A. B. 1999c N-acylhomoserine lactones and quorum sensing in the proteobacteria. In *Cell–cell signaling in bacteria* (ed. G. M. Dunny and S. C. Winans), pp. 291–313. Washington, DC: American Society of Microbiology Press.

Tan, M.-W., Rahme, L. G., Sternberg, J. A., Tompkins, R. G. & Ausubel, F. M. 1999 *Pseudomonas aeruginosa* killing of *Caenorhabditis elegans* used to identify *P. aeruginosa* virulence factors. *Proc. Natl Acad. Sci. USA* **96**, 2408–2413.

Tang, H. B., Dimango, E., Bryan, R., Gambello, M. J., Iglewski, B. H., Goldberg, J. B. & Prince, A. 1996 Contribution of specific *Pseudomonas aeruginosa* virulence factors to pathogenesis of pneumonia in a neonatal mouse model of infection. *Infect. Immun.* **64**, 37–43.

Telford, G., Wheeler, D., Williams, P., Tomkins, P. T., Appleby, P., Sewell, H., Stewart, G. S. A. B., Bycroft, B. W. & Pritchard, D. I. 1998 The *Pseudomonas aeruginosa* quorum sensing signal molecule,

N-(3-oxododecanoyl)-L-homoserine lactone has immunomodulatory activity. *Infect. Immun.* **66**, 36–42.

Thornley, J. P., Shaw, J. G., Gryllos, I. A. & Eley, A. 1997 Virulence properties of clinically significant *Aeromonas* species: evidence for pathogenicity. *Rev. Med. Microbiol.* **8**, 61–72.

Throup, J., Camara, M., Briggs, G., Winson, M. K., Chhabra, S. R., Bycroft, B. W., Williams, P. & Stewart, G. S. A. B. 1995 Characterisation of the *yenI/yenR* locus from *Yersinia enterocolitica* mediating the synthesis of two *N*-acylhomoserine lactone signal molecules. *Mol. Microbiol.* **17**, 345–356.

Van Delden, C., Pesci, E. C., Pearson, J. P. & Iglewski, B. H. 1998 Starvation selection restores elastase and rhamnolipid production in a *Pseudomonas aeruginosa* quorum sensing mutant. *Infect. Immun.* **66**, 4499–4502.

Wang, J., Mushegian, A., Lory, S. & Jin, S. 1996 Large-scale isolation of candidate virulence genes of *Pseudomonas aeruginosa* by *in vivo* selection. *Proc. Natl Acad. Sci. USA* **93**, 10434–10439.

Wang, X. D., de Boer, P. A. J. & Rothfield, L. I. 1991 A factor that positively regulates cell division by activating transcription of the major cluster of essential cell division genes of *Escherichia coli*. *EMBO J.* **10**, 3363–3372.

Williams, P. 1988 Role of the cell envelope in bacterial adaptation to growth *in vivo* in infections. *Biochimie* **70**, 987–1011.

Williams, P., Stewart, G. S. A. B., Camara, M., Winson, M. K., Chhabra, S. R., Salmond, G. P. & Bycroft, B. W. 1996 Signal transduction through quorum sensing in *Pseudomonas aeruginosa*. In *Pseudomonas: molecular biology and biotechnology* (ed. T. Nakazawa, K. Furukawa, D. Haas and S. Silver), pp. 195–205. Washington, DC: American Society of Microbiology Press.

Winson, M. K. (and 12 others) 1995 Multiple *N*-acyl-L-homoserine lactone signal molecules regulate production of virulence determinants and secondary metabolites in *Pseudomonas aeruginosa*. *Proc. Natl Acad. Sci. USA* **92**, 9427–9431.

Winson, M. K., Swift, S., Fish, L., Throup, J. P. Jorgensen, F., Chhabra, S. R., Bycroft, B. W., Williams, P. & Stewart, G. S. A. B. 1998 Construction and analysis of *luxCDABE*-based plasmid sensors for investigating *N*-acylhomoserine lactone-mediated quorum sensing. *FEMS Microbiol. Lett.* **163**, 185–192.

Withers, H. L. & Nordstrm, K. 1999 Quorum sensing acts at initiation of chromosomal replication in *Escherichia coli*. *Proc. Natl Acad. Sci. USA* **96**, 4832–4837.

Zhu, J. & Winans, S. C. 1999 Autoinducer binding by the quorum-sensing regulator TraR increases affinity for target promoters *in vitro* and decreases TraR turnover rates in whole cells. *Proc. Natl Acad. Sci. USA* **96**, 4832–4837.

TYPE III SECRETION: A BACTERIAL DEVICE FOR CLOSE COMBAT WITH CELLS OF THEIR EUKARYOTIC HOST

GUY R. CORNELIS*

Microbial Pathogenesis Unit,
Christian de Duve Institute of Cellular Pathology (ICP),
and Faculté de Médecine, Université Catholique de Louvain,
B-1200 Brussels, Belgium

Salmonella, Shigella, Yersinia, Pseudomonas aeruginosa, enteropathogenic *Escherichia coli* and several plant-pathogenic Gram-negative bacteria use a new type of systems called 'type III secretion' to attack their host. These systems are activated by contact with a eukaryotic cell membrane and they allow bacteria to inject bacterial proteins across the two bacterial membranes and the eukaryotic cell membrane to reach a given compartment and destroy or subvert the target cell. These systems consist of a secretion apparatus made up of about 25 individual proteins and a set of proteins released by this apparatus. Some of these released proteins are 'effectors' that are delivered by extracellular bacteria into the cytosol of the target cell while the others are 'translocators' that help the 'effectors' to cross the membrane of the eukaryotic cell. Most of the 'effectors' act on the cytoskeleton or on intracellular signalling cascades. One of the proteins injected by the enteropathogenic *E. coli* serves as a membrane receptor for the docking of the bacterium itself at the surface of the cell.

Keywords: bacterial pathogenesis; *Salmonella*; *Shigella*; *Yersinia*; enteropathogenic *E. coli*; translocation

1. INTRODUCTION

For millions of years, eukaryotes gradually built up their multicellular complexity and some cells, such as the epithelial cells and the phagocytes, specialized in the exclusion and clearing of intruding micro-organisms. The latter, on the other hand, remained unicellular but developed an impressive pool of genes that they exchanged more or less freely and, by doing so, they acquired an extraordinary adaptative potential, which perpetuates conflicts and equilibria that date from the era when the unicellular state was the rule.

*Address for correspondence: Microbial Pathogenesis Unit, Université Catholique de Louvain, Avenue Hippocrate 74, UCL 74.49, B-1200 Brussels, Belgium; (cornelis@mipa.ucl.ac.be).

For a rather long period of time, it was assumed that Gram-negative bacteria do not 'secrete' proteins in their environment; they were only supposed to export proteins in their strategical periplasm. However, research of the past two decades revealed that Gram-negative bacteria do indeed transfer proteins across their sophisticated outer membrane. They do this by a variety of systems that are now classified in four major types and several minor ones (Salmond & Reeves 1993). Type I, exemplified by the haemolysin secretion system of *Escherichia coli* is a rather simple system based on only three proteins that belong to the universal multidrug resistance (MDR) type of efflux pumps. Type II is a very complex apparatus that extends the general secretory pathway and transfers fully folded enzymes or toxins from the periplasm to the extracellular medium, across the outer membrane. Type IV, another complex system that transfers pertussis toxin among others, is related to the apparatus of *Agrobacterium*, which transfers DNA to plant cells (Covacci *et al.* 1999). Finally, type III, the object of this review, is a sophisticated apparatus that allows bacteria adhering at the membrane of a eukaryotic host cell or of an intracellular organelle, to inject specialized proteins across this membrane. The injected proteins subvert the functioning of the aggressed cell or destroy its communications, favouring the entry or survival of the invading bacteria. Type III is thus not a secretion apparatus in the strict sense of the term but rather a complex weapon for close combat. It is used by a growing number of animal pathogens but also by a number of plant pathogens. The type of intercellular communication this device allows is not restricted to pathogenesis: it is also used to initiate symbiosis by *Rhizobium* spp. (Viprey *et al.* 1998). The rule seems to be that the communication event follows close contact between the players.

'Type III secretion' contributes to a number of totally different diseases with different symptoms and severities, going from a fatal septicaemia to a mild diarrhoea or from a fulgurant diarrhoea to a chronic infection of the lung. Among the animal pathogens, type III systems have been extensively studied in *Yersinia* spp. (reviewed by Cornelis 1998; Cornelis *et al.* 1998), *Salmonella* spp. (reviewed by Galan 1998), *Shigella* spp. (reviewed by Van Nhieu & Sansonnetti 1999) and pathogenic *E. coli* (enteropathogenic *E. colis* (EPECs) and enterohaemorrhagic *E. colis* (EHECs)) (Jarvis *et al.* 1995; Elliott *et al.* 1998; reviewed by Frankel *et al.* 1998; Goosney *et al.* 1999). It has also been described in *Pseudomonas aeruginosa*,

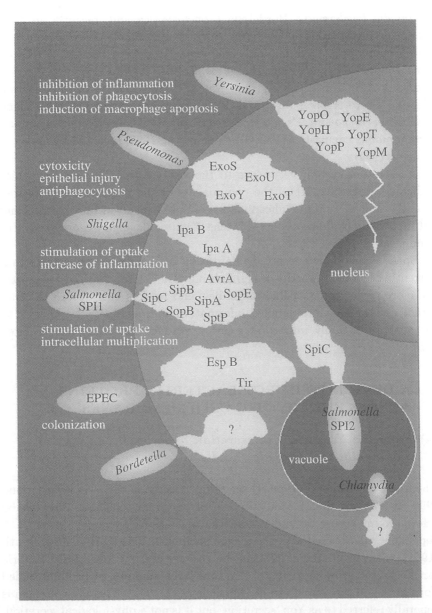

Fig. 1. Illustration of the various bacterial pathogens endowed with type III secretion, injecting effectors into the cytosol of a eukaryotic target cell. See Table 2 for references.

Chlamydia trachomatis, Bordetella bronchiseptica and, recently, in *Bordetella pertussis* (Kerr *et al.* 1999). Surprisingly, *Salmonella typhimurium* and *Yersinia* spp. have not only one type III system but two (Ochman *et al.* 1996; Shea *et al.* 1996; Hensel *et al.* 1997, 1998; Carlson & Pierson, GenBank AF005744), playing their role at different stages of the infection (Fig. 1). There is also substantial documentation in plant pathogens such as *Erwinia amylovora*, *Pseudomonas syringae*, *Ralstonia solanacearum* and *Xanthomonas campestris* (Van Gijseghem *et al.* 1995; reviewed by Galan & Collmer 1999).

A secretion system very close to the various type III systems is also dedicated to the export of the components of the flagellum. It appeared recently that a *Yersinia* phospholipase (YplA) involved in virulence and not in motility could be exported by the *Yersinia* flagellar export apparatus (Young *et al.* 1999).

Space restrictions necessitate limitation of this review to aspects that currently seem most worthy of highlighting. The references cited are mostly the more recent or lesser-known studies, and readers are referred to other recent reviews for more detailed information and references. I apologize to the authors of many important relevant studies that could not be cited here.

2.   A DEVICE TO INJECT BACTERIAL PROTEINS ACROSS EUKARYOTIC CELL MEMBRANES

(a) *From the Yersinia Ysc secretion apparatus... to the Salmonella and Shigella 'needle'*

The first observation of 'type III secretion' was made with *Yersinia* around 1990. It was the first major outcome of long and tenacious research by a few groups trying to understand the mysterious phenomenon of Ca^{2+} dependency discovered with the plague bacillus in 1961 by Higushi & Smith (1961): when incubated at $37°C$ in the absence of Ca^{2+} ions, these bacteria can no longer grow, instead they release large amounts of proteins called Yops in the culture supernatant (Michiels *et al.* 1990). This phenomenon is generally referred to as Yop 'secretion' but it is not a physiological secretion: it is rather a massive leakage resulting from the artificial opening of a tightly controlled delivery apparatus. In spite of the fact that it is presumably artefactual, this observation turned out to be of paramount importance because it allowed one to carry out a genetic analysis, which led to the identification of 29 genes involved in this process of Yop release and called '*ysc*' for Yop secretion. Only a minority of the Ysc proteins have been characterized so far but the Ysc system remains the archetype of the type III secretons. The *ysc* gene nomenclature has been transposed in EPECs (*esc* genes), as well as in plant pathogens for the type III genes that are conserved (*hrc* genes for hypersensitive response (HR)-conserved): the *hrc* and *esc* genes thus carry the same letter code as their *ysc* homologues. For the sake of clarity and consistency, we will first describe the various Ysc proteins (see Cornelis *et al.* (1998) for details) and mention afterwards the elements that are different or better-known in the other systems.

YscC is one of the best known Ysc proteins. It belongs to the family of secretins, a group of outer-membrane proteins involved in the transport of various macromolecules and filamentous phages across the outer membrane (Genin & Boucher 1994). As the other secretins, it exists as a very stable multimeric complex of about 600 kDa that forms a ring-shaped structure with an external diameter of about 200 Å and an apparent central pore of about 50 Å (Koster *et al.* 1997). As a matter of comparison, the PIV secretin of phage f1 has an internal diameter of about 80 Å, allowing the passage of the filamentous capsid with a diameter of 65 Å (Russel 1994). Lipoprotein YscW is ancillary to YscC in the sense that it is required for the proper insertion of YscC in the outer membrane (Koster *et al.* 1997). The Ysc apparatus also contains another lipoprotein called YscJ. Four proteins (YscD, -R, -U and -V, formerly called LcrD) have been shown, and two other (YscS and -T) proteins have been predicted to span the inner membrane. YscN is a 47.8 kDa protein with ATP-binding motifs (Walker boxes A and B) resembling the β-catalytic subunit of F_oF_1 proton translocase and related ATPases (Woestyn *et al.* 1994). It probably energizes the secretion process. Surprisingly, the two proteins YscO and YscP which are also necessary for Yop secretion are themselves released on Ca^{2+}-chelation, suggesting that they belong to the external part of the apparatus (Payne & Straley 1998).

Eight Ysc proteins (YscJ, -N, -O, -Q, -R, -S, -T, -U and -V) have counterparts in almost every type III secreton, including the one serving the flagellum. In the flagellum, the corresponding proteins belong to the most internal part of the basal body, i.e. the MS ring, the C ring and the ATPase.

The YscC secretin also has counterparts in almost any type III system but not in the flagellum. The similarity between the inner parts of the type III secreton and the basal body of the flagellum prompted the groups of I. Aizawa and J. Galan to apply the well-established extraction and purification procedures of the basal body to the *Salmonella* type III secreton. This allowed them to visualize under the electron microscope a supramolecular structure that strikingly resembles a needle (Kubori *et al.* 1998). This needle complex is a hollow structure about 1200 Å long composed of two clearly identifiable domains: a needle-like portion projecting outwards from the surface of the bacterial cell and a cylindrical base that anchors the structure to the inner and outer membranes. The base closely resembles the flagellar basal body, further supporting the evolutionary relationship between flagella and type III secretion systems. An immunoblot analysis of the purified needle complex revealed that it is composed of at least three proteins: the secretin homologous to YscC and two lipoproteins, one of which resembles YscJ. More recently, a similar needle-like structure could be seen on the surface of plasmolysed *Shigella* (Blocker *et al.* 1999).

Little is known about the actual mechanism of export. The structure of the needle-like complex suggests that it serves as a hollow conduit through which the 'exported' proteins travel to cross the two membranes and the peptidoglycan barrier. The type III secreton would thus be operating in one step, taking its energy from the hydrolysis of ATP. Whether proteins travel folded or unfolded has not been demonstrated yet but, given the size of the channel, it is likely that they travel at least partially unfolded.

(b) *Other structural components of the type III systems*

Apart from the needle, described in *Salmonella* and in *Shigella*, other structural components have been found to be associated with type III machineries. *P. syringae* pv. tomato produces a filamentous surface appendage 6–8 nm in diameter, called the Hrp pilus, that is dependent on at least two type III system genes, *hrpS* and *hrcC*, encoding the secretin (Roine *et al.* 1997). The major structural protein of this Hrp pilus is encoded by *hrpA*, another essential gene for the type-III-mediated hypersensitive response and pathogenicity.

A filamentous organelle is also associated to the type III system of EPECs (Knutton *et al.* 1998). It has a diameter of about about 7–8 nm and a length of up to 2 m. It contains EspA (Knutton *et al.* 1998), one of the proteins secreted by the Esc type III secreton of EPECs (Kaper 1998). It seems likely that these EspA filaments would play a role in the translocation process and the authors speculate that they may act as 'molecular go-betweens' transporting proteins from the bacterium to the host cell. However, this has not been demonstrated yet.

Finally, Ginocchio *et al.* (1994) have reported that contact with cultured epithelial cells results in the formation of filamentous appendages on the surface of *S. typhimurium*, but the significance of this observation and its relation to the more recently discovered 'needle' are not clear.

(c) *Translocation of effectors across eukaryotic cell membranes*

Purified secreted Yops have no cytotoxic effect on cultured cells, although live extracellular *Yersinia* have such an activity. Cytotoxicity was nevertheless found to depend on the capacity of the bacterium to secrete YopE and YopD. However, YopE alone was found to be cytotoxic when microinjected into the cells. This observation led to the hypothesis that YopE is a cytotoxin that needs to be injected into the eukaryotic cell's cytosol by a mechanism involving YopD, in order to exert its effect (Rosqvist *et al.* 1991). In 1994, this hypothesis was demonstrated by two different approaches. The group of Hans Wolf-Watz used immunofluorescence and confocal laser scanning microscopy examinations (Rosqvist *et al.* 1994) while the group of Guy Cornelis introduced a reporter enzyme strategy based on the calmodulin-activated adenylate cyclase (Sory & Cornelis 1994). Infection of a monolayer of eukaryotic cells by a recombinant *Y. enterocolitica* producing a hybrid protein made of the N-terminus of YopE and the catalytic domain of the adenylate cyclase of *Bordetella pertussis* (YopE-Cya protein) led to an accumulation of cyclic AMP in the cells. Since the cyclase is not functional in the bacterial cell and in the culture medium because of a lack of calmodulin, this accumulation of cAMP signified the internalization of YopE-Cya into the cytosol of eukaryotic cells (Sory & Cornelis 1994). Thus extracellular *Yersinia* inject YopE into the cytosol of eukaryotic cells by a mechanism that involves at least one other Yop protein, YopD. YopH was later demonstrated to be also injected into the target cells cytosol (Persson *et al.* 1995; Sory *et al.* 1995) and YopB was shown to be required for delivery of YopE and YopH, like YopD. These observations led to the present concept that Yops are a collection of intracellular effectors (including YopE and YopH) and proteins required for translocation of these effectors across the plasma membrane of eukaryotic cells (including YopB and YopD) (Cornelis & Wolf-Watz 1997). Delivery of effector Yops into eukaryotic cells appears to be a directional phenomenon in the sense that the majority of the Yop effector molecules produced are directed into the cytosol of the eukaryotic cell and not to the outside environment (Rosqvist *et al.* 1994; Persson *et al.* 1995).

This model of intracellular delivery of Yop effectors by extracellular adhering bacteria is now largely supported by a number of other results, including immunological observations. While antigens processed in phago-cytic vacuoles of phagocytes are cleaved and presented by MHC class II molecules, epitope 249–257 of YopH produced by *Y. enterocolitica* during a mouse infection is presented by MHC class I molecules, such as cytosolic proteins (Starnbach & Bevan 1994). This prompted some authors to convert *Salmonella* (Russmann *et al.* 1998) or *Y. enterocolitica* (Chaux *et al.* 1999) into antigen-presenting vectors.

As already mentioned, translocation across the cell membrane requires other secreted proteins, including YopB and YopD (Rosqvist *et al.* 1994; Sory & Cornelis 1994; Persson *et al.* 1995; Boland *et al.* 1996). These two Yops contain hydrophobic domains suggesting that they could act as transmembrane proteins (Håkansson *et al.* 1993). In agreement with this, *Yersinia* has a contact-dependent lytic activity on sheep erythrocytes, depending on YopB and YopD (Håkansson *et al.* 1996*b*; Neyt & Cornelis 1999). All this suggests that the translocation apparatus involves some kind of a pore in the target cell membrane by which the Yop effectors pass through into the cytosol. This YopB- and YopD-dependent lytic activity is higher when the effector yop genes are deleted suggesting that the pore is nor-mally filled with effectors (Håkansson *et al.* 1996*b*). Osmoprotectants can inhibit YopB- and YopD-mediated sheep erythrocyte lysis, provided they are large enough so that they cannot traffic through the pore. This allowed Håkansson *et al.* (1996*b*) to estimate the inner diameter of the putative pore to be between 1.2 nm and 3.5 nm. The idea of a translocation pore was further documented in macrophages: infection of PU5–1.8 macrophages with an effector polymutant *Y. enterocolitica* leads to complete flattening of the cells, similar to treatment with the pore-forming streptolysin O from *Streptococcus pyogenes* (Neyt & Cornelis 1999). When the macrophages are pre-loaded with the low molecular weight fluorescent marker BCECF (623 Da), prior to the infection, they release the fluorescent marker but not cytosolic proteins, indicating that there is no membrane lysis but rather insertion of a pore of small size into the macrophage plasma membrane. Macrophages infected with the same polymutant strain also become permeable to extracellular lucifer yellow CH (443 Da) but not to Texas red-X phalloidin (1490 Da), supporting further the hypothesis of a pore. The hypothesis of a channel was recently reinforced by the observation that artificial liposomes that have been incubated with *Yersinia* also contain channels detectable by electrophysiology (Tardy *et al.* 1999). The observed channel has a conductance of $105 + 5$ pS and no ion selectivity. In agreement with the findings on translocation, all these events are dependent on translocators YopB and YopD. These two hydrophobic Yops seem thus to be central for the translocation of the effectors and for the formation of a channel in lipid membranes. Whether the two events are linked is very likely but not formally proven, so far.

Translocation of the effectors also requires the secreted LcrV protein, which interacts with YopB and YopD and is surface-exposed before target-cell contact (Sarker *et al.* 1998; Petterson *et al.* 1999). Finally, the 11 kDa LcrG protein is also required for efficient translocation of *Yersinia* Yop effector proteins into the eukaryotic cells but it is not required for pore formation. LcrG was shown to bind to heparan sulphate proteoglycans, suggesting that it could play a role in the control of release by contact but its exact localization in the bacterium remains elusive (Boyd *et al.* 1998). These four proteins are encoded by the same large operon *lcrGVsycDyopBD*, which also encodes SycD, the chaperone of YopB and YopD (Wattiau *et al.* 1994; Neyt & Cornelis 1999). This genetic organization reinforces the idea that YopB, YopD, LcrV and LcrG act together as 'translocators'. This does not necessarily exclude that some of them could themselves end up in the eukaryotic cytoplasm as was shown for YopD (Francis & Wolf-Watz 1998). *P. aeruginosa* has a translocation apparatus consisting of PcrG, -V, PopB and -D, that is very similar to the LcrG, -V, YopB, -D apparatus of *Yersinia*. However, the other type III systems diverge somehow at this level. *Shigella* and SPI1 have very similar apparatus made of IpaB, -C, -D and SipB,-C, -D, respectively. IpaB and SipB could be considered as the counterparts of YopB, but IpaC, -D and SipC, -D are not similar to either YopD or LcrV. The translocators of the EPECs are called EspB and EspD. The latter could be considered as a counterpart of YopB (Wachter *et al.* 1999).

A central question is, of course, do the translocators belong to the 'needle' or the 'pilus'? It is difficult to answer for *Yersinia* since their 'needle' or 'pilus' has not been seen yet but there are clues in *Shigella*. *Shigella* also has a contact-dependent haemolytic activity and this activity requires IpaB, a secreted protein that has similarities with YopB. Blocker *et al.* (1999) examined under the electron microscope the 'needle' of a mutant deficient in IpaB and found it to be undistinguishible from that of the wild-type, suggesting that the needle probably does not comprise the translocators or, at least, that the translocators are not an abundant element of the 'needle'. *Salmonella* secretes a homologue to IpaB, called SipB. Surprisingly, IpaB and SipB were the two first proteins to be found to have an apoptotic activity by reacting with the cytosolic macrophage protein ICE (Chen *et al.* 1996; Hersh *et al.* 1999). Similarly, EspB, which somehow resembles YopD was also found to be translocated, in an Esp-dependent manner, into eukaryotic cells (Wolff *et al.* 1998). These observations indicate that the translocators are not restricted to the area of contact between bacteria and eukaryotic cells but that they are themselves trafficking in the eukaryotic cell, possibly associated with membranes, but this has not been determined yet. Thus, although there is a general agreement on the fact that the hydrophobic secreted proteins of the YopB, YopD family are involved in the translocation of the effectors, the understanding of their situation in the structure and of their exact role still deserves a long-standing effort.

(d) *The cytosolic chaperones*

A hallmark of type III secretion is that normal secretion of some substrate proteins requires the presence of small cytosolic chaperones of a new type (Wattiau *et al.* 1994, 1996; Ménard *et al.* 1994). Generally, these chaperones are encoded by a gene located close to the gene encoding the protein they serve and this is a useful indication for recognizing such chaperones. However, there are examples of gene reshuffling such as in *Y. pseudotuberculosis* where the gene encoding the chaperone of YopH was separated from the *yopH* gene by a large inversion. The latest observations suggest that these chaperones may not form a single homogeneous group but rather could belong to two different subfamilies.

SycE, the chaperone of YopE, is the archetype of the first family (Wattiau & Cornelis 1993). There are four typical representatives of this family

Table 1. Type III cytosolic chaperones.

	Protein	kDa	pI	Assisted protein	Strong simiarities	References
SycE family	SycE	14.7	4.55	YopE (*Yersinia*) binds to amino acids 15–50	ORF1 (*P. aeruginosa*) Sccl (*C. psittaci*)	Wattiau & Cornelis 1993
	SycH	14.7	4.88	TopH (*Yersinia*) binds to amino acids 20–70		Wattiau *et al.* 1994
	SycT	15.7	4.4	YopT (*Yersinia*)		Iriarte & Cornelis 1998
	SycN	15.1	5.2	YopN (*Yersinia*)	Pcr2 (*P. aeruginosa*)	Day & Plano 1998 / Iriarte & Cornelis 1999
	YscB	15.4	9.3	YopN (*Yersinia*) (co-chaperone)		Jackson & Plano 1998 / Day & Plano 1998
	SicP	13.6	4.0	SptP (*Salmonella*)		Fu & Galan 1998
	SpcU		4.4	ExoU (*P. aeruginosa*)		Finck-Barbançon *et al.* 1998
	CesT		7.1	Tir (EPECs and EHECs)		Abe *et al.* 1999 / Elliott *et al.* 1999
	FlgN	16.5		FlgK and FlgL (*Proteus flagellum*)		Fraser *et al.* 1999
	FliT	14		HAP2 (*Proteus flagellum*)		Fraser *et al.* 1999
SycD family	SycD	19.0	4.53	YopB and YopD (*Yersinia*)	PcrH (*P. aeruginosa*)	Wattiau *et al.* 1994
	CesD			(EPECs)		Wainwright & Kaper 1998
	IpgC	18.0		IpaB and IpaC (*Shigella*)		Ménard *et al.*
	SicA	19	4.61	SipB and SipC (*Salmonella*)		Kaniga *et al.*

Table 2. Type III effectors.

	Effector	Enzymatic activity	Target	Similarity	Effect	Reference
Yersinia	YopE	unknown	unknown	ExoS, SptP	cytotoxin, actin filaments disruption, antiphagocytic	Michiels et al. 1990; Rosqvist et al. 1994; Sory & Cornelis 1994
	YopH	PTPase	P130cas FAK	SptP	disruption of peripheral focal complexes, antiphagocytic	Persson et al. 1997
	YopM	unknown	unknown	IpaH	migrates to the nucleus	Skrzypek et al. 1998; Boland et al. 1996
	YpkaA/YopO	serine, threonine kinase	unknown	—	unknown	Håkansson et al. 1996a
	YopP/Yop	unknown	unknown	AvrA AvrRxv	inhibition of TNFα release, apoptotic	Mills et al. 1997; Monack et al. 1997; Schesser et al. 1998
	YopT	unknown	RhoA	—	cytotoxin, actin filaments disruption, antiphagocytic	Iriarte & Cornelis 1998
Salmonella SPI1	AvrA	unknown	unknown	YopP/YopJ AvrRxv		Hardt & Galan 1997; Schesser et al. 2000
	SipA	unknown	actin	SipA	enhances actin polymerization, macropinocytosis	Zhou et al. 1999
	SipB		caspase-1		induction of apoptosis	Hersh et al. 1999
	SopB/SigD	InsP phosphatase	various		intestinal chloride secretion	Norris et al. 1998; Jones et al. 1998

Table 2. (*Continued*)

	Effector	Enzymatic activity	Target	Similarity	Effect	Reference
	SopE	GDP-GTP exchange factor	CDC42, Rac			Hardt *et al.* 1998
	SptP	PTPase	unknown	YopE, YopH		Kaniga *et al.* 1996
	SopD					Jones *et al.* 1998
Salmonella SPI2	SpiC				inhibition of fusion between phagosomes and lysosomes	Uchiya *et al.* 1999
Pseudomonas aeruginosa	ExoS	ADP-ribosyl-transferase	Ras	YopE, ExoT		Finck-Barbançon *et al.* 1997; Frithz-Lindsten *et al.* 1997; McGuffie *et al.* 1998; Pederson *et al.* 1999
	ExoT	ADP-ribosyl-transferase	unknown	ExoS		Finck-Barbançon *et al.* 1997
	ExoU/PepA				cytotoxin	Finck-Barbançon *et al.* 1997; Hauser *et al.* 1998
	ExoY	adenylate cyclase				Yahr *et al.* 1998
Shigella	IpaA	unknown	vinculin	SipA		Van Nhieu *et al.* 1997
	IpaB		caspase-1	SipB	induction of apoptosis	Hilbi *et al.* 1998
	IpaC				activation of Cde42, entry of *Shigella*	Van Nhieu *et al.* 1999
EPEC/EHEC	Tir/Esp E	receptor			receptor for intimin	Kenny *et al.* 1997; Deibel *et al.* 1998

in *Yersinia*: SycE, SycH, SycT and SycN, one in *Salmonella* (SicP), one in *P. aeruginosa* (SpcU), one in EPECs (CesT) and two in the *Proteus flagellum* assembly system (Table 1). One could also add to this list the less typical YscB from *Yersinia*, acting as a co-chaperone for YopN (Day & Plano 1998; Jackson *et al.* 1998). All these chaperones are small (14–15 kDa) proteins with a putative C-terminal amphiphilic α-helix and most of them are acidic (pI 4.4–5.2). They specifically bind only to their partner Yop. The main feature is that, in the absence of these chaperones, secretion of their cognate protein is severely reduced, if not abolished. However, the exact role of these chaperones remains elusive. Research on Syc chaperones focused first on SycE and SycH. They both bind to their partner Yop at a unique site spanning roughly residues 20 to 70 (Sory *et al.* 1995). Surprisingly, when this site is removed, the cognate Yop is still secreted — though maybe in reduced amounts — and the chaperone becomes dispensible for secretion (Woestyn *et al.* 1996). This suggests that it is the binding site itself that creates the need for the chaperone and thus that the chaperone somehow protects this site from premature associations which would lead to degradation. In agreement with this hypothesis, SycE has indeed an anti-degradation role: the half-life of YopE is longer in wild-type bacteria than in *sycE* mutant bacteria (Frithz-Lindsten *et al.* 1995; Cheng *et al.* 1997). In addition to this putative role of bodyguard, SycE also acts as a secretion pilot leading the YopE protein to the secretion locus (see §2(e)). Finally, both SycE and SycH are required for efficient translocation of their partner Yop into eukaryotic cells (Sory *et al.* 1995). However, when YopE is delivered by a *Yersinia* polymutant strain that synthesizes an intact secretion and translocation apparatus but no other effector, it appears that YopE is delivered without the chaperone and the chaperone-binding site (Boyd *et al.* 2000). Thus, the SycE chaperone appears to be needed, only when YopE competes with other Yops for delivery. This suggests that the Syc chaperones could be involved in some kind of a hierarchy for delivery. This new hypothesis about the role of the Syc chaperones fits quite well with the observation that only a subset of the effectors seem to have a chaperone. Little is known about the role of SycT and SycN. However, there is an unexpected complexity for the latter: SycN apparently requires YscB working as a co-chaperone (Day & Plano 1998).

SycD is the archetype of the second group of 'type III chaperones'. It serves both YopB and YopD (Wattiau *et al.* 1994; Neyt & Cornelis 1999) and in its absence, YopD and YopB are less detectable inside the bacterial cell. SycD appears to be different from SycE and SycH in the sense that it binds to several domains on YopB, which evokes SecB, a molecular chaperone in *E. coli* that is dedicated to the export of newly synthesized proteins and also has multiple binding sites on its targets (reviewed by Fekkes & Driessen 1999). IpgC, the related chaperone from *S. flexneri* has been shown to prevent the intrabacterial association between translocators IpaB and IpaC (Ménard *et al.* 1994). The similarity between IpgC and SycD suggested that SycD could play a similar role and would thus prevent the intrabacterial association of YopB and YopD. However, Neyt & Cornelis (1999) observed that intrabacterial YopB and YopD are associated even in the presence of SycD. Since YopB and YopD also have the capacity to bind to LcrV, one could speculate that SycD prevents a premature association, not between YopB and YopD but rather between YopB, YopD and LcrV, but this hasn't been shown yet. CesD, the homologue from the EPECs, has also been shown to be required for full secretion of the translocators EspB and EspD, but it was only shown to bind to EspD, the translocator that is closest from YopB and IpaB. Like SycD and IpgC, CesD is present in the bacterial cytosol, but a substantial amount of this protein was also found to be associated with the inner membrane of the bacterium (Wainwright & Kaper 1998).

(e) *Recognition of the transported proteins*

Effectors delivered by the type III secretion systems have no classical cleaved N-terminal signal sequence (Michiels *et al.* 1990). However, it appeared very clearly that Yops are recognized by their N-terminus but that no classical signal sequence is cleaved off during Yop secretion (Michiels *et al.* 1990). The minimal region shown to be sufficient for secretion was gradually reduced to 17 residues of YopH (Sory *et al.* 1995), 15 residues of YopE (Sory *et al.* 1995) and 15 residues of YopN (Anderson & Schneewind 1997).

There is no similarity between the secretion domains of the Yops, which suggested recognition of a conformational motif of the nascent protein (Michiels *et al.* 1990). To explain that proteins with no common signal could be recruited by the same secretion apparatus, Wattiau & Cornelis (1993) suggested that the Syc chaperones could serve as pilots. However, this hypothesis was questioned when it appeared that YopE could be secreted even if its chaperone-binding domain had been deleted (Woestyn *et al.* 1996). It was then concluded that secretion was dependent only on the short N-terminal signal but secretion of a Yop lacking only this N-terminal signal had never been tested.

A systematic mutagenesis of the secretion signal by Anderson & Schneewind (1997, 1999*b*) led to doubts about the proteic nature of this signal. No point mutation could be identified that specifically abolished secretion of YopE, YopN and YopQ. Moreover, some frameshift mutations that completely altered the peptide sequences of the YopE, YopN signals also failed to prevent secretion. Anderson & Schneewind (1997, 1999) concluded from these observations that the signal that leads to the secretion of these Yops could be in the 5-end of the messenger RNA rather than in the peptide sequence. Translation of *yop* mRNA might be inhibited by a property either of its own RNA structure or as a result of its binding to other regulatory elements. If this is correct, one would expect that no Yop could be detected inside bacteria. However, while this is reported to be true for YopQ (Anderson & Schneewind 1999), it is certainly not true for other Yops such as YopE.

To determine whether this N-terminal (or 5-terminal) signal is absolutely required for YopE secretion, Cheng *et al.* (1997) deleted codons 2–15 and they observed that 10% of the hybrid proteins deprived of the N-terminal secretion signal were still secreted. They inferred that there is a second secretion signal and they showed that this second, and weaker, secretion signal corresponds to the SycE-binding site. Not surprisingly, this secretion signal is only functional in the presence of the SycE chaperone (Cheng *et al.* 1997), rejuvenating the pilot hypothesis of Wattiau & Cornelis (1993). The Syc chaperone could ensure stability and proper conformation of the protein and target it to the secretion channel. At the moment of secretion, the chaperone must be released from the partner Yop to allow secretion.

Thus, the effectors that have a chaperone, such as YopE, YopH, YopN and YopT, are likely to have two secretion signals that operate during *in vitro* secretion, one linked to translation and one post-translational. What the relative importance of these two systems is *in vivo* remains to be elucidated. For the other Yops, for instance YopQ, the N-terminal secretion signal would be the only one. This non-cleavable N-terminal or 5mRNA signal seems to be a hallmark of type III secretion systems.

(f) *Control of the injection*

We have seen that 'type III secretons' can secrete their substrate *in vitro* under given conditions, such as Ca^{2+}-chelation for instance. What is the triggering signal *in vivo*? Most probably contact with a eukaryotic cell. Several reports in *Yersinia* have shown that Yops delivery is a 'directional' phenomenon in the sense that most of the load is delivered inside the eukaryotic cell and that there is little leakage (Persson *et al.* 1995). According to the assays used, there is some discrepancy on the degree of 'directionality' (Boland *et al.* 1996) but there is no doubt that the majority of the released Yops load ends up in the eukaryotic cell and thus that contact must be the signal. Pettersson *et al.* (1996) provided a nice visual demonstration of the phenomenon. By expressing luciferase under the control of a *yop* promoter, they showed indeed that active transcription of *yop* genes is limited to bacteria that are in close contact with eukaryotic cells. Release of Ipa proteins from *Shigella* was also shown to depend on contact between bacteria and epithelial cells (Watarai *et al.* 1995).

3.   EFFECTOR PROTEINS AND HOST RESPONSES

(a) *A panoply of enzyme activities*

Delivery of effectors across the plasma or vacuolar membrane appears to be the object of type III secretion. We have seen in §2 that some of the translocators, namely IpaB, SipB, EspB (Wolff *et al.* 1998) and YopD (Francis & Wolf-Watz 1998), have been shown to be delivered themselves into the eukaryotic cell. In addition to these proteins, 18 effectors have been described in the various animal pathogen systems and this relatively large list is increasing very fast. The effectors and their activity are detailed in Table 2. Six effectors have been characterized in *Yersinia*: YopE, YopH, YopM, YopJ/P, YopO/YpkA and YopT (Cornelis *et al.* 1998). Five effectors are delivered by the *Salmonella* SPI1-encoded apparatus: AvrA, SipA, SopB, SopE and SptP, and one, SpiC, has been identified for SPI2 (Uchiya *et al.* 1999). Four are delivered by the Psc apparatus of *Pseudomonas aeruginosa*: ExoS (Frithz-Lindsten *et al.* 1997), ExoT (Vallis *et al.* 1999), ExoU (Finck-Barbencon *et al.* 1998) and ExoY (Yahr *et al.* 1998). *Shigella* delivers IpaA and IpaC (Van Nhieu *et al.* 1997, 1999). Finally, EPECs or EHECs deliver their own receptor, Tir (Kenny *et al.* 1997) or EspE (Deibel *et al.* 1998). No effector has been characterized yet for the other systems.

Five different enzymatic activities could be identified so far in the panoply of type III effectors: phosphotyrosine phosphatases (YopH, SptP), serine-threonine kinase (YpkA/YopO), inositol phosphate phosphatase (SopB) (Norris *et al.* 1998), ADP-ribosyltransferases (ExoS, ExoT) and an adenylate cyclase (ExoY). It is worthwhile noticing that the two latter activities are classical in A-B toxins. However, although ExoY resembles the toxins of *Bordetella pertussis* and *Bacillus anthracis*, it does not require calmodulin for its activity. The similarity between activities of type III effectors and A-B toxins suggest that these type III effectors could be considered as some kind of toxins that need a very sophisticated apparatus for their delivery. Some of the type III effectors are hybrid proteins composed of two domains that display different activities. SptP from *S. enterica* appears to be a hybrid between YopE and YopH from *Yersinia*: the C-terminal part is a phosphotyrosine phosphatase homologous to YopH while the N-terminal part is homologous to YopE (Kaniga *et al.* 1996). This YopE-like domain also occurs in the N-terminal part of ExoS from *P. aeruginosa* (Frithz-Lindsten *et al.* 1997).

(b) *The cytoskeleton is a major target*

There is also a great diversity among the targets and the effects induced by the effectors. However, two major themes emerge. The first one is the cytoskeleton. Several effectors stimulate the cytoskeleton activity, which leads to macropinocytosis of *Salmonella* and *Shigella* (SipA, SopE, IpaA), while others disrupt the actin filaments, which leads to cytotoxicity and inhibition of phagocytosis of *Yersinia* and *P. aeruginosa* (YopE, YopH, YopT, ExoS). Small GTP-binding proteins such as Rho, Rac and CDC42 are essential in the control of the cytoskeleton movements. These GTP-binding proteins can cycle between two states: a GDP-bound (inactive) and a GTP-bound (active) form capable of engaging different effector molecules. Not surprisingly, several of the effectors that affect the cytoskeleton have been shown to act on such small G proteins. SopE is a GDP–GTP exchange factor acting on Cdc42 and Rac-1 (Hardt *et al.* 1998*a*); the ADP-ribosyltransferase domain of ExoS acts on Ras (McGuffie *et al.* 1998) and YopT has just been shown to act on RhoA

(Zumbihl *et al.* 1999). SipA has been reported to act directly on actin and to decrease its critical concentration for polymerization while IpaA has been shown to bind to vinculin, which initiates the formation of focal adhesion-like structures required for *Shigella* invasion (Van Nhieu *et al.* 1997).

(c) *Signalling interference*

The second theme for the action of type III effectors is inflammation and cell signalling. Key elements in the induction of the inflammatory response are some cytokines. Central to their synthesis are the transcriptional activator NFkB and the mitogen-activated protein kinases ERK, JNK and p38. Several type III effectors downregulate the inflammatory response. The best example of this is YopP (YopJ in *Y. pestis* and *Y. pseudotuberculosis*). Injection of YopP/YopJ into macrophages leads to a significant reduction in the release of TNF, a pro-inflammatory cytokine, and to apoptosis (Mills *et al.* 1997; Monack *et al.* 1997; Boland & Cornelis 1998). The two events are probably the consequence of the same early event in a common signalling cascade but reduction in the release of TNF is not simply the consequence of apoptosis since it occurs even if apoptosis is prevented by caspase inhibitors. Concomitantly with these two events, one can observe the inhibition of NF-KB activation and the inhibition of the ERK1/2, p38 and JNK mitogen-activated protein kinases (MAPKs) activities (Ruckdeschel *et al.* 1998; Schesser *et al.* 1998; Boland *et al.* 1998). Interestingly, YopP and YopJ share a high level of similarity with an Avr protein from *Xanthomonas campestris* and a protein from the nitrogen-fixing *Rhizobium*. Because of this similarity, the *S. enterica* counterpart of YopP/J was called AvrA (Hardt & Galan 1997) but so far, no activity described for YopP/J could be assigned to AvrA.

In contrast to the *Yersinia* Ysc system, the *S. enterica* SPI1 system tends to induce a profound inflammatory response in the intestinal epithelium. The exact effector(s) responsible for this have not been identified yet, but again MAPKs are involved (Hobbie *et al.* 1997). JNK MAPK is also activated as a consequence of SopE-induced activation of Rho (Hardt *et al.* 1998a).

(d) *Intracellular trafficking of the effectors*

Not very much is known so far on the intracellular traffic of the effectors. Most of them are presumed to be cytosolic but two of them have been shown to follow a different route. YopM is a strongly acidic protein containing LRRs whose action and target remain unknown. However, it has been shown to traffick to the cell's nucleus by means of a vesicle-associated pathway that is strongly inhibited by brefeldin A, perturbed by monensin or bafilomycin (Skrzypek *et al.* 1998). Tir from EPECs (EspE in EHECs) is particularly interesting in the sense that it inserts in the plasma membrane of the target enterocytes and serves as a receptor for intimin, a powerful adhesin of EPECs. Thus EPECs and EHECS insert their own receptor into mammalian cell surfaces, to which they then adhere to trigger additional host signalling events and actin nucleation (Kenny *et al.* 1997; Deibel *et al.* 1998).

(e) *Intracellular action of translocators*

The *Shigella* IpaB and its *Salmonella* counterpart SipB, bind caspase 1 (Casp-1) and, by doing this, they induce apoptosis (Chen *et al.* 1996; Hilbi *et al.* 1998; Hersh *et al.* 1999), bypassing signal transduction events and caspases upstream of Casp-1. *Shigella*-induced apoptosis is thus distinct from other forms of apoptosis and seems uniquely dependent on Casp-1. Binding studies show that SipB associates with the pro-apoptotic protease Casp-1. This interaction results in the activation of Casp-1, as seen in its proteolytic maturation and the processing of its substrate interleukin-1 beta. Functional inhibition of Casp-1 activity by acetyl-Tyr-Val-Ala-Asp-chloromethyl ketone blocks macrophage cytotoxicity, and macrophages lacking Casp-1 are not susceptible to *Salmonella*-induced apoptosis. Taken together, the data demonstrate that the *Shigella* IpaB and the *Salmonella* SipB function not only as translocators but also as effectors inducing apoptosis. Thus, type III systems of *Yersinia, Salmonella* and *Shigella* all induce apoptosis but it must be stressed that they do it by two totally distinct pathways.

Finally, IpaB and SipB are not the only bifunctional translocators: purified IpaC was recently shown to nucleate and to bundle actin filaments (Hayward & Koronakis 1999). How this discovery correlates with the role of SipA (Zhou *et al.* 1999) remains to be clarified.

4. COMPARISON OF THE VARIOUS TYPE III SYSTEMS

(a) *Three major groups of systems among the animal pathogens*

A superficial comparison of the sequences of the secretion–translocation systems encountered in the animal pathogens suggests the existence of at least three families: the Psc system of *Pseudomonas aeruginosa* is extremely close to the Ysc system of *Y. enterocolitica*, which is quite surprising given the long evolutionary distance between these two bacterial species. The system of *S. typhimurium* encoded by centisome 63 and the Mxi/Spa system of *Shigella*, both involved in bacterial invasion of epithelial cells, are also very similar. Finally, the second system of *S. typhimurium*, encoded by centisome 30 (Ochman *et al.* 1996; Shea *et al.* 1996; Hensel *et al.* 1997) seems to be rather related to the system found in EPECs and EHECs. Several attempts have been made to trans-complement mutations in secretion genes using the homologue from another but these were generally unsuccessful. However, the *pcrV* gene from *P. aeruginosa* can complement an *lcrV* mutation in *Y. pseudotuberculosis* (Petterson *et al.* 1999). It seems thus that, apart from the couple *Yersinia–Pseudomonas*, it is impossible to mix the pieces of various injectisomes.

(b) *Exchangeability between the effectors of the different systems*

Are the various type III systems functionally interchangeable in the sense that effectors from one system could be secreted or even delivered intracellularly by another system? The N-terminal domain (217 residues) of ADP-ribosyltransferase ExoS from *P. aeruginosa* (453 residues total) is 54% similar to the entire YopE (see §3) and the protein encoded by the gene next to ExoS (ORF1) is very similar to SycE (Wattiau *et al.* 1996). These observations prompted Frithz-Lindsten *et al.* (1997) to introduce the two genes from *P. aeruginosa*, transcribed from the P_{lac} promoter, into *Y. pseudotuberculosis*. Since they observed that the recombinant *Y. pseudotuberculosis* could secrete ExoS, they pursued their investigation and they wondered whether ExoS would not be delivered by a recombinant *Y. pseudotuberculosis* into

HeLa cells, just like YopE. They introduced the *exoS* gene and ORF1 in a non-cytotoxic double *yopE*, *yopH* mutant of *Y. pseudotuberculosis* and they infected HeLa cells. The result was clear cytotoxicity, indicating that ExoS is translocated across the HeLa cell plasma membrane and also that ExoS has a cytotoxic activity. Repeating the experiment with a mutated form of ExoS that has a 2000-fold reduced ADP-ribosyltransferase activity, they still observed cytotoxicity, which indicated that ExoS is a bifunctional protein endowed with a YopE-like cytotoxic activity. These experiments demonstrated that the closely related *Yersinia* and *Pseudomonas* type III systems are functionally interchangeable. Given the taxonomic distance between these two species, the observation is of importance because it strengthens the idea of a horizontal spread of these type III systems.

Wolf-Watz's group also observed that *Y. pseudotuberculosis* can secrete IpaB from *S. flexneri* and that *S. typhimurium* can secrete YopE (Rosqvist *et al.* 1995). The latter recombinant *Salmonella* is also cytotoxic for HeLa cells, suggesting that YopE could even be translocated across the cell plasma membrane.

5.   GENETIC SUPPORT

Comparison of the systems and the phylogeny analyses suggest that these systems must have been transferred hortizontally during evolution. Not surprisingly, the genes that encode these systems have been found to be part of elements that are more mobile than most of the other bacterial genes. In *Yersinia* and *Shigella*, the whole systems are plasmid-borne, while they are on pathogenicity islands in *Salmonella* (SPI1 and SPI2) and in EPECs. In general, the genes encoding the secretion–translocation systems appear to be part of large, compact operons, while the genes encoding the effectors are more scattered. Pathogenicity islands are sometimes considered as vestigial phages. Interestingly, in *S. typhimurium*, Hardt *et al.* (1998a) observed that SopE, one of the substrates of the system encoded by SPI1 is encoded outside the SPI, but on a cryptic P2-like phage. This observation tends to suggest that the effectors could be horizontally transferred independently from the secretion–translocation systems. This hypothesis is consistent with the observation that the effectors from one system can generally be delivered by another one, provided there is no limitation in their synthesis.

6. PROSPECTS

Since its discovery in 1994, type III secretion has expanded very rapidly, to become a whole field. Study of the type III systems allowed a better understanding of the pathogenesis of Gram-negative bacteria and discoveries with the different pathogens benefited from a constant cross-feed. The recent very fast progress made with *P. aeruginosa*, taking advantage of its similarity with *Yersinia*, is a spectacular example of such cross-feeding. The *Yersinia* lesson was not limited to the understanding of the fate of the known *P. aeruginosa* exotoxins but it extended to promising vaccination attempts. The *Yersinia* translocator LcrV was known to represent a protective antigen against plague since the mid-1950s (Burrows & Bacon 1958). Because of its extensive similarity to LcrV, one could guess that PcrV from *P. aeruginosa* could also act as a protective antigen. This was indeed shown recently by Sawa *et al.* (1999), using lung infection in mice as a model. As appealing as it may be, the development of new vaccines is not the only spin-off of this exciting new field. From a medical point of view, it could lead to the development of 'antipathogenicity molecules'. From a more basic point of view, it could also be beneficial to eukaryotic cell biology, by bringing in new tools if not new concepts.

I acknowledge M. Monteforte for handling the reference files. Work in my laboratory is supported by the Belgian Fonds National de la Recherche Scientifique Médicale (Convention 3.4595.97), the Direction Générale de la Recherche Scientifique-Communauté Française de Belgique (Action de Recherche Concertée 99/04–236) and by the Interuniversity Poles of Attraction Program–Belgian State, Prime Minister's Office, Federal Office for Scientific, Technical and Cultural affairs (PAI 4/03).

REFERENCES

Abe, A., de Grado, M., Pfuetzner, R. A., Sanchez-SanMartin, C., DeVinney, R., Puente, J. L., Strynadka, N. C. J. & Finlay, B. B. 1999 Enteropathogenic *Escherichia coli* translocated intimin receptor, Tir, requires a specific chaperone for stable secretion. *Mol. Microbiol.* **33**, 1162–1175.

Anderson, D. M. & Schneewind, O. 1997 A mRNA signal for the type III secretion of Yop proteins by *Yersinia enterocolitica*. *Science* **278**, 1140–1143.

Anderson, D. M. & Schneewind, O. 1999a Type III machines of gram-negative pathogens: injecting virulence factors into host cells and more. *Curr. Opin. Microbiol.* **2**, 18–24.

Anderson, D. M. & Schneewind, O. 1999b *Yersinia enterocolitica* type III secretion: an mRNA signal that couples translation and secretion of YopQ. *Mol. Microbiol.* **31**, 1139–1148.

Blocker, A., Gounon, P., Larquet, E., Allaoui, A., Niebuhr, K., Cabiaux, V., Parsot, C. & Sansonetti, P. 1999 Role of *Shigella*'s type III secretion system in insertion of IpaB and IpaC into the host membrane. *J. Cell Biol.* **147**, 683–693.

Boland, A. & Cornelis, G. R. 1998 Role of YopP in suppression of tumour necrosis factor alpha release by macrophages during *Yersinia* infection. *Infect. Immun.* **66**, 1878–1884.

Boland, A., Sory, M. P., Iriarte, M., Kerbourch, C., Wattiau, P. & Cornelis, G. R. 1996 Status of YopM and YopN in the *Yersinia* Yop virulon: YopM of *Y. enterocolitica* is internalized inside the cytosol of PU5–1.8 macrophages by the YopB, D, N delivery apparatus. *EMBO J.* **15**, 5191–5201.

Boyd, A. P., Sory, M. P., Iriarte, M. & Cornelis, G. R. 1998 Heparin interferes with translocation of Yop proteins into HeLa cells and binds to LcrG, a regulatory component of the *Yersinia* Yop apparatus. *Mol. Microbiol.* **27**, 425–436.

Boyd, A. P. *et al.* 2000 Competition between the Yops of *Yersinia enterocolitica* for delivery into eukaryotic cells: role of the SycE chaperone binding domain. (In preparation.)

Burrows, T. W. & Bacon, G. A. 1958 The effects of loss of different virulence determinants on the virulence and immunogenicity of strains of *Pasteurella pestis*. *Br. J. Exp. Pathol.* **39**, 278–291.

Chaux, P. (and 10 others) 1999 Identification of five Mage-A1 epitopes recognized by cytolytic T lymphocytes obtained by *in vitro* stimulation with dendritic cells transduced with Mage-A1. *J. Immunol.* **163**, 2928–2936.

Chen, Y., Smith, M. R., Thirumalai, K. & Zychlinsky, A. 1996 A bacterial invasin induces macrophage apoptosis by binding directly to ICE. *EMBO J.* **15**, 3853–3860.

Cheng, L. W., Anderson, D. M. & Schneewind, O. 1997 Two independent type III secretion mechanisms for YopE in *Yersinia enterocolitica*. *Mol. Microbiol.* **24**, 757–765.

Cornelis, G. R. 1998 The *Yersinia* deadly kiss. *J. Bacteriol.* **180**, 5495–5504.

Cornelis, G. R. & Wolf-Watz, H. 1997 The *Yersinia* Yop virulon: a bacterial system for subverting eukaryotic cells. *Mol. Microbiol.* **23**, 861–867.

Cornelis, G. R., Boland, A., Boyd, A. P., Geuijen, C., Iriarte, M., Neyt, C., Sory, M.-P. & Stainier, I. 1998 The virulence plasmid of *Yersinia*, an antihost genome. *Microbiol. Mol. Biol. Rev.* **62**, 1315–1352.

Covacci, A., Telford, J. L., Del Giudice, G., Parsonnet, J. & Rappuoli, R. 1999 *Helicobacter pylori* virulence and genetic geography. *Science* **284**, 1328–1333.

Day, J. B. & Plano, G. V. 1998 A complex composed of SycN and YscB functions as a specific chaperone for YopN in Yersinia pestis. *Mol. Microbiol.* **30**, 777–788.

Deibel, C., Krämer, S., Chakraborty, T. & Ebel, F. 1998 EspE, a novel secreted protein of attaching and effacing bacteria, is directly translocated into infected host cells, where it appears as a tyrosine-phosphorylated 90 kDa protein. *Mol. Microbiol.* **28**, 463–474.

Elliott, S. J., Wainwright, L. A., McDaniel, T. K., Jarvis, K. G., Deng, Y., Lai, L., McNamara, B. P., Donnenberg, M. S. & Kaper, J. B. 1998 The complete sequence of the locus of enterocyte effacement (LEE) from enteropathogenic *Escherichia coli* E2348/69. *Mol. Microbiol.* **28**, 1–4.

Elliott, S. J., Hutcheson, S. W., Dubois, M. S., Mellies, J. L., Wainwright, L. A., Batchelor, M., Frankel, G., Knutton, S. & Kaper, J. B. 1999 Identification of CesT, a chaperone for the type III secretion of Tir in enteropathogenic *Escherichia coli*. *Mol. Microbiol.* **33**, 1176–1189.

Fekkes, P. & Driessen, A. J. M. 1999 Protein targeting to the bacterial cytoplasmic membrane. *Microbiol. Mol. Biol. Rev.* **63**, 161–173.

Finck-Barbançon, V., Goranson, J., Zhu, L., Sawa, T., Wiener Kronish, J. P., Fleiszig, S. M. J., Wu, C., Mend Mueller, L. & Frank, D. W. 1997 ExoU expression by *Pseudomonas aeruginosa* correlates with acute cytotoxicity and epithelial injury. *Mol. Microbiol.* **25**, 547–557.

Finck-Barbançon, V., Yahr, T. L. & Frank, D. W. 1998 Identification and characterization of SpcU, a chaperone required for efficient secretion of the ExoU cytotoxin. *J. Bacteriol.* **180**, 6224–6231.

Francis, M. S. & Wolf-Watz, H. 1998 YopD of *Yersinia pseudotuberculosis* is translocated into the cytosol of HeLa epithelial cells: evidence of a structural domain necessary for translocation. *Mol. Microbiol.* **29**, 799–813.

Frankel, G., Phillips, A. D., Rosenshine, I., Dougan, G., Kaper, J. B. & Knutton, S. 1998 Enteropathogenic and enterohaemorrhagic *Escherichia coli*: more subversive elements. *Mol. Microbiol.* **30**, 911–921.

Fraser, G. M., Bennett, J. C. Q. & Hughes, C. 1999 Substrate-specific binding of hook-associated proteins by FlgN and FliT, putative chaperones for flagellum assembly. *Mol. Microbiol.* **32**, 569–580.

Frithz-Lindsten, E., Rosqvist, R., Johansson, L. & Forsberg, A. 1995 The chaperone-like protein YerA of *Yersinia pseudotuberculosis* stabilizes YopE in the cytoplasm but is dispensible for targeting to the secretion loci. *Mol. Microbiol.* **16**, 635–647.

Frithz-Lindsten, E., Du, Y., Rosqvist, R. & Forsberg, A. 1997 Intracellular targeting of exoenzyme S of *Pseudomonas aeruginosa* via type III-dependent translocation induces phago-cytosis resistance, cytotoxicity and disruption of actin microfilaments. *Mol. Microbiol.* **25**, 1125–1139.

Fu, Y. X. & Galan, J. E. 1998 Identification of a specific chaperone for SptP, a substrate of the centisome 63 type III secretion system of *Salmonella typhimurium*. *J. Bacteriol.* **180**, 3393–3399.

Galan, J. E. 1998 Interactions of *Salmonella* with host cells: encounters of the closest kind. *Proc. Natl Acad. Sci. USA* **95**, 14006–14008.

Galan, J. E. & Collmer, A. 1999 Type III secretion machines: bacterial devices for protein delivery into host cells. *Science* **284**, 1322–1328.

Genin, S. & Boucher, C. A. 1994 A superfamily of proteins involved in different secretion pathways in gram-negative bacteria: modular structure and specificity of the N-terminal domain. *Mol. Gen. Genet.* **243**, 112–118.

Ginocchio, C. C., Olmsted, S. B., Wells, C. L. & Galan, J. E. 1994 Contact with epithelial cells induces the formation of surface appendages on *Salmonella typhimurium*. *Cell* **76**, 717–724.

Goosney, D. L., de Grado, M. & Finlay, B. B. 1999 Putting *E. coli* on a pedestal: a unique system to study signal transduction and the actin cytoskeleton. *Trends Cell Biol.* **9**, 11–14.

Håkansson, S., Galyov, E. E., Rosqvist, R. & Wolf-Watz, H. 1996*a* The *Yersinia* YpkA Ser/Thr kinase is translocated and subsequently targeted to the inner surface of the HeLa cell plasma membrane. *Mol. Microbiol.* **20**, 593–603.

Håkansson, S., Schesser, K., Persson, C., Galyov, E. E., Rosqvist, R., Homble, F. & Wolf-Watz, H. 1996*b* The YopB protein of *Yersinia pseudotuberculosis* is essential for the translocation of Yop effector proteins across the target cell plasma membrane and displays a contact dependent membrane disrupting activity. *EMBO J.* **15**, 5812–5823.

Hardt, W. D. & Galan, J. E. 1997 A secreted *Salmonella* protein with homology to an avirulence determinant of plant pathogenic bacteria. *Proc. Natl Acad. Sci. USA* **94**, 9887–9892.

Hardt, W. D., Chen, L. M., Schuebel, K. E., Bustelo, X. R. & Galan, J. E. 1998*a S. typhimurium* encodes an activator of Rho GTPases that induces membrane ruffling and nuclear responses in host cells. *Cell* **93**, 815–826.

Hardt, W. D., Urlaub, H. & Galan, J. E. 1998*b* A substrate of the centisome 63 type III protein secretion system of *Salmonella typhimurium* is encoded by a cryptic bacteriophage. *Proc. Natl Acad. Sci. USA* **95**, 2574–2579.

Hauser, A. R., Kang, P. J. & Engel, J. N. 1998 PepA, a secreted protein of *Pseudomonas aeruginosa*, is necessary for cytotoxicity and virulence. *Mol. Microbiol.* **27**, 807–818.

Hayward, R. D. & Koronakis, V. 1999 Direct nucleation and bundling of actin by the SipC protein of invasive *Salmonella*. *EMBO J.* **18**, 4926–4934.

Hensel, M., Shea, J. E., Raupach, B., Monack, D. M., Falkow, S., Gleeson, C., Kubo, T. & Holden, D. W. 1997 Functional analysis of *ssaJ* and the *ssaK/U* operon, 13 genes encoding components of the type III secretion apparatus of *Salmonella pathogenicity* island 2. *Mol. Microbiol.* **24**, 155–167.

Hensel, M., Shea, J. E., Waterman, S. R., Mundy, R., Nikolaus, T., Banks, G., Vazquez-Torres, A., Gleeson, C., Fang, F. C. & Holden, D. W. 1998 Genes encoding putative effector proteins of the type III secretion system of *Salmonella* pathogenicity island 2 are required for bacterial virulence and proliferation in macrophages. *Mol. Microbiol.* **30**, 163–174.

Hersh, D., Monack, D. M., Smith, M. R., Ghori, N., Falkow, S. & Zychlinsky, A. 1999 The *Salmonella* invasin SipB induces macrophage apoptosis by binding to caspase-1. *Proc. Natl Acad. Sci. USA* **96**, 2396–2401.

Higushi, K. & Smith, J. L. 1961 Studies on the nutrition and physiology of *Pasteurella pestis*. VI. A differential plating medium for the estimation of the mutation rate to avirulence. *J. Bacteriol.* **81**, 605–608.

Hilbi, H., Moss, J. E., Hersh, D., Chen, Y. J., Arondel, J., Banerjee, S., Flavell, R. A., Yuan, J. Y., Sansonetti, P. J. & Zychlinsky, A. 1998 *Shigella*-induced apoptosis is dependent on caspase-1 which binds to IpaB. *J. Biol. Chem.* **273**, 32895–32900.

Hobbie, S., Chen, L. M., Davis, R. J. & Galan, J. E. 1997 Involvement of mitogen-activated protein kinase pathways in the nuclear responses and cytokine production induced by *Salmonella typhimurium* in cultured intestinal epithelial cells. *J. Immunol.* **159**, 5550–5559.

Iriarte, M. & Cornelis, G. R. 1998 YopT, a new *Yersinia* Yop effector protein, affects the cytoskeleton of host cells. *Mol. Microbiol.* **29**, 915–929.

Iriarte, M. & Cornelis, G. R. 1999 Identification of SycN, YscX, and YscY, three new elements of the *Yersinia* Yop virulon. *J. Bacteriol.* **181**, 675–680.

Jackson, M. W., Day, J. B. & Plano, G. V. 1998 YscB of *Yersinia pestis* functions as a specific chaperone for YopN. *J. Bacteriol.* **180**, 4912–4921.

Jarvis, K. G., Giron, J. A., Jerse, A. E., McDaniel, T. K., Donnenberg, M. S. & Kaper, J. B. 1995 Enteropathogenic *Escherichia coli* contains a putative type III secretion system necessary for the export of proteins involved in attaching and effacing lesion formation. *Proc. Natl Acad. Sci. USA* **92**, 7996–8000.

Jones, M. A., Wood, M. W., Mullan, P. B., Watson, P. R., Wallis, T. S. & Galyov, E. E. 1998 Secreted effector proteins of *Salmonella dublin* act in concert to induce enteritis. *Infect. Immun.* **66**, 5799–5804.

Kaniga, K., Tucker, S., Trollinger, D. & Galan, J. E. 1995 Homologs of the *shigella* IpaB and IpaC invasins are required for *Salmonella* entry into host cells. *J. Bacteriol.* **177**, 3965–3971.

Kaniga, K., Uralil, J., Bliska, J. B. & Galan, J. E. 1996 A secreted protein tyrosine phosphatase with modular effector domains in the bacterial pathogen *Salmonella typhimurium*. *Mol. Microbiol.* **21**, 633–641.

Kaper, J. B. 1998 EPEC delivers the goods. *Trends Microbiol.* **6**, 169–172.

Kenny, B., DeVinney, R., Stein, M., Reinscheid, D. J., Frey, E. A. & Finlay, B. B. 1997 Enteropathogenic *E. coli* (EPEC) transfers its receptor for intimate adherence into mammalian cells. *Cell* **91**, 511–520.

Kerr, J. R., Rigg, G. P., Matthews, R. C. & Burnie, J. P. 1999 The Bpel locus encodes type III secretion machinery in *Bordetella pertussis*. *Microb. Pathogen.* **27**, 349–367.

Knutton, S., Rosenshine, I., Pallen, M. J., Nisan, I., Neves, B. C., Bain, C., Wolf, C., Dougan, G. & Frankel, G. 1998 A novel EspA-associated surface organelle of enteropathogenic *Escherichia coli* involved in protein translocation into epithelial cells. *EMBO J.* **17**, 2166–2176.

Koster, M., Bitter, W., de Cock, H., Allaoui, A., Cornelis, G. R. & Tommassen, J. 1997 The outer membrane component, YscC, of the Yop secretion machinery of *Yersinia enterocolitica* forms a ring-shaped multimeric complex. *Mol. Microbiol.* **26**, 789–798.

Kubori, T., Matsushima, Y., Nakamura, D., Uralil, J., Lara-Tejero, M., Sukhan, A., Galan, J. E. & Aizawa, S.-I. 1998 Supramolecular structure of the *Salmonella typhimurium* type III protein secrection system. *Science* **280**, 602–605.

McGuffie, E. M., Frank, D. W., Vincent, T. S. & Olson, J. C. 1998 Modificiation of Ras in eukaryotic cells by *Pseudomonas aeruginosa* exoenzyme S. *Infect. Immun.* **66**, 2607–2613.

Ménard, R., Sansonetti, P. J., Parsot, C. & Vasselon, T. 1994 Extracellular association and cytoplasmic partitioning of the IpaB and IpaC invasins of *S. flexneri*. *Cell* **79**, 515–525.

Michiels, T., Wattiau, P., Brasseur, R., Ruysschaert, J. M. & Cornelis, G. R. 1990 Secretion of Yop proteins by *Yersiniae*. *Infect. Immun.* **58**, 2840–2849.

Mills, S. D., Boland, A., Sory, M. P., Van der Smissen, P., Kerbourch, C., Finlay, B. B. & Cornelis, G. R. 1997 *Yersinia enterocolitica* induces apoptosis in macrophages by a process requiring functional type III secretion and translocation mechanisms and involving YopP, presumably acting as an effector protein. *Proc. Natl Acad. Sci. USA* **94**, 12638–12643.

Monack, D. M., Mecsas, J., Ghori, N. & Falkow, S. 1997 *Yersinia* signals macrophages to undergo apoptosis and YopJ is necessary for this cell death. *Proc. Natl Acad. Sci. USA* **94**, 10385–10390.

Neyt, C. & Cornelis, G. R. 1999 Insertion of a Yop translocation pore into the macrophage plasma membrane by *Yersinia enterocolitica*: requirement for translocators YopB and YopD, but not LcrG. *Mol. Microbiol.* **33**, 971–981.

Norris, F. A., Wilson, M. P., Wallis, T. S., Galyov, E. E. & Majerus, P. W. 1998 SopB, a protein required for virulence of *Salmonella dublin*, is an inositol phophate phosphatase. *Proc. Natl Acad. Sci. USA* **95**, 14057–14059.

Ochman, H., Soncini, F. C., Solomon, F. & Groisman, E. A. 1996 Identification of a pathogenicity island required for *Salmonella* survival in host cells. *Proc. Natl Acad. Sci. USA* **93**, 7800–7804.

Payne, P. L. & Straley, S. C. 1998 YscO of *Yersinia pestis* is a mobile core component of the Yop secretion system. *J. Bacteriol.* **180**, 3882–3890.

Pederson, K. J., Vallis, A. J., Aktories, K., Frank, D. W. & Barbieri, J. T. 1999 The amino-terminal domain of *Pseudomonas aeruginosa* ExoS disrupts actin filaments via small-molecular-weight GTP-binding proteins. *Mol. Microbiol.* **32**, 393–401.

Persson, C., Nordfelth, R., Holmstrm, A., Hakansson, S., Rosqvist, R. & Wolf-Watz, H. 1995 Cell-surface-bound *Yersinia* translocate the protein tyrosine phosphatase YopH by a polarized mechanism into the target cell. *Mol. Microbiol.* **18**, 135–150.

Persson, C., Carballeira, N., Wolf-Watz, H. & Fllman, M. 1997 The PTPase YopH inhibits uptake of *Yersinia*, tyrosine phosphorylation of p130^{Cas} and FAK, and the associated accumulation of these proteins in peripheral focal adhesions. *EMBO J.* **16**, 2307–2318.

Pettersson, J., Nordfelth, R., Dubinina, E., Bergman, T., Gustafsson, M., Magnusson, K. E. & Wolf-Watz, H. 1996 Modulation of virulence factor expression by pathogen target cell contact. *Science* **273**, 1231–1233.

Pettersson, J., Holmstrm, A., Hill, J., Leary, S., Frithz-Lindsten, E., von Euler-Matell, A., Carlsson, E., Titball, R., Forsberg, A. & Wolf-Watz, H. 1999 The V-antigen of *Yersinia* is surface exposed before target cell contact and involved in virulence protein translocation. *Mol. Microbiol.* **32**, 961–976.

Roine, E., Wei, W., Yuan, J., Nurmiaho-Lassila, E. L., Kalkkinen, N., Romantschuk, M. & He, S. Y. 1997 Hrp pilus: an *hrp*-dependent bacterial surface appendage produced by *Pseudomonas syringae* pv. tomato DC3000. *Proc. Natl Acad. Sci. USA* **94**, 3459–3464.

Rosqvist, R., Forsberg, A. & Wolf-Watz, H. 1991 Intracellular targeting of the *Yersinia* YopE cytotoxin in mammalian cells induces actin microfilament disruption. *Infect. Immun.* **59**, 4562–4569.

Rosqvist, R., Magnusson, K. E. & Wolf-Watz, H. 1994 Target cell contact triggers expression and polarized transfer of *Yersinia* YopE cytotoxin into mammalian cells. *EMBO J.* **13**, 964–972.

Rosqvist, R., Håkansson, S., Forsberg, A. & Wolf-Watz, H. 1995 Functional conservation of the secretion and translocation machinery for virulence proteins of yersiniae, salmonellae and shigellae. *EMBO J.* **14**, 4187–4195.

Ruckdeschel, K., Harb, S., Roggenkamp, A., Hornef, M., Zumbihl, R., Kohler, S., Heesemann, J. & Rouot, B. 1998 *Yersinia enterocolitica* impairs activation of transcription factor NF-B: involvement in the induction of programmed cell death and in the suppression of the macrophage TNF- production. *J. Exp. Med.* **187**, 1069–1079.

Russel, M. 1994 Phage assembly: a paradigm for bacterial virulence factor export? *Science* **265**, 612–614.

Russmann, H., Shams, H., Poblete, F., Fu, Y. X., Galan, J. E. & Donis, R. O. 1998 Delivery of epitopes by the *Salmonella* type III secretion system for vaccine development. *Science* **281**, 565–568.

Salmond, G. P. & Reeves, P. J. 1993 Membrane traffic wardens and protein secretion in gram-negative bacteria. *Trends Biochem. Sci.* **18**, 7–12.

Sarker, M. R., Neyt, C., Stainier, I. & Cornelis, G. R. 1998 The *Yersinia* Yop virulon: LcrV is required for extrusion of the translocators YopB and YopD. *J. Bacteriol.* **180**, 1207–1214.

Sawa, T., Yahr, T. L., Ohara, M., Kurahashi, K., Gropper, M. A., Wiener-Kronish, J. P. & Frank, D. W. 1999 Active and passive immunization with the *Pseudomonas* V antigen protects against type III intoxication and lung injury. *Nat. Med.* **5**, 392–398.

Schesser, K., Spiik, A.-K., Dukuzumuremyi, J.-M., Neurath, M. F., Pettersson, S. & Wolf-Watz, H. 1998 The *yopJ* locus is required for *Yersinia*-mediated inhibition of NF-KB activation and cytokine expression: YopJ contains a eukaryotic SH2-like domain that is essential for its repressive activity. *Mol. Microbiol.* **28**, 1067–1079.

Schesser, K., Dukuzumuremyi, J. M., Cilio, C., Borg, S., Wallis, T. S., Pettersson, S. & Galyov, E. E. 2000 The *Salmonella* YopJ-homologue AvrA does not possess YopJ-like activity. *Microb. Pathogen.* **28**, 59–70.

Shea, J. E., Hensel, M., Gleeson, C. & Holden, D. W. 1996 Identification of a virulence locus encoding a second type III secretion system in *Salmonella typhimurium*. *Proc. Natl Acad. Sci. USA* **93**, 2593–2597.

Skrzypek, E., Cowan, C. & Straley, S. C. 1998 Targeting of the *Yersinia pestis* YopM protein into HeLa cells and intracellular trafficking to the nucleus. *Mol. Microbiol.* **30**, 1051–1065.

Sory, M. P. & Cornelis, G. R. 1994 Translocation of a hybrid YopE-adenylate cyclase from *Yersinia enterocolitica* into HeLa cells. *Mol. Microbiol.* **14**, 583–594.

Sory, M. P., Boland, A., Lambermont, I. & Cornelis, G. R. 1995 Identification of the YopE and YopH domains required for secretion and internalization into the cytosol of macrophages, using the *cyaA* gene fusion approach. *Proc. Natl Acad. Sci. USA* **92**, 11998–12002.

Starnbach, M. N. & Bevan, M. J. 1994 Cells infected with *Yersinia* present an epitope to class I MHC-restricted CTL. *J. Immunol.* **153**, 1603–1612.

Tardy, F., Homblé, F., Neyt, C., Wattiez, R., Cornelis, G. R., Ruysschaert, J.-M. & Cabiaux, V. 1999 *Yersinia enterocolitica* type III secretion-translocation system: channel formation by secreted Yops. *EMBO J.* **18**, 6793–6799.

Uchiya, K., Barbieri, M. A., Funato, K., Shah, A. H., Stahl, P. D. & Groisman, E. A. 1999 A *Salmonella* virulence protein that inhibits cellular trafficking. *EMBO J.* **18**, 3924–3933.

Vallis, A. J., Finck-Barbançon, V., Yahr, T. L. & Frank, D. W. 1999 Biological effects of *Pseudomonas aeruginosa* type III-secreted proteins on CHO cells. *Infect. Immun.* **67**, 2040–2044.

Van Gijsegem, F., Gough, C. L., Zischek, C., Niqueux, E., Arlat, M., Genin, S., Barberis, P., German, S., Castello, P. & Boucher, C. 1995 The *hrp* gene locus of *Pseudomonas solanacearum*, which controls the production of a type III secretion system, encodes eight proteins related to components of the bacterial flagellar biogenesis complex. *Mol. Microbiol.* **15**, 1095–1114.

Van Nhieu, G. T. & Sansonetti, P. J. 1999 Mechanism of *Shigella* entry into epithelial cells. *Curr. Opin. Microbiol.* **2**, 51–55.

Van Nhieu, G. T., BenZeev, A. & Sansonetti, P. J. 1997 Modulation of bacterial entry into epithelial cells by association between vinculin and the *Shigella* IpaA invasin. *EMBO J.* **16**, 2717–2729.

Van Nhieu, G. T., Caron, E., Hall, A. & Sansonetti, P. J. 1999 IpaC induces actin polymerization and filopodia formation during *Shigella* entry into epithelial cells. *EMBO J.* **18**, 3249–3262.

Viprey, V., Del Greco, A., Golinowski, W., Broughton, W. J. & Perret, X. 1998 Symbiotic implications of type III protein secretion machinery in *Rhizobium. Mol. Microbiol.* **28**, 1381–1389.

Wachter, C., Beinke, C., Mattes, M. & Schmidt, M. A. 1999 Insertion of EspD into epithelial target cell membranes by infecting enteropathogenic *Escherichia coli. Mol. Microbiol.* **31**, 1695–1707.

Wainwright, L. A. & Kaper, J. B. 1998 EspB and EspD require a specific chaperone for proper secretion from enteropathogenic *Escherichia coli. Mol. Microbiol.* **27**, 1247–1260.

Watarai, M., Tobe, T., Yoshikawa, M. & Sasakawa, C. 1995 Contact of *Shigella* with host cells triggers release of Ipa invasins and is an essential function of invasiveness. *EMBO J.* **14**, 2461–2470.

Wattiau, P. & Cornelis, G. R. 1993 SycE, a chaperone-like protein of *Yersinia enterocolitica* involved in the secretion of YopE. *Mol. Microbiol.* **8**, 123–131.

Wattiau, P., Bernier, B., Deslee, P., Michiels, T. & Cornelis, G. R. 1994 Individual chaperones required for Yop secretion by *Yersinia. Proc. Natl Acad. Sci. USA* **91**, 10493–10497.

Wattiau, P., Woestyn, S. & Cornelis, G. R. 1996 Customized secretion chaperones in pathogenic bacteria. *Mol. Microbiol.* **20**, 255–262.

Woestyn, S., Allaoui, A., Wattiau, P. & Cornelis, G. R. 1994 YscN, the putative energizer of the *Yersinia* Yop secretion machinery. *J. Bacteriol.* **176**, 1561–1569.

Woestyn, S., Sory, M. P., Boland, A., Lequenne, O. & Cornelis, G. R. 1996 The cytosolic SycE and SycH chaperones of *Yersinia* protect the region of YopE and YopH involved in translocation across eukaryotic cell membranes. *Mol. Microbiol.* **20**, 1261–1271.

Wolff, C., Nisan, I., Hanski, E., Frankel, G. & Rosenshine, I. 1998 Protein translocation into host epithelial cells by infecting enteropathogenic *Escherichia coli. Mol. Microbiol.* **28**, 143–155.

Yahr, T. L., Vallis, A. J., Hancock, M. K., Barbieri, J. T. & Frank, D. W. 1998 ExoY, an adenylate cyclase secreted by the *Pseudomonas aeruginosa* type III system. *Proc. Natl Acad. Sci. USA* **95**, 13 899–13 904.

Young, G. M., Schmiel, D. H. & Miller, V. L. 1999 A new pathway for the secretion of virulence factors by bacteria: the flagellar export apparatus functions as a protein-secretion system. *Proc. Natl Acad. Sci. USA* **96**, 6456–6461.

Zhou, D., Mooseker, M. S. & Galan, J. E. 1999 Role of the *S. typhimurium* actin-binding protein SipA in bacterial internalization. *Science* **283**, 2092–2095.

Zumbihl, R., Aepfelbacher, M., Andor, A., Jacobi, C. A., Ruckdeschel, H., Rouot, B. & Heesemann, J. 1999 The cytotoxin YopT of *Yersinia enterocolitica* induces modification and cellular redistribution of the small CTP-binding protein RhoA. *J. Biol. Chem.* **274**, 29 289–29 293.

EVOLUTION OF MICROBIAL PATHOGENS

JOACHIM MORSCHHÄUSER[†], GERWALD KÖHLER[†],
WILMA ZIEBUHR[‡], GABRIELE BLUM-OEHLER[‡],
ULRICH DOBRINDT[‡] and JORG HACKER[*,‡]

[†] *Zentrum fur Infektionsforschung, and*
[‡] *Institut fur Molekulare Infektionsbiologie der Universität Wurzburg,*
Röntgenring 11, D-97070 Wurzburg, Germany

Various genetic mechanisms including point mutations, genetic rearrangements and lateral gene transfer processes contribute to the evolution of microbes. Long-term processes leading to the development of new species or subspecies are termed macroevolution, and short-term developments, which occur during days or weeks, are considered as microevolution. Both processes, macro- and microevolution need horizontal gene transfer, which is particularly important for the development of pathogenic microorganisms. Plasmids, bacteriophages and so-called pathogenicity islands (PAIs) play a crucial role in the evolution of pathogens. During microevolution, genome variability of pathogenic microbes leads to new phenotypes, which play an important role in the acute development of an infectious disease. Infections due to *Staphylococcus epidermidis*, *Candida albicans* and *Escherichia coli* will be described with special emphasis on processes of microevolution. In contrast, the development of PAIs is a process involved in macro-evolution. PAIs are especially important in processes leading to new pathotypes or even species. In this review, particular attention will be given to the fact that the evolution of pathogenic microbes can be considered as a specific example for microbial evolution in general.

Keywords: evolution; pathogenicity; genomic island; biofilm; codon usage; resistance

1. INTRODUCTION: MICROEVOLUTION AND MACROEVOLUTION

The scientific observations on the evolution of organisms made by Charles Darwin (1809–1882) in the 19th century are not only true for eukaryotic organisms, they also seem to be valid for prokaryotes. The key processes of Darwinian evolution can be described by four different termini: genetic variability, phenotype formation, selection and isolation (Fig. 1). There is no doubt that the permanent development of new genetic variants represents the main requisite for the development of life. As many genetic alterations

---

*Author for correspondence (j.hacker@mail.uni-wuerzburg.de).

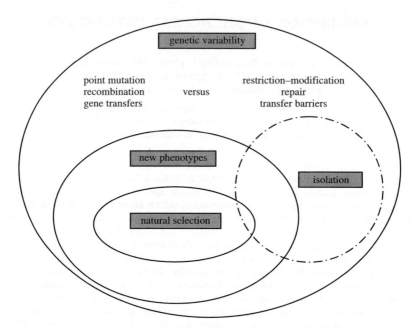

Fig. 1. Darwinian principles of evolution.

(mutations) do not lead to new phenotypes it is crucial to point out that only the alterations that do produce new phenotypes are decisive for evolution. New phenotypic variants are immediately subject to selection by biological and non-biological forces. As already described by Darwin, the geographical isolation of certain species or groups of organisms may be useful for evolutionary development. In addition, a genetic isolation due to transfer barriers, strong restriction-modification systems or changes in the codon usage may also contribute to the speed of evolutionary development.

The first micro-organisms appeared on Earth more than three billion years ago. As humans have developed in the last 1.5 million years, strictly human pathogens can be considered as very young microbes. Nevertheless, evolutionary development is also seen in these microbes. An evolutionary process that occurs within a longer period of time and that leads to the formation of new species or subspecies is considered as macroevolution. According to Ernst Mayr, macroevolution is the key developmental process for the evolution of life.

In contrast, processes of microevolution take days or weeks. As a consequence of microevolution, new variants of a certain species or sub-species are generated. During evolution, 'variability generators' such as insertion sequence (IS) elements, switching DNA fragments or transposons have appeared, which play a key role in microevolution. Microevolution is extremely important for the pathogenesis of infectious diseases. The expression or non-expression of particular genes (phase variation) or the alteration of microbial structures, especially surface structures such as pili or outer membrane proteins (antigenic variation) *in vivo* represent paradigms of micro-evolution. In addition, these processes are virulence mechanisms, important in many infectious diseases.

In this article we will first describe the general genetic mechanisms involved in the development of micro-organisms. Particular attention, however, will be given to pathogenic microbes. It is our view that pathogens produce many virulence or pathogenicity factors, which directly or indirectly contribute to the development of an infectious disease. These factors and the underlying gene clusters are of particular importance for the evolution of pathogens. In addition, pathogenic micro-organisms express resistance factors, which render the organisms resistant to antimicrobial drugs. The genetic basis of antimicrobial drug resistance is also an example for 'evolution under the microscope'. In this review we will present examples for microevolution of virulence factors as well as resistance mechanisms; both are important in various microbe host systems. Finally, the present knowledge on processes of macroevolution of pathogens with special emphasis on the occurrence and structure of pathogenicity islands (PAIs) will be summarized.

2. MECHANISMS OF GENETIC VARIABILITY

The evolution of organisms relies on heritable changes that are transmitted to the offspring. Phenotypic alterations that confer a selective advantage therefore must have a genetic basis to be relevant in evolutionary terms. There are several different mechanisms by which micro-organisms produce genetic variability: the accumulation of point mutations, genetic rearrangements and the acquisition of new genetic material by horizontal gene transfer.

(a) *Point mutations*

Point mutations occur more or less randomly throughout the genome and are generated by replication errors or incorrect repair following DNA damage. They can be silent; for example, when they are introduced into the coding sequence of a gene without altering the amino-acid sequence of the gene product. However, if the mutation results in an amino-acid exchange, a modified protein will be the consequence. Alternatively, such mutations may also occur in regulatory regions, thereby affecting the expression pattern of the respective gene(s). Single point mutations usually affect one specific trait that may confer an advantage in a changing environment (Musser 1995). This can generate new variants of a clone within relatively short periods of time (micro-evolution). On a larger scale, i.e. the generation of new species, evolution by accumulation of point mutations is a very slow process. This is especially true for mutations in essential species-specific housekeeping genes, e.g. loci encoding ribosomal RNAs, particular enzymes (for example ATPases) or structural proteins.

(b) *Genetic rearrangements*

Microbes may alter their genome also by rearrangement of existing parts. For example, gene amplifications result in an increase in the number of blueprints available for production of the corresponding gene product, which would be of advantage in an environment that demands a constantly higher expression. Such amplifications may be caused by recombination between repetitive DNA elements flanking the amplified sequence. Gene duplications also provide the material for further mutational evolution of the redundant copies without destroying the function of the product encoded by the original gene. In this way, variants can be generated with fine-tuned functional and/or regulatory properties. Recombination between individual members of gene families generated by duplications provides an additional level of variability, leading to new variants with altered properties by shuffling modules from pre-existing copies. The same genetic rearrangements that cause amplifications can also result in the loss of parts of the genome that are not essential any longer when a microbe has adapted to a new ecological niche, thereby freeing the organism of this genetic burden. Mobile genetic elements like IS elements are frequently substrates of such recombination events; however, they can also generate variability by their insertion into new sites in the genome, which may activate or inactivate genes located at those sites.

(c) *Gene transfer*

Microbes may alter their characteristics much more rapidly by acquisition of new genetic material from other organisms. Three basic mechanisms of gene transfer are known: transformation, transduction and conjugation (the mating between bacteria). Some bacteria, like *Neisseria gonorrhoeae* or *Haemophilus influenzae*, have a high natural competence to take up free DNA from lysed cells of their own species, which may subsequently recombine with corresponding genomic sequences within the cell to generate new variants. Within a given species there is often a great variety of individual organisms, especially in pathogenic bacteria that may or may not possess genes encoding virulence factors or antibiotic resistance determinants. Such genes may be located on the chromosome, on transposable elements, conjugative plamids or on phages that have integrated these genetic elements into their genome, and can be transferred to members of the same species or even to other species by conjugation or by phage transduction. These mechanisms are very important for the evolution of pathogenic bacteria since, instead of the slow adaptation of their own molecules to new functions, they allow the rapid generation of new variants with considerably altered properties by the exploitation of 'ready to use' genetic material from other organisms.

3.   EXAMPLES FOR MICROEVOLUTION OF PATHOGENS

(a) *Microevolution in* Candida albicans

The opportunistic fungal pathogen *C. albicans* is a harmless commensal in many healthy humans, residing on the mucosal surfaces of the gastrointestinal and urogenital tract. In immunocompromised patients, however, *C. albicans* can cause superficial as well as life-threatening disseminated infections. AIDS patients with oropharyngeal candidiasis or women with *Candida vaginitis* frequently suffer from recurrent infections, which are in most cases caused by the same *C. albicans* strain (Lischewski *et al.* 1995). Sensitive fingerprinting methods have shown that these strains undergo micro-evolution, as demonstrated by subtle changes in the DNA finger-print pattern between individual isolates from different infection episodes (Lockhart *et al.* 1996; Schröppel *et al.* 1994).

Oropharyngeal candidiasis in AIDS patients is usually treated effectively with azole antifungals, especially fluconazole, which inhibit the biosynthesis of ergosterol, the major sterol in the fungal plasma membrane. Since the introduction of fluconazole and its broad use during the past decade, resistant *C. albicans* strains have emerged in patients receiving long-term fluconazole treatment, resulting in therapy failure (White *et al.* 1998). As compared with previous isolates from earlier episodes, the resistant isolates have frequently undergone several genomic changes that contribute to their reduced drug susceptibility (Fig. 2). For example, point mutations within the *ERG11* gene, encoding the drug target enzyme sterol 14α-demethylase, result in the production of an enzyme with lowered affinity for the drug (Franz *et al.* 1998; Sanglard *et al.* 1998; White 1997*b*). Additionally, chromosomal

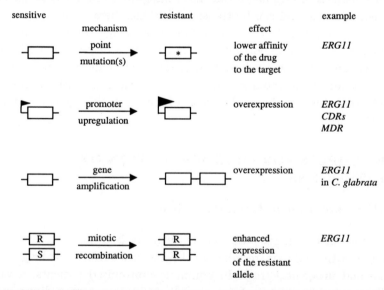

Fig. 2. Molecular mechanisms of azole resistance in *Candida*. The schematic representation shows the genetic mechanisms which lead to phenotypes with reduced susceptibility. Examples of the genes involved in azole resistance are indicated (see §3(a)).

rearrangements may contribute to enhanced resistance, because mitotic recombination in the diploid *C. albicans* can lead to homozygosity for a mutated *ERG11* allele (Franz *et al.* 1998; White 1997*b*), such that the cell produces only the resistant version of the enzyme. Similarly, constitutive overexpression of the target gene has also been observed, and this in turn may result in enhanced enzyme activity within the cell (Franz *et al.* 1998; White 1997*a*). In *C. albicans*, *ERG11* overexpression is probably caused by enhanced promoter activity, whereas in *C. glabrata* gene amplification has also been demonstrated (Marichal *et al.* 1997).

In addition, *C. albicans* possesses efflux pumps that enable the fungus to transport drugs actively out of the cell. In many fluconazole-resistant isolates a reduced intracellular accumulation of the drug correlating with stable, constitutive overexpression of these membrane transport proteins was observed (Sanglard *et al.* 1995, 1997). The ATP-binding cassette transporters Cdr1p and Cdr2p transport other azoles in addition to fluconazole. Consequently, *CDR1* and/or *CDR2* overexpression confers cross-resistance to several different azoles. In contrast, the substrate specificity of the Mdr1 protein, which belongs to the major facilitator superfamily, is more limited. *C. albicans* strains in which fluconazole resistance correlates with activation of the *MDR1* gene are still susceptible to ketoconazole and itraconazole, although several other unrelated drugs are transported by this efflux pump (Goldway *et al.* 1995; Sanglard *et al.* 1995, 1996). The exact nature of the mutations leading to overexpression of membrane transport proteins has not yet been clarified, but it seems that, similar to the baker's yeast *Saccharomyces cerevisiae*, mutations in regulatory proteins controlling the expression of the transporter genes are involved (Carvajal *et al.* 1997; Nourani *et al.* 1997; Wirsching *et al.* 2000).

The genomic alterations described are probably rare events, but once they have occurred the selective pressure exerted by the presence of the drug favours the overgrowth of *C. albicans* cells with increased resistance in a previously susceptible population. The constant micro-evolution of *C. albicans* results in the generation of strains in which multiple mechanisms have contributed to produce highly resistant strains. Under normal circumstances, the mutations would presumably reduce the fitness of the strains. For example, it has been shown that a mutation in sterol 14α-demethylase lowering the affinity of the enzyme to fluconazole also reduces enzyme activity (Kelly *et al.* 1999) and therefore might affect the growth rate of the cells when ergosterol biosynthesis is a limiting factor. Similarly, unregulated expression of biosynthesis enzymes or efflux pumps might be unfavourable for variant cells in many host niches as compared with cells that maintained the ability to fine-tune gene expression in response to environmental signals. The selective conditions in the clinical situation, however, give an advantage to cells with mutations conferring drug resistance Therefore, it can be concluded that genomic alterations, especially point mutations and rearrangements, which occur during the *in vivo* process of infection result in microevolutionary development of *C. albicans*.

(b) *Variation of biofilm formation in* Staphylococcus epidermidis *and* S. aureus

In aquatic, nutrient-limited ecosystems many bacteria develop a marked tendency to attach to surfaces and to initiate their organization in biofilms (Fig. 3). Bacterial biofilms are characterized by the production of slimy matrix substances that enclose the bacterial cells and mediate their adherence to each other and to solid surfaces (Costerton *et al.* 1995). Bacterial

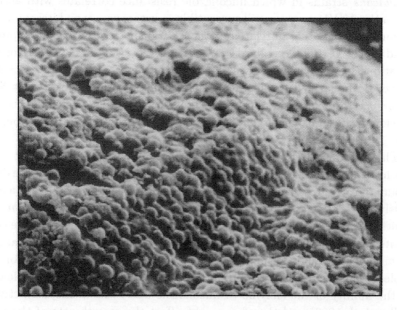

Fig. 3. Biofilm formation of an *S. epidermidis* strain on a polystyrene surface.

biofilms are widespread in nature and, in a sense, their organization can be considered to resemble that of multicellular organisms (Shapiro 1998). This point of view is supported by the fact that bacterial cells in biofilms differ considerably from their planktonic counterparts in terms of metabolic activity, gene expression and an inherent higher resistance to antibiotics.

Numerous pathogens, which are common sources of persistent and recurrent infections, e.g. *Pseudomonas aeruginosa*, *Escherichia coli*, streptococci and staphylococci, have been shown to generate biofilms (Costerton *et al.* 1999). In the course of an infection caused by biofilm-forming bacteria, planktonic cells are constantly released from the sessile population and they obviously are not efficiently eliminated by host defence mechanisms. In this respect, it has been hypothesized that specific detachment programmes exist that, once activated, abolish the extracellular matrix substance expression (Costerton *et al.* 1995). However, the genetic mechanisms of bacterial biofilm detachment are poorly understood. Recently, the molecular basis of biofilm formation in *S. epidermidis* and *S. aureus* has been elucidated and it has been demonstrated that biofilm expression undergoes strong phenotypic and genetic variations, which are believed to contribute to the detachment programmes mentioned above. Phase variation in biofilm formation indeed represents a good example for microevolution of staphylococcal isolates.

S. aureus and the coagulase-negative species *S. epidermidis* are the most common causes of nosocomial infections. Most of these infections are associated with indwelling medical devices. The formation of staphylococcal biofilms on plastic material was shown to be mediated by the *ica*ADBC-operon (Heilmann *et al.* 1996). Upon activation of this operon, a polysaccharide intercellular adhesin (PIA) is synthesized that mediates biofilm production. The evolutionary origin of the *ica* genes is not known. It appears, however, that this genetic information is widespread in clinical *S. epidermidis* isolates (Ziebuhr *et al.* 1997). In contrast, it is rarely observed in saprophytic strains from the healthy mucosa. With regard to *S. aureus* it has been found that all isolates, regardless of their origin, contain the *ica* genes (Cramton *et al.* 1999). However, only very few strains indeed express the operon and, consequently, most *S. aureus* strains are biofilm negative *in vitro*. Apart from these genetic differences, the PIA synthesis has been reported to undergo a phase variation process in biofilm-producing strains which is, in a substantial number of variants, caused by the alternating insertion and excision of the mobile genetic element IS*256* into (from) different

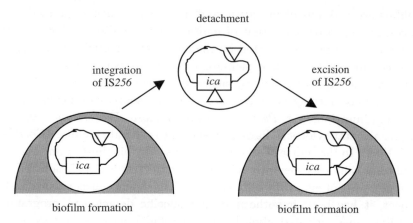

Fig. 4. Phase variation in staphylococci. Biofilm-forming bacteria are detached after insertional inactivation of the *ica* genes by IS*256*. Excision of IS*256* enables the cells to recolonize and form another biofilm.

sites of the *ica* gene cluster (Ziebuhr *et al.* 1999a) (Fig. 4). Especially, the *ica*C gene seems to represent a preferred target for IS*256* insertions. This study also revealed the reversible nature of this transposition. Thus, following repeated passages of PIA-negative insertional mutants, the biofilm-forming phenotype could be restored. Nucleotide sequence analyses of the revertants confirmed the complete excision of IS*256*. Recently, *ica*C::IS*256* insertional mutants were detected in clinical isolates from patients with indwelling medical devices and therefore it is assumed that the process might play a significant role during an infection. It is conceivable that the switch-off of the PIA production enables single bacterial cells to detach from the biofilm and to disseminate into novel habitats. A possible back-switch to the biofilm-producing phenotype, at a later stage of infection, would again render them capable of forming new biofilms on suitable surfaces (Fig. 4). In addition to phase variation, biofilm formation can also be irreversibly affected by the complete loss of the *ica* gene cluster or large chromosomal rearrangements (W. Ziebuhr *et al.* 2000). Chromosomal *ica* deletions were detected in different clinical isolates and comprised DNA fragments of *ca.* 70–100 kb in size. Additionally, chromosomal rearrangements were demonstrated during infections by biofilm-forming *S. epidermidis* strains.

Interestingly, both deletions and rearrangements resulted in altered IS*256*-specific hybridization patterns. The data suggest that the genetic and phenotypic flexibility of this pathogen contributes to its successful adaptation to changing environmental conditions and therefore, might be involved in the persistence and also the relapse of an infection. However, more experimental work is needed to investigate the exact molecular mechanisms of these processes.

(c) *Genomic deletions in pathogenic* E. coli

Some strains of *E. coli* are able to cause intestinal as well as extra-intestinal infectious diseases. Urinary tract infections represent the main diseases due to extra-intestinal *E. coli*. Uropathogenic *E. coli* produce a number of virulence factors such as adherence factors (P-fimbriae, S-fimbriae), toxins (haemolysin, cytotoxic necrotizing factor (CNF) I), capsules and particular iron uptake systems (aerobactin, yersinabactin). Recent studies revealed that uropathogenic *E. coli* producing α-haemolysin on blood agar plates undergo a switch from a haemolytic to a non-haemolytic phenotype. This was independently shown for different uropathogenic isolates such as strain 536 (O6:K15) and strain J96 (O4:K6). The modulation of virulence properties was also demonstrated *in vivo* in a rat pyelonephritis model. As the genetic rearrangements of uropathogenic *E. coli* occur with relatively high frequency during days and weeks, they can be considered microevolutionary processes.

Southern hybridization experiments and other molecular methods indicated that the non-haemolytic phenotype was due to deletion of large fragments of DNA, a mechanism also detected in *S. epidermidis* (see §3(b)). The deletion in *E. coli* comprises DNA fragments of 70 kb up to 190 kb (Fig. 5). Genes encoding α-haemolysin and gene clusters responsible for P-fimbriae and CNF are located on this unstable region in the genome of uropathogenic *E. coli* strains. The deletions are *recA* independent and occur with frequencies of 10^{-4} to 10^{-5}. It was shown recently that they represent site-specific events. The analysis of 40 independently generated deletion mutants of strain 536 showed a unique pattern, which argues for the action of a site-specific integrase–exicase, which may direct this process of virulence modulation. Short direct repeats are involved in the generation of the deletions in both uropathogenic *E. coli* strains J96 and 536. In strain 536 the direct repeats are 16 and 18 bp long. Following the deletion process one copy of the repeated DNA retained in the genome of the deletion

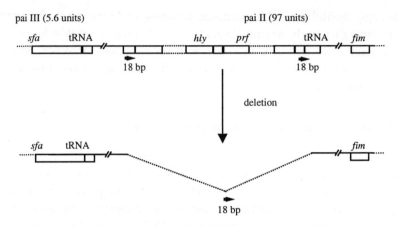

Fig. 5. Deletion of a PAI. Spontaneous deletion of PAI II in uropathogenic *E. coli* can occur by recombination between the flanking 18 bp direct repeats. Figure not drawn to scale.

mutant (Blum *et al.* 1994). We have indications that deletion formation does not only occur *in vitro* or in animal *in vivo* tests, but also during episodes of chronic urinary tract infections (M. Maibaum, G. Blum-Oehler and J. Hacker, unpublished data). The advantage to generate deletion mutants by microevolutionary processes may be that less pathogenic variants may have a better chance to survive in a late stage of urinary tract infection compared with fully virulent bacteria.

4. MECHANISMS OF LONG-TERM EVOLUTIONARY DEVELOPMENT

(a) *On pathogenicity and genomic islands*

In principle, PAIs belong to the repertoire of virulence gene carriers such as plasmids and bacteriophages. These three genetic elements strongly contribute to the macro-evolution of pathogens. Plasmids and bacteriophages as well as PAIs, however, may also be involved in processes of microevolution (Finlay & Falkow 1989; Karaolis *et al.* 1999; Ratti *et al.* 1997; Waldor & Mekalanos 1996). The microevolutionary processes of deletion formation in uropathogenic *E. coli* (see §3(c)) gave excellent indications of macroevolutionary developments: the formation of PAIs. It was demonstrated that the deleted DNA fragments from strain 536 fulfil the criteria of PAIs. PAIs represent large fragments of genomic DNA which are present in pathogenic microbes but absent in less pathogenic or apathogenic strains of the same species or related species. PAIs often carry more than one virulence gene and additional mobility genes such as integrases and parts of IS elements. PAIs represent distinct sections of DNA with repeats or IS elements at their boundary and they are frequently associated with tRNA genes. In addition, PAIs are highly unstable. They may represent former transferable elements, which have been fixed in the genome of certain species and subspecies during evolution (Dobrindt & Hacker 2000; Hacker *et al.* 1997; Hacker & Kaper 1999).

Strain 536 carries characteristic PAIs with genes encoding P-fimbriae and α-haemolysin. The PAIs are associated with tRNA genes: PAI I is accompanied by the selenocysteine-specific tRNA gene *sel*C, PAI II is located in the vicinity of the tRNA gene *leu*X, which encodes for the minor tRNA$_5^{Leu}$. One should mention here that the *leu*X-specific tRNA seems to act as a specific modulator of gene expression in pathogenic *E. coli*. The absence of the *leu*X-specific tRNA following deletion of PAI II leads to a decrease in expression of a number of virulence factors such as flagella, haemolysin or type I fimbriae (Dobrindt *et al.* 1998; Ritter *et al.* 1995, 1997). Additionally, the tRNA$_5^{Leu}$-specific gene, *leu*X, is part of a regulatory network, because it is regulated by certain global response regulators such as alternative σ-factors (Dobrindt & Hacker 2000). This example illustrates the complex genetic interplay between the core chromosome, PAIs and particular regulators involved in expression of virulence-associated genes.

PAIs seem to be established during long-term processes, which are macroevolutionary. It was shown that certain PAIs specific for a pathogenic *E. coli* strain such as the yersiniabactin-specific island or the locus for enterocyte effacement element of enteropathogenic *E. coli* have been part of the *E. coli* genome for more than one million years. The *Salmonella enterica* genome carries at least five different *Salmonella* PAIs (SPI). SPI I seems to have been present in the genome for more than 100 million years, because it is not only present in the genome of *Salmonella enterica*, but also in the genome of *Salmonella bongeri*. Both species, however, diverged about 100 million years ago. Therefore, it seems that PAIs represent former mobile DNA elements, which became fixed in the genome of certain pathogenic bacteria. If a pathogenic variant is successful, evolutionary pressure leads to the 'homing' of the island in the genome. This may promote the establishment of new pathotypes such as uropathogenic or enteropathogenic *E. coli* or certain serotypes of *Salmonella enterica*.

The increasing number of fully sequenced bacterial genomes reveals that PAIs do not represent unique DNA elements specific for pathogenic bacteria (Buchrieser *et al.* 1998; Censini *et al.* 1996; Lindsay *et al.* 1998). Similar DNA elements though encoding different functions were also found in non-pathogenic bacteria (Table 1). The *mec*A gene responsible for methicillin- and oxacillin-resistance encodes an alternative penicillin-binding protein of pathogenic staphylococci. It is also located on a large and unstable DNA fragment that fulfils the criteria for a PAI with the only difference that the fragment harbours a resistance determinant instead of virulence genes. This element could therefore be termed a 'resistance island' (Ito *et al.* 1999). Following integration, a large 500 kb plasmid, which carries the *nif* genes specific for nitrogen fixation in particular strains of *Mesorhizobium loti* was designated as a 'symbiosis island' (Sullivan & Ronson 1998). In addition, conjugative transposons of *Salmonella senftenberg* were denominated as 'metabolic islands', because they carry genes involved in sugar uptake (Hochhut *et al.* 1997). Various bacteria harbour genes for type III secretion (Bonas 1994) or iron uptake on plasmids or distinct sections of the genome. In order to bring all these different structures into a category we propose to use the term 'genomic islands' for such additional elements of DNA with the capacity to encode particular functions of microbes. These genomic

Table 1. Examples of genomic islands

organism	property	type of island	genetic feature
Mesorhizobium loti	N$_2$ fixation	symbiosis	500 kb plasmid
S. aureus	MecA protein	resistance	51 kb element
Pseudomonas putida	phenol degradation	degradation	104 kb plasmid
Salmonella senftenberg	sucrose uptake	metabolism	45 kb conjugative transposon
various bacteria	type III secretion	secretion	part of chromosome or plasmid
various bacteria	iron uptake	fitness	part of chromosome or plasmid

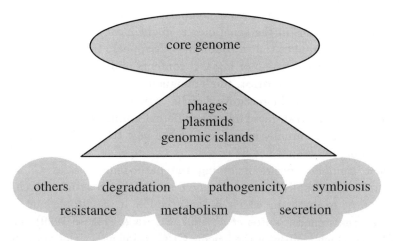

Fig. 6. Genome structure of prokaryotes. Various genetic elements like genomic islands, plasmids and phages can integrate in the core genome and deliver new traits to the micro-organism.

islands are not present in all isolates of a species, are unstable, represent a particular fragment of DNA and are very often associated with tRNA genes. It is our opinion that PAIs represent a subgroup of genomic islands. Genomic islands are not restricted to pathogens, rather they are present in the whole microbial world. Figure 6 depicts the possible contributions of the various genetic elements to the core genome of prokaryotes.

(b) *The role of codon usage*

As already mentioned, isolation of organisms or genetic barriers, which may lead to a separation of the genetic material between different groups of organisms, may also play an important role in the evolution of species. One example of 'genetic isolation' is the establishment of an alternative codon-usage programme as found in some organisms. After the discovery that the genetic code is universal in all organisms from bacteria to man it was a long-held view that codon changes would not be allowed during evolution because, by affecting so many genes, this would be a lethal event for the cells (Crick 1968). However, many exceptions concerning specific codons have since been discovered in bacteria, mitochondria and lower eukaryotes

(Dybvig & Voelker 1996; Osawa *et al.* 1992). There are two theories of how such alterations in the genetic code might have occurred. The 'codon capture theory' implies that a codon first disappears from the genome due to a drift to extremely high or low GC contents and reappears later with a different meaning. In contrast, the 'ambiguous intermediate theory' postulates that a codon can be recognized by its cognate as well as a near-cognate tRNA and thus be translated into two different amino acids within the same cell before it is eventually taken over by the new tRNA. This implies that codon ambiguity would not be lethal.

Most *Candida* species including the pathogenic species *C. albicans* translate the normally leucine-specific codon CUG to serine, which is the only known example in which a sense codon in a cytoplasmic mRNA has been reassigned (Santos *et al.* 1997; Sugita & Nakase 1999). The ser-tRNA$_{CAG}$ which translates the leucine CUG codon as serine has an unusual structure which lowers its decoding efficiency, thus allowing cells to survive low-level serine CUG translation (Santos *et al.* 1996). At least in some *Candida* species the CUG codon is still ambiguous, i.e. it is read both as serine and as leucine. It has been shown that the codon ambiguity generates a general stress response, which allows cells to grow under normally lethal conditions. This could have been the selective force that provided an advantage of codon ambiguity under stress conditions and enabled *Candida* species to adapt to new ecological niches. The constitutive tolerance of *C. albicans* to stress conditions like high temperature and oxidants might have evolved as a consequence of CUG reassignment, which could therefore also have played a role in the evolution of *Candida* pathogenicity (Santos *et al.* 1999).

(c) *Bacterial IS elements as 'variation generators'*

Bacterial IS elements are small mobile DNA units that encode only features necessary for their own mobilization. In general, they consist of the genetic information for a transposase protein and inverted repeat sequences that exactly define the borders of the element. IS elements can be regarded as repetitive DNA sequences that are randomly distributed on the bacterial chromosome. They also occur on plasmids, phages and in composite transposons where ISs often form the ends of the element. IS elements are known to play an important role in the microevolution of the bacterial genome

(Arber 1993; Mahillon & Chandler 1998; Ziebuhr *et al.* 1999*b*), their 'genetic programme', however, has been developed over millions of years. Therefore, the existence of such 'variation generators' (Arber 1993) is a matter of macro-evolutionary processes.

IS elements are also involved in the variation of gene expression during infection. Thus, IS elements have the capacity to cause irreversible inactivation of genes by random and in some cases also by site-specific transposition. Additionally, in *Neisseria meningitidis* and *S. epidermidis* it was shown that distinct elements (i.e. IS*1301* and IS*256*, respectively) can also contribute to reversible inactivation of virulence genes. The *ica* genes of *S. epidermidis* (see §3(b)) encoding the production of the polysaccharide adhesin PIA represent the target sites for integration and precise excision of IS*256*, leading to repression and reactivation of the *ica*-specific genes (Ziebuhr *et al.* 1999*a*). In *N. meningitidis*, the IS*1301* element mediates phase variation of capsule synthesis by a similar mechanism (Hammerschmidt *et al.* 1996). Both the findings in *Neisseria* and in *S. epidermidis* suggest a general role of bacterial IS elements in the modulation of gene expression in pathogenic bacteria.

Another interesting property of IS elements is their capacity to control the expression of adjacent genes. Numerous elements were shown to contain outwardly directed −35 promoter boxes in their terminal inverted repeats. When such an element transposes at the correct distance from a resident −10 promoter hexamer, new promoters capable of activating genes located downstream are created. These effects have been shown, for example, for the expression of the aminoglycoside resistance genes in Tn*4001* or for the expression of methicillin-resistance-associated genes in *S. aureus* (Maki & Murakami 1997; Rouch *et al.* 1987).

In addition to simple transposition, IS elements also give rise to complex DNA rearrangements including deletions, inversions, gene amplifications and the fusion of two DNA molecules by co-integrate formation (Arber 1993). In addition, the acquisition of genetic material or its distribution to other strains or species is often mediated by IS elements. Recent studies in enterococci indicate that different IS elements contribute to the evolution of large composite transposons that encode antibiotic resistance genes and carry integrated copies of plasmids (Bonafede *et al.* 1997; Rice & Carias 1998). The harboured genes are then mobilized to susceptible strains by horizontal gene transfer.

In general, the effects described above are mediated by specific actions of IS transposases. However, IS elements are also passively involved in recombination, since their nucleotide sequences represent homologous DNA stretches which might serve as recombinational cross-over points. Apparently, these IS-mediated mechanisms play a major role in genome flexibility. They are involved in the creation of new antigenic variants and since many IS elements are associated with antibiotic resistance genes, they also contribute considerably to the spread of resistance among bacterial populations.

(d) *Gene expression: the right gene at the right time*

All organisms must have the capacity to respond to changing environmental conditions by adapting the expression pattern of their genes. In pathogenic microbes, there are two fundamentally different important principles of changing the expression of genes: the regulated activation and repression of genes in response to external stimuli, and the transition between expressed and non-expressed states, which is usually random and results in phase or antigenic variation (Deitsch *et al.* 1997; Miller *et al.* 1989). The latter is often observed with structures on the cell surface that interact with host cells, like capsules or adhesins (Ölschläger *et al.* 1997). In certain situations, expression of these structures may be advantageous for the cell, whereas in others it may also represent a handicap. For example, the presence of a capsule renders *N. meningitidis* resistant to host defence mechanisms, but it hinders invasion into epithelial cells. Similarly, adhesins are necessary for tissue colonization, but they delay the crossing of the bacteria through mucus and can also mediate binding to phagocytic cells that destroy the bacteria. In addition, surface structures are often immunogenic, and the expression of variants with altered antigenic properties helps the cells to escape from the immune system. Micro-organisms have therefore evolved mechanisms that allow them to switch the expression state of such structures, thereby generating variants from a homogeneous population that gain access to new host niches or evade the host's immune response (Pouttu *et al.* 1999; Sokurenko *et al.* 1999). The genetic loci involved are often hypermutable, which allows the bacteria to rapidly produce the necessary diversity from a clonal population of infecting cells (Moxon & Tang, this issue).

As discussed above, gene families generated by duplication of an ancestral gene allow the evolution of variants which may be better adapted to specialized functions or environmental niches. Correspondingly, regulatory mechanisms have also evolved that allow for the expression of the proper variant under different environmental conditions, which is induced by external signals. *C. albicans* secretes aspartic proteinases that seem to be involved in the pathogenicity of the fungus. The role of this virulence factor during infection has not yet been clarified, but it may include the degradation of tissue barriers, the evasion of host defence mechanisms, or simply the supply of nitrogen sources. *C. albicans* possesses a family of at least ten homologous genes encoding secreted aspartic proteinases, and the individual members of this gene family are differentially regulated *in vitro* (Hube *et al.* 1994). Such a differential regulation is also ensured by host signals during infection, suggesting that the various proteinase isoenzymes might each have a specialized function in different host niches and at different stages of the infection (Staib *et al.* 1999).

5. CONCLUSIONS

In this review we intended to demonstrate that the general mechanisms of Darwinian evolution, generation of genetic variability, phenotypic expression of new variants and selection as well as physical and genetic 'isolation' are also valid for the evolution of microbial pathogens. Therefore, the development of pathogens can be considered as an example of Darwinian evolution. The generation of new variants of microbes also leads to the generation of new pathogens, which may represent new pathotypes belonging to the same species or forming new pathogenic species of microbes. The formation of PAIs is paradigmatic for the generation of new pathogenic variants. In addition, isolation of certain groups of organisms is also a key process of Darwinian evolution. The change of codon usage in pathogenic microbes such as *C. albicans* or *Mycoplasma* spp. illustrates these processes of genetic isolation. Isolated variants have the capacity of rapid development in particular ecological niches.

The mechanisms of microbial evolution elucidated in pathogenic bacteria might not be limited to this group of micro-organisms. While recent data illustrate the evolution of pathogenic bacteria, many processes inherent in the evolution of eukaryotic pathogens are still undiscovered. It will be one major future goal in molecular pathogenesis research to study evolution of pathogenic protozoan and fungal organisms to get further insights into the evolution of eukaryotes.

Our own work related to the topic of this review was supported by the Bundesministerium für Bildung, Wissenschaft, Forschung und Technologie (BMBF grant O1 K18906–0), the BMBF Stipendienprogramm Infektionsforschung (G.K.), the Deutsche Forschungsgemeinschaft, and the Fonds der Chemischen Industrie. We thank Hilde Merkert for help with preparation of the figures.

REFERENCES

Arber, W. 1993 Evolution of prokaryotic genomes. *Gene* **135**, 49–56.

Blum, G., Ott, M., Lischewski, A., Ritter, A., Imrich, H., Tschäpe, H. & Hacker, J. 1994 Excision of large DNA regions termed pathogenicity islands from tRNA-specific loci in the chromosome of an *Escherichia coli* wild-type pathogen. *Infect. Immun.* **62**, 606–614.

Bonafede, M. E., Carias, L. L. & Rice, L. B. 1997 Enterococcal transposon Tn5384: evolution of a composite transposon through cointegration of enterococcal and staphylococcal plasmids. *Antimicrob. Agents Chemother.* **41**, 1854–1858.

Bonas, U. 1994 Hrp genes of phytopathogenic bacteria. *Curr. Top. Microbiol. Immunol.* **192**, 79–98.

Buchrieser, C., Brosch, R., Bach, S., Guiyoule, A. & Carniel, E. 1998 The high-pathogenicity island of *Yersinia pseudotuberculosis* can be inserted into any of the three chromosomal tRNA genes. *Mol. Microbiol.* **30**, 965–978.

Carvajal, E., Van den Hasel, H. B., Cybularz-Kalaczkowska, A., Balzi, E. & Goffeau, A. 1997 Molecular and phenotypic characterization of yeast *PDR1* mutants that show hyperactive transcription of various ABC multidrug transporter genes. *Mol. Gen. Genet.* **256**, 406–415.

Censini, S., Lange, C., Xiang, Z., Crabtree, J. E., Ghiara, P., Borodovsky, M., Rappuoli, R. & Covacci, A. 1996 Cag, a pathogenicity island of *Helicobacter pylori*, encodes type I-specific and disease-associated virulence factors. *Proc. Natl Acad. Sci. USA* **10**, 14 648–14 653.

Costerton, J. W., Lewandowski, Z., Caldwell, D. E., Korber, D. R. & Lappin-Scott, H. M. 1995 Microbial biofilms. *A. Rev. Microbiol.* **49**, 711–745.

Costerton, J. W., Stewart, P. S. & Greenberg, E. P. 1999 Bacterial biofilms: a common cause of persistent infections. *Science* **284**, 1318–1322.

Cramton, S. E., Gerke, C., Schnell, N. F., Nichols, W. W. & Götz, F. 1999 The intercellular adhesion (*ica*) locus is present in *Staphylococcus aureus* and is required for biofilm formation. *Infect. Immun.* **67**, 5427–5433.

Crick, F. H. 1968 The origin of the genetic code. *J. Mol. Biol.* **38**, 367–379.

Deitsch, K. W., Moxon, E. R. & Wellems, T. E. 1997 Shared themes of antigenic variation and virulence in bacterial, protozoan, and fungal infections. *Microbiol. Mol. Biol. Rev.* **61**, 281–293.

Dobrindt, U. & Hacker, J. 1999 Plasmids, phages and pathogenicity islands in relation to bacterial protein toxins: impact on the evolution of microbes. In *The comprehensive sourcebook of bacterial protein toxins* (ed. J. E. Alouf and J. H. Freer), pp. 3–23. London: Academic Press.

Dobrindt, U. & Hacker, J. 2000 Regulation of the pathogenicity island associated tRNA$_5^{Leu}$-encoding gene LeuX in the uropathogenic *Escherichia coli* strain 536. (Submitted.)

Dobrindt, U., Cohen, P. S., Utley, M., Mhldorfer, I. & Hacker, J. 1998 The *leuX*-encoded tRNA$_5^{Leu}$ but not the pathogenicity islands I and II influence the survival of the uropathogenic *Escherichia coli* strain 536 in CD-1 mouse bladder mucus in the stationary phase. *FEMS Microbiol. Lett.* **162**, 135–141.

Dybvig, K. & Voelker, L. L. 1996 Molecular biology of mycoplasmas. *A. Rev. Microbiol.* **50**, 25–27.

Finlay, B. & Falkow, S. 1989 Common themes in microbial pathogenicity. *Microb. Rev.* **53**, 210–230.

Franz, R., Kelly, S. L., Lamb, D. C., Kelly, D. E., Ruhnke, M. & Morschhäuser, J. 1998 Multiple molecular mechanisms contribute to a stepwise development of fluconazole resistance in clinical *Candida albicans* strains. *Antimicrob. Agents Chemother.* **42**, 3065–3072.

Goldway, M., Teff, D., Schmidt, R., Oppenheim, A. B. & Koltin, Y. 1995 Multidrug resistance in *Candida albicans*: disruption of the *BEN^r* gene. *Antimicrob. Agents Chemother.* **39**, 422–426.

Hacker, J. & Kaper, J. 1999 The concept of pathogenicity islands. In *Pathogenicity islands and other mobile virulence elements* (ed. J. Kaper and J. Hacker), pp. 1–11. Washington, DC: American Society for Microbiology Press.

Hacker, J., Blum-Oehler, G., Mhldorfer, I. & Tschäpe, H. 1997 Pathogenicity islands of virulent bacteria: structure, function and impact of microbial evolution. *Mol. Microbiol.* **23**, 1089–1097.

Hammerschmidt, S., Hilse, R., Van Putten, J. P., Gerardy-Schahn, R., Unkmeir, A. & Frosch, M. 1996 Modulation of cell surface sialic acid expression in *Neisseria meningitidis* via a transposable genetic element. *EMBO J.* **15**, 192–198.

Heilmann, C., Schweitzer, O., Gerke, C., Vanittanakom, N., Mack, D. & Götz, F. 1996 Molecular basis of intercellular adhesion in the biofilm-forming *Staphylococcus epidermidis*. *Mol. Microbiol.* **20**, 1083–1091.

Hochhut, B., Jahreis, K., Lengeler, J. W. & Schmid, K. 1997 CTnscr94, a conjugative transposon found in enterobacteria. *J. Bacteriol.* **179**, 2097–2102.

Hube, B., Monod, M., Schofield, D. A., Brown, A. J. P. & Gow, N. A. R. 1994 Expression of seven members of the gene family encoding secretory aspartyl proteinases in *Candida albicans. Mol. Microbiol.* **14**, 87–99.

Ito, T., Katayama, Y. and Hiramatsu, K. 1999 Cloning & nucleotide sequence determination of the entire *mec* DNA of pre-methicillin-resistant *Staphylococcus aureus* N315. *Antimicrob. Agents Chemother.* **43**, 1449–1458.

Karaolis, D. K., Somara, S., Manevil Jr, D. R., Johnson, J. A. & Kaper, J. B. 1999 A bacteriophage encoding a pathogenicity island, a type IV pilus and a phage receptor in cholera bacteria. *Nature* **399**, 375–379.

Kelly, S. L., Lamb, D. C., Loeffler, J., Einsele, H. & Kelly, D. E. 1999 The G464S amino acid substitution in *Candida albicans* sterol 14 alpha-demethylase causes fluconazole resistance in the clinic through reduced affinity. *Biochem. Biophys. Res. Commun.* **262**, 174–179.

Lindsay, J. A., Ruzin, A., Ross, H. F., Kurepina, N. & Novick, R. P. 1998 The gene for toxic shock toxin is carried by a family of mobile pathogenicity islands in *Staphylococcus aureus. Mol. Microbiol.* **29**, 527–543.

Lischewski, A., Ruhnke, M., Tennagen, I., Schönian, G., Morschhäuser, J. & Hacker, J. 1995 Molecular epidemiology of *Candida* isolates from AIDS patients showing different fluconazole resistance profiles. *J. Clin. Microbiol.* **33**, 769–771.

Lockhart, S. R., Reed, B. D., Pierson, C. L. & Soll, D. R. 1996 Most frequent scenario for recurrent *Candida vaginitis* is strain maintenance with 'substrain shuffling': demonstration by sequential DNA fingerprinting with probes Ca3, C1, and CARE2. *J. Clin. Microbiol.* **34**, 767–777.

Mahillon, J. & Chandler, M. 1998 Insertion sequences. *Microbiol. Mol. Biol. Rev.* **62**, 725–774.

Maki, H. & Murakami, K. 1997 Formation of potent hybrid promoters of the mutant *llm* gene by IS*256* transposition in methicillin-resistant *Staphylococcus aureus. J. Bacteriol.* **179**, 6944–6948.

Marichal, P., Van den Bosche, H., Odds, F. C., Nobels, G., Warnock, D. W., Timmerman, V., Van Broeckhoven, C., Fay, S. & Mose Larsen, P. 1997 Molecular-biological characterization of an azole-resistant *Candida glabrata* isolate. *Antimicrob. Agents Chemother.* **41**, 2229–2237.

Miller, J. F., Mekalanos, J. J. & Falkow, S. 1989 Coordinate regulation and sensory transduction in the control of bacterial virulence. *Science* **243**, 916–922.

Musser, J. M. 1995 Antimicrobial agent resistance in mycobacteria: molecular genetic insights. *Clin. Microbiol. Rev.* **8**, 496–514.

Nourani, A., Papajova, D., Delahodde, A., Jacq, C. & Subik, J. 1997 Clustered amino acid substitutions in the yeast transcription regulator Pdr3p increase pleiotropic drug resistance and identify a new central regulatory domain. *Mol. Gen. Genet.* **256**, 397–405.

Ölschläger, T. A., Khan, A. S., Meier, C. & Hacker, J. 1997 Receptors and ligands in adhesion and invasion of *Escherichia coli. Nova Acta Leopoldina* **301**, 195–205.

Osawa, S., Jukes, T. H., Watanabe, K. & Muto, A. 1992 Recent evidence for evolution of the genetic code. *Microbiol. Rev.* **5**, 229–264.

Pouttu, R., Puustinen, T., Virkola, R., Hacker, J., Klemm, P. & Korhonen, T. K. 1999 Amino acid residue Ala-62 in the FimH fimbrial adhesin is critical for the adhesiveness of meningitis-associated *Escherichia coli* to collagens. *Mol. Microbiol.* **31**, 1747–1757.

Ratti, G., Covacci, A. & Rappuoli, R. 1997 A tRNA$_2^{Arg}$ gene of *Corynebacterium diphtheriae* is the chromosomal integration site for toxinogenic bacteriophages. *Mol. Microbiol.* **25**, 1179–1181.

Rice, L. B. & Carias, L. L. 1998 Transfer of Tn5385, a composite, multiresistance chromosomal element from *Enterococcus faecalis*. *J. Bacteriol.* **180**, 714–721.

Ritter, A., Blum, G., Emödy, L., Kerenyi, M., Böck, A., Neuhierl, B., Rabsch, W., Scheutz, F. & Hacker, J. 1995 tRNA genes and pathogenicity islands: influence on virulence and metabolic properties of uropathogenic *Escherichia coli*. *Mol. Microbiol.* **17**, 109–121.

Ritter, A., Gally, D. L., Olsen, P. B., Dobrindt, U., Friedrich, A., Klemm, P. & Hacker, J. 1997 The Pai-associated *leu*X-specific tRNA$_5^{Leu}$ affects type 1 fimbriation in pathogenic *Escherichia coli* by control of FimB recombinase expression. *Mol. Microbiol.* **25**, 871–872.

Rouch, D. A., Byrne, M. E., Kong, Y. C. & Skurray, R. A. 1987 The aacA-aphD gentamicin and kanamycin resistance determinant of Tn4001 from *Staphylococcus aureus*: expression and nucleotide sequence analysis. *J. Gen. Microbiol.* **133**, 3039–3052.

Sanglard, D., Kuchler, K., Ischer, F., Pagani, J. L., Monod, M. & Bille, J. 1995 Mechanisms of resistance to azole antifungal agents in *Candida albicans* isolates from AIDS patients involve specific multidrug transporters. *Antimicrob. Agents Chemother.* **39**, 2378–2386.

Sanglard, D., Ischer, F., Monod, M. & Bille, J. 1996 Susceptibilities of *Candida albicans* multidrug transporter mutants to various antifungal agents and other metabolic inhibitors. *Antimicrob. Agents Chemother.* **40**, 2300–2305.

Sanglard, D., Ischer, F., Monod, M. & Bille, J. 1997 Cloning of *Candida albicans* genes conferring resistance to azole antifungal agents: characterization of *CDR2*, a new multidrug ABC transporter gene. *Microbiology* **143**, 405–416.

Sanglard, D., Ischer, F., Koymans, L. & Bille, J. 1998 Amino acid substitutions in the cytochrome P-450 lanosterol 14α-demethylase (CYP51A1) from azole-resistant *Candida albicans* clinical isolates contribute to resistance to azole antifungal agents. *Antimicrob. Agents. Chemother.* **42**, 241–253.

Santos, M. A. S., Perreau, V. M. & Tuite, M. F. 1996 Transfer RNA structural change is a key element in the reassignment of the CUG codon in *Candida albicans*. *EMBO J.* **15**, 5060–5068.

Santos, M. A. S., Ueda, T., Watanabe, K. & Tuite, M. F. 1997 The non-standard genetic code of *Candida* spp.: an evolving genetic code or a novel mechanism for adaptation? *Mol. Microbiol.* **26**, 423–431.

Santos, M. A. S., Cheesman, C., Costa, V., Moradas-Ferreira, P. & Tuite, M. F. 1999 Selective advantages created by codon ambiguity allowed for the evolution of an alternative genetic code in *Candida* spp. *Mol. Microbiol.* **31**, 937–947.

Schröppel, K., Rotman, M., Galask, R. & Soll, D. R. 1994 Evolution and replacement of *Candida albicans* strains during recurrent vaginitis demonstrated by DNA fingerprinting. *J. Clin. Microbiol.* **32**, 2646–2654.

Shapiro, J. A. 1998 Thinking about bacterial populations as multicellular organisms. *A. Rev. Microbiol.* **52**, 81–104.

Sokurenko, E. V., Hasty, D. L. & Dykhuizen, D. E. 1999 Pathoadaptive mutations: gene loss and variation in bacterial pathogens. *Trends Microbiol.* **7**, 191–195.

Staib, P., Kretschmar, M., Nichterlein, T., Köhler, G., Michel, S., Hof, H., Hacker, J. & Morschhäuser, J. 1999 Host-induced, stage-specific virulence gene activation in *Candida albicans* during infection. *Mol. Microbiol.* **32**, 533–546.

Sugita, T. & Nakase, T. 1999 Non-universal usage of the leucine CUG codon and the molecular phylogeny of the genus *Candida*. *Syst. Appl. Microbiol.* **22**, 79–86.

Sullivan, J. T. & Ronson, C. W. 1998 Evolution of rhizobia by acquisition of a 500-kb symbiosis island that integrates into a phe-tRNA gene. *Proc. Natl Acad. Sci. USA* **95**, 5145–5149.

Waldor, M. K. & Mekalanos, J. J. 1996 Lysogenic conversion by a filamentous phage encoding cholera toxin. *Science* **272**, 1910–1914.

White, T. C. 1997*a* Increased mRNA levels of *ERG16*, *CDR*, and *MDR1* correlate with increases in azole resistance in *Candida albicans* isolates from a patient infected with human immunodeficiency virus. *Antimicrob. Agents Chemother.* **41**, 1482–1487.

White, T. C. 1997*b* The presence of an R467K amino acid substitution and loss of allelic variation correlate with an azole-resistant lanosterol 14α-demethylase in *Candida albicans*. *Antimicrob. Agents Chemother.* **41**, 1488–1494.

White, T. C., Marr, K. A. & Bowden, R. A. 1998 Clinical, cellular, and molecular factors that contribute to antifungal drug resistance. *Clin. Microbiol. Rev.* **11**, 382–402.

Wirsching, S., Michel, S., Köhler, G. & Morschhäuser, J. 2000 Activation of the multiple drug resistance gene *MDR1* in fluconazole-resistant, clinical *Candida albicans* strains is caused by mutations in a *trans*-regulatory factor. *J. Bacteriol.* **182**, 400–404.

Ziebuhr, W., Heilmann, C., Götz, F., Meyer, P., Wilms, K., Straube, E. & Hacker, J. 1997 Detection of the intercellular adhesion gene cluster (ica) and phase variation in *Staphylococcus epidermidis* blood culture strains and mucosal isolates. *Infect. Immun.* **65**, 890–896.

Ziebuhr, W., Krimmer, V., Rachid, S., Lößner, I., Götz, F. & Hacker, J. 1999*a* A novel mechanism of phase variation of virulence in *Staphylococcus* *epidermidis*: evidence for control of the polysaccharide intercellular adhesin synthesis by alternating insertion and excision of the insertion sequence element IS*256*. *Mol. Microbiol.* **32**, 345–356.

Ziebuhr, W., Ohlsen, K., Karch, H., Korhonen, T. & Hacker, J. 1999*b* Evolution of bacterial pathogenesis. *Cell. Mol. Life Sci.* **56**, 719–728.

Ziebuhr, W., Dietrich, U., Trautmann, M. & Wilhelm, M. 2000 Chromosomal rearrangements affecting biofilm production and antibiotic resistance in a *Staphylococcus epidermidis* strain causing shunt-associated ventriculitis. *Int. J. Med. Microbiol.* **290**, 115–120.

THE IMMUNE RESPONSES TO BACTERIAL ANTIGENS ENCOUNTERED *IN VIVO* AT MUCOSAL SURFACES

GORDON DOUGAN*, MARJAN GHAEM-MAGHAMI, DEREK PICKARD,
GAD FRANKEL, GILL DOUCE, SIMON CLARE, SARAH DUNSTAN
and CAMERON SIMMONS

*Department of Biochemistry, Imperial College of Science,
Technology and Medicine, London SW7 2Z, UK*

Mammals have evolved a sophisticated immune system for handling antigens encountered at their mucosal surfaces. The way in which mucosally delivered antigens are handled influences our ability to design effective mucosal vaccines. Live attenuated derivatives of pathogens are one route towards the development of mucosal vaccines. However, some molecules, described as mucosal immunogens, are inherently immunogenic at mucosal surfaces. Studies on mucosal immunogens may facilitate the identification of common characteristics that contribute to mucosal immunogenicity and aid the development of novel, non-living mucosal vaccines and immunostimulators.

Keywords: mucosal vaccines; IgA; *Citrobacter*; enterotoxins

1. INTRODUCTION

During a lifetime individuals ls continuously encounter antigens at the mucosal suraces (respiratory, gut, urinogenital, corneal) of the body. Most of these are harmless environmental antigens, whereas others can be components of potentially life-threatening pathogens. Mounting an immune response to environmental antigens can be hazardous in terms of energy expenditure and the danger of autoimmunity and allergy. Since mammals have been continuously exposed to both environmentally derived and pathogen-associated antigens they have evolved mechanisms that enable them to tightly regulate immune responses to mucosally encountered antigens. In this there is a contradiction. There is a need to respond weakly to non-hazardous antigens but vigorously to pathogens. Mammals are further compromised by their dependence on the exchange of nutrients and gases with the environment. Rather than cover themselves in an impregnable shell they have evolved to exchange essential molecules with the environment via mucosal surfaces. These sites of nutrient exchange are the most vulnerable points for infection. Perhaps in response to these

---

*Author for correspondence (g.dougan@ic.ac.uk).

evolutionary pressures mammals have evolved a sophisticated mucosal-associated immune system that is integrated closely with the systemic immune system.

2. THE MUCOSAL IMMUNE SYSTEM: A BARRIER TO DELIVERING THERAPEUTIC AGENTS AND VACCINE ANTIGENS

The mucosal immune system presents a number of practical problems to the immunologist and vaccinologist. Most antigens are apparently poorly immunogenic when processed through mucosal surfaces (Levine & Dougan 1998). In this definition they are poor inducers of serum IgG or secretory IgA. However, early investigations demonstrated that animals can display systemic tolerance following a mucosal encounter with an antigen even in the absence of antigen-specific antibody production (Garside *et al.* 1999). Thus, an immune phenotype was present if the correct readout was measured. The physical location of the mucosal immune cells within the body also limited experimental access to immune inductive sites forcing investigators to use invasive or systemic techniques to measure immunological changes. In spite of these hurdles the interest in mucosal immunology has increased in recent years. This is partly because of the potential practical benefits of mucosally targeted therapies. These include mucosally deliverable vaccines against infectious agents or mucosal-tolerizing agents to treat autoimmune disease. The fear of needle contamination and the spread of infection (HIV or hepatitis) have also favoured oral or nasal delivery of antigens and drugs.

3. THE CONCEPT OF MUCOSAL IMMUNOGENS

Some antigens are clearly more immunogenic than others when delivered to mucosal surfaces. Antigens that have inherent immunogenicity when delivered to the mucosa can be described as mucosal immunogens. Mucosal immunogens fall into two classes, those that are alive and those that are non-living.

4.   THE IMMUNOGENICITY OF LIVE MUCOSAL VACCINES

It makes sense that the immune surveillance system should be able to identify the presence of live pathogens. Exactly how this recognition works may
be very complex but attenuated live micro-organisms are clearly a potential route towards vaccine development. Many effective vaccines have been
based on this approach, the caveat being that attenuated strains must not
be able to revert to virulence (Levine *et al.* 1997). Live vaccines have proven
particularly effective when delivered through the mucosal route. Examples
of mucosally deliverable vaccines include those based on attenuated mycobacteria, *Vibrio cholerae*, *Salmonella typhi* and polio virus. Attenuated
live micro-organisms are likely to follow a similar preliminary route of colonization of the host compared to the fully virulent parent. This will include
potential encounters with M and immune cells. In addition, direct contacts
with eukaryotic cells and ligands (possibly leading to an intracellular phase)
may be closely paralleled. Thus, live vaccines may be treated in a similar
manner to virulent pathogens. Factors that may enhance the immunogenicity of live vaccines may include (i) the ability to produce antigens expressed
only in the host and not on laboratory media, (ii) the ability to adhere to
or colonize immune cells and deposit antigen directly into particular intracellular processing pathways, (iii) the ability to activate innate surveillance
mechanisms through generic molecules such as lipopolysaccharides (LPS),
(iv) the ability to produce metabolites that can activate surveillance cells,
and (v) the ability to reach immune-inductive sites.

A number of pathogens (see above) have been attenuated to create
live vaccines that can be delivered via mucosal routes. This approach was
first explored soon after Pasteur had generated vaccines based on passaged
micro-organisms. Indeed, BCG was extensively exploited as a mucosal (oral)
vaccine. Mucosal vaccines based on live attenuated micro-organisms rely on
the ability of the pathogen to target the mucosa and penetrate mucosal associated lymphoid tissues. Such vaccines, for example the Sabin polio vaccine, can be extremely effective. The underlying basis of attenuation does
not have to be known in order for a live vaccine to be successful. However,
with the improving knowledge of the molecular basis of infection and an
emphasis on safety above all else, it would currently be difficult to regis-
ter a live human vaccine without knowing the basis of attenuation. This

knowledge can greatly simplify the process of vaccine quality control and can allow licensing agencies to evaluate the likelihood of reversion to virulence. Perhaps the biggest challenge facing new live vaccines is obtaining the balance between optimal immunogenicity with an absence of reactogenicity in the vaccine. This is a window of acceptability that has proved very difficult to hit using genetically characterized live vaccines. Much recent work on live mucosal vaccine development has focused on enteric bacteria. Work on shigella has been extensive, but so far no vaccine has been brought forward for registration. Although several candidate shigella vaccines have been developed, so far, all promising candidates have had problems with reactogenicity in the clinic (Formal *et al.* 1989; Klee *et al.* 1997; Levine *et al.* 1997; Coster *et al.* 1999). Work in *V. cholerae* has focused on cholera enterotoxin (CT) defective strains. One CT-negative candidate, CVD103, has been extensively evaluated in the clinic and in the field (Tacket *et al.* 1992; Kaper *et al.* 1995). This vaccine performed very impressively in the clinic during phase I and II immunogenicity and challenge studies. The vaccine was immunogenic in a single dose, eliciting serum and local anti-vibrio antibodies in most volunteers and was impressively protective. After initial successes in the field this vaccine encountered some problems. A recent efficacy study in Indonesia generated disappointing protection against cholera. This may, in part, be due to the different intestinal environment in many individuals in developing compared to developed countries (Lagos *et al.* 1999). The normal flora and even gut architecture can differ enormously between the two groups, and higher doses of CVD103 vaccine were required in locals from the tropics (Thailand) to elicit similar levels of immunity to Westerner (in this case from the USA) (Tacket *et al.* 1992, 1999; Su-Arehawaratana *et al.* 1992). This could be due to competitive exclusion by the normal flora or activated or primed immunity in individuals from developing countries. Whatever the causes, this is an area of major interest to vaccine developers and must be better understood if we are to improve mucosal vaccine delivery.

Arguably, the most intensively studied area of live enteric vaccine development has involved *Salmonella*. This area has been driven by work using the murine model to identify virulence-associated genes in pathogenic *Salmonella* such as *Salmonella typhimurium*. In the last several years over 100 *Salmonella* genes have been implicated in virulence using the murine model (e.g. see Hensal *et al.* 1995). Many of these attenuating mutations have been proposed as potential components of live salmonella vaccines either in veterinary species or in man (usually as typhoid vaccines on a *S. typhi* background; Dougan *et al.* 1994). Out of these attenuating lesions, relatively few have been evaluated systematically in the murine model as components of live vaccines. In order to be useful in a live vaccine, an attenuating mutation must not over-attenuate (leading to poor protection) or under-attenuate (leading to reactogenicity). Many candidate attenuating mutations cannot be used in salmonella vaccines because they either over- or under-attenuate the vaccine strain (O'Callaghan *et al.* 1988). Of course this is a simplification because the mouse is not the ideal model for selecting candidate mutations for salmonella derivatives to be used as vaccines in other mammalian species. This we know, as some mutations that attenuate salmonella in the mouse do not attenuate in other mammals such as the cow (Tsolis *et al.* 1999). Nevertheless, the murine model has been used to underpin a number of candidate attenuating lesions for use in veterinary salmonellae and *S. typhi*. For example, an *S. typhi* Ty2 derivative based on *aroC*, *aroD* and *htrA* attenuation is currently in phase II clinical trial as a candidate human typhoid vaccine (Tacket *et al.* 1997). This strain, known as CVD908 *htrA* is highly attenuated in humans, not reaching the bloodstream in detectable numbers after oral immunization with as many as 109 viable bacteria. CVD908 *htrA* is also immunogenic, generating circulating B cells producing anti-*Salmonella* LPS antibodies. It is too early to say if this vaccine will perform well in the field.

Why are live salmonella and other pathogen-based vaccines effective mucosal immunogens? Like their wild-type parents, these candidate vaccine strains are likely to target the mucosal surface via M cells or epithelia and interact with antigen-presenting cells, such as dendritic cells and monocytes. We know *Salmonella* can invade mammalian cells and localize within a vacuole in both phagocytic and non-phagocytic cells (Finlay & Falkow 1997). From these sites there is plenty of opportunity to interact with the immune system and at the same time avoid the attention of antibodies and other immune effectors. The expression of generic activating molecules such as peptidoglycan, lipoproteins and LPS, recognized by innate immune effectors, may help increase the immune potential of these live vaccines. Unfortunately, these are the same molecules that can activate the fever response potentially leading to reactogenicity.

5.   THE MUCOSAL IMMUNOGENICITY OF NON-LIVING ANTIGENS

It is relatively easy to picture how a live pathogen may be immunogenic when administered via a mucosal surface. What about non-living antigenic preparations? Why are some more immunogenic than others? What molecular features of antigens might contribute to their mucosal immunogenicity? There are relatively few antigens (mucosal immunogens) that in a purified form are able to elicit significant levels of secretory IgA and serum IgG when administered to mucosal surfaces. Furthermore, some mucosal immunogens can only stimulate local IgA responses whereas others can stimulate both local and systemic immunity. In addition, these properties may, in part, be host dependent.

We know that the immunogenicity of antigens can be improved by using generic methods that protect the antigen from denaturation or degradation (e.g. encapsulation) (O'Hagen 1990). Antigens differ in their ability to resist the harsh conditions they are likely to encounter close to host body surfaces. Thus, antigens from pathogens that have evolved to retain biological function *in vivo* in body tissues might be more adapted to persist in the host and resist degradation and consequently be more immunogenic (Dougan 1994). Antigen persistence in the host may also be enhanced by an ability to bind mammalian cells and target mucosal surfaces. Aizapurua & Russell-Jones (1988) attempted to define classes of molecules that could act as mucosal immunogens by screening different antigens in a model oral immunization model. They were able to show that some, but not all, antigens that targeted mucosal surfaces had enhanced mucosal immunogenicity. Polymerized molecules such as flagella or fimbriae fell into the immunogenic class. This requirement for binding is not surprising as an antigen that can target the mucosa is likely to reach immune inductive sites and cells at a higher concentration than non-binding antigens. However, it is important to recognize that not all antigens that bind mucosal surfaces are necessarily mucosal immunogens as the nature of the binding site, coupled with antigen persistence, may be critical. For example, binding to enterocytes or epithelial mucus may not enhance antigen translocation to immune inductive sites in some tissues. Unfortunately, most mucosal immunogens are only moderately immunogenic in uncapsulated form and this factor has limited our ability to design and perform serious comparative experiments. Thus, data in this area are limited.

6.   ENTEROTOXINS AS MUCOSAL IMMUNOGENS AND ADJUVANTS

Perhaps the best known example of a class of mucosal immunogens are the bacterial enterotoxins. Indeed, CT and *Escherichia coli* (LT) entero-toxins are recognized as the most potent of all known mucosal immuno-gens. They are so immunogenic at mucosal surfaces that they can activate immune responses to co-administered, non-coupled, bystander molecules that are normally poorly immunogenic at mucosal surfaces (Elson & Eald-ing 1984; Lycke & Holmgren 1986). For example, mice will not normally mount significant secretory or systemic antibody responses to tetanus tox-oid administered orally or intranasally. However, if tetanus toxoid is mixed with microgram quantities of CT or LT, mice will readily sero-convert and produce anti-tetanus toxoid IgA at the local mucosal surfaces as well as serum IgG (Jackson *et al.* 1994; Douce *et al.* 1995). As a consequence both LT and CT are referred to as mucosal adjuvants. This attractive property of these molecules is compromised by the fact that both LT and CT, al-though relatively weak toxins for mice, are highly toxic for humans and some other animals. This factor alone limits their value as practical mu-cosal adjuvants but does not preclude their use as experimental adjuvants in mice. The structures of both LT and CT have been defined using crys-tallography and the elegant structure of these enterotoxins has provided some clues as to why they are effective mucosal immunogens (Rappuoli *et al.* 1999). They have a very compact structure and the holotoxins are quite resistant to denaturation and degradation by proteases. Furthermore, they have the ability to target receptors (gangliosides and glycosylated proteins) at the surface of both epithelial and immune cells. Thus, their ability to target different types of cells may be an important characteris-tic. The influence of the cell-binding activities of LT (Nashar *et al.* 1998) and other ADP-ribosylating toxins such as pertussis toxin (Cropley *et al.* 1995; Roberts *et al.* 1995) have been studied using site-directed mutants or chemical inactivation. These studies have confirmed cell binding as a critical property contributing mucosal adjuvanticity and immunogenicity. However, it is important to note that there may be some situations in which non-binding enterotoxin derivatives may retain mucosal adjuvant activity.

How is adjuvanticity linked to other biological activities of the entero-toxins? Both CT and LT have a sophisticated tertiary structure. They belong to the AB class of bacterial enterotoxins and are composed of a pentameric B oligomer that binds receptors on the surface of eukaryotic cells and an enzymatically active A-subunit that is an ADP-ribosyltransferase responsible for toxicity. The A-subunit is associated with the B-subunit and together they form a tight complex which is highly protease resistant. The structure of both CT and LT has been probed by introducing site-directed amino-acid substitutions into both the A- and B-subunits. Mutations in the A-subunit have been identified that fully or partially inactivate the ADP-ribosyltransferase activity and hence reduce the toxicity of the molecule (Pizza *et al.* 1994). Careful studies using different A-subunit mutant derivatives of both LT and CT have been used to elucidate the contribution of holotoxin formation and enzymatic activity and/or toxicity to mucosal immunogenicity and adjuvanticity. This work has recently been reviewed in detail (Rappuoli *et al.* 1999) and will be described briefly here.

LT and CT mutants that have a destabilized AB structure in terms of subunit association or ability to resist degradation or denaturation are less immunogenic. Thus, the compact structure of the molecule may be required to enhance resistance and persistence in tissues. Some non-toxic derivatives of LT retain significant mucosal adjuvant activity, e.g. LTK63 (Douce *et al.* 1995). Thus, toxicity and ADP-ribosyltransferase activity are not essential for adjuvanticity. Although the ADP-ribosyltransferase activity is not essential for mucosal adjuvant activity it can enhance this activity. For example, mutants that retain partial ADP-ribosyltransferase activity are better mucosal adjuvants than LTK63, e.g. LTR72 (Douce *et al.* 1997; Giuliani *et al.* 1998). Hence, by studying these mutant derivatives it has proved possible to identify combinations of important features that contribute to both immunogenicity and adjuvanticity at mucosal surfaces. What is the importance of this information for vaccine design? By breaking down the individual features of LT and CT that contribute to mucosal adjuvant activity it has been possible to use this information to design completely novel mucosal adjuvants. For example, by combining the antibody-binding domain of *Staphylococcus aureus* protein A with the A-subunit of CT, a novel mucosal adjuvant CTA1-DD has been design as a prototype of a new class of artificial mucosal adjuvants (Agren *et al.* 1999).

7.   UNDERSTANDING THE REGULATION OF MUCOSAL IMMUNE RESPONSES

If we are to improve methods for mucosal therapies and mucosal vaccination, we need to understand the fundamentals of how the immune system regulates the immune response to mucosally delivered antigens (Bienenstock *et al.* 1978; McGhee *et al.* 1992). What are the molecular mechanisms that initiate an active (IgA, serum IgG) compared to a tolerant immune response? Indeed, what is the basis of responsiveness compared to non-responsiveness and how is this balance maintained or changed? It could be argued that there are two fundamental elements interacting to maintain this balance. These elements are the formulation of the antigens derived exogenously from the environment (either environmental or pathogen derived) together with the mucosal immune regulator systems. Evidence has accumulated that T-cell responses to antigens presented via mucosal cells are tightly regulated. Both CD4$^+$ and CD8$^+$ T cells have been implicated in this form of regulation in experimental approaches involving vaccination and tolerance induction. A new class of regulatory T cell, producing high levels of interleukin 10 that proliferate poorly in response to antigen have been identified in mucosal tissues (Groux *et al.* 1997). The establishment of particular cytokine expression patterns by these and other regulatory cells may be critical in this phenomenon. It is also likely that the regulation of antigen presentation by antigen-presenting cells at mucosal surfaces is a critical regulatory step. Indeed, dendritic cells associated with mucosal surfaces may possess significantly different properties from other dendritic cell populations. If tight regulatory networks are operating *in vivo*, this will complicate any attempts to use *in vitro* methods of assessment of the function of mucosal immune cells since key signals may be missing. Thus, *in vivo* studies are essential if we are to understand how the regulation of mucosal immune responses is achieved.

8.   *IN VIVO* MODELS OF MUCOSAL IMMUNE DYSFUNCTION

The genetic manipulation of pathogens has provided a rich source of information on the role of bacterial gene products in infection and pathogenesis.

Murine models have proved to be particularly fruitful and they now have the added value that both the pathogen and the host can be genetically manipulated. A murine model for analysing mucosal immunity and immune dysfunction could be particularly attractive. An *in vivo* approach to understanding mucosal T-cell regulation is evolving through the study of the pathogenic mechanisms of a family of bacterial pathogens that cause attaching and effacing lesions on gut enterocytes. The interaction of these pathogens with cells of the gastrointestinal tract is complex, but new studies (summarized below) have suggested that these bacteria can profoundly modulate mucosal T-cell responses. These bacteria may therefore represent useful tools with which to dissect elements of the intricate immune regulatory network present in the mucosa.

9. *CITROBACTER RODENTIUM*, AN EXPERT MUCOSAL IMMUNOMODULATOR

Citrobacter rodentium colonizes the distal colon of susceptible inbred mouse strains via the formation of attaching and effacing (AE) lesions on colonic enterocytes (Schauer *et al.* 1993a). Ultrastructurally, these AE lesions are indistinguishable from those caused by enteropathogenic *E. coli* (EPEC) infection in humans. AE lesions are characterized by the intimate attachment of bacteria to cup-like pedestals on the luminal side of the enterocyte cell membrane and the subsequent destruction of host cell microvilli. The bacterial virulence determinants required for AE lesion formation have been most extensively described in EPEC. Formation of AE lesions is dependent on expression of several bacterial proteins, which are encoded by genes located on a chromosomal pathogenicity island called the locus for enterocyte effacement (LEE). The *eae* gene, which encodes intimin, an outer membrane protein adhesin, was the first gene in the LEE to be associated with AE lesion formation (Jerse *et al.* 1990). In addition to intimin, the LEE encodes a type III secretion system, a translocated intimin receptor (Tir) and three EPEC secreted proteins required for protein translocation (reviewed in Frankel *et al.* 1998). Characterization of these LEE-encoded virulence determinants has been performed primarily by examining the interaction of EPEC bacteria with continuous human epithelial cell lines or human tissue explants. While these studies have helped facilitate a dissection of the events leading to AE lesion formation, they have not fully revealed the extent to which individual LEE-encoded proteins contribute to bacterial pathogenesis *in vivo*. Furthermore, these *in vitro* studies have not provided information on the extent, type or specificity of the infected host's response to EPEC antigens; information which is likely to be important in the rational design of vaccines or therapeutics to prevent EPEC infections in humans. While *in vivo* studies of EPEC pathogenesis and immunity are clearly desirable, the inability of human isolates of EPEC to colonize small rodents has meant alternative animal models have been investigated. One of these models, *C. rodentium* infection of mice, has several features which make it attractive for furthering research into EPEC and also mucosal T-cell regulation.

10. IMMUNOBIOLOGY OF *CITROBACTER RODENTIUM* INFECTION

The *C. rodentium* chromosome has been shown to contain genes with homology to those located in the LEE pathogenicity island found in human EPEC strains (Schauer *et al.* 1993). The identification of LEE homologues in *C. rodentium* has facilitated the construction of defined mutants and their subsequent characterization in mice. An *eae* mutant of *C. rodentium* (strain DBS255) was shown to be avirulent. However, virulence could be restored by complementation with either the *eae* gene from *C. rodentium* (Schauer *et al.* 1993b), or the *eae* gene from the prototype EPEC strain E2348/69 (strain DBS255 (pCVD438) (Frankel *et al.* 1996). The fact that an *eae* mutant of *C. rodentium* is attenuated in mice is consistent with human studies demonstrating that an *eae* mutant of EPEC strain E2348/69 is attenuated in humans (Donnenberg *et al.* 1993). In addition to its role as an adhesin, intimin also contributes to the induction of colonic epithelial cell hyperplasia, the second characteristic feature of *C. rodentium* infection in mice. The colonic hyperplasia that occurs during *C. rodentium* colonization is associated with the expression of inflammatory cytokines (IL-1, tumour necrosis factor-alpha (TNFα)) and a strong T-cell infiltrate consisting predominantly of CD4$^+$ T cells with a Th1 phenotype (Higgins *et al.* 1999). Interestingly, the characteristics of the T-cell infiltrate which occurs in the mouse colon during *C. rodentium* infection share striking similarity to the cellular events occurring in murine models of inflammatory bowel disease (IBD). Epithelial cell hyperplasia in mice can also be induced via intrarectal administration of formalin-killed *C. rodentium* or EPEC, but not by *eae* mutants of *C. rodentium* or EPEC, indicating a critical role for intimin in this effect (Higgins *et al.* 1999). Indeed, intimin-bearing bacterial cells appear to be sufficient for this effect, since intrarectal administration of formalin-killed *E. coli* K12 expressing intimin, but not *E. coli* K12 alone, also evokes colonic hyperplasia (Higgins *et al.* 1999). The mechanism through which intimin-associated bacteria promotes hyperplasia is not clear, but clues are provided by the ability of the C-terminal 280 amino acids of intimin to co-stimulate T cells *in vitro*, which potentially occurs as a result of intimin binding to cell-surface receptors (Higgins *et al.* 1999). If intimin does indeed bind resident or infiltrating T cells in the gut, then this may promote unregulated proliferation of these mucosal T cells, which under normal situations are hyporesponsive and require strong co-stimulatory signals before cytokine production occurs. Following mucosal T-cell activation, cytokine production has been shown to trigger a cascade of events, including production of the epithelial cell mitogen keratinocyte growth factor (KGF) by mesenchymal cells (Bajaj-Elliott *et al.* 1998). The uncontrolled epithelial cell proliferation characteristically seen during *C. rodentium* infection and in other models of IBD may result from inflammatory cytokine-driven overexpression of KGF. Although the colonic hyperplasia observed in *C. rodentium* infected mice has many similarities with other murine models of IBD, *C. rodentium* elicited hyperplasia represents a unique model of mucosal T-cell unregulation. This syndrome is caused by a specific microbial agent and is critically dependent in the presence of one well-defined molecule, intimin. Thus, intimin can be regarded as having a dual role as both a cell adhesin and as a molecule which the pathogen uses to modulate the function of immune cells *in vivo*. The epithelial cell hyperplasia observed during *C. rodentium* infection may be of benefit to the bacterium by providing the pathogen with a greater surface area to colonize and thereby increase shedding. Conversely, the peak of the hyperplastic response in mice occurs when *C. rodentium* numbers in the colon begin to subside. This may suggest that hyperplasia represents a component of a protective immune response. Although hyperplasia has not been regarded as a feature of EPEC infection there are reports describing hyperplasia and villus atrophy in small intestinal biopsies from EPEC-infected children. This suggests that hyperplasia resulting from the unregulation of mucosal T cells may occur in some individuals (Fagundes Neto *et al.* 1989; Hill *et al.* 1991).

The results summarized here highlight the importance of intimin in *C. rodentium* colonization and infection and form the rationale behind current studies that are designed to determine whether pre-existing immune responses to intimin can prevent *C. rodentium* infection of mice.

11. *CITROBACTER RODENTIUM* INFECTION OF MICE AS A MODEL FOR EVALUATING CANDIDATE EPEC VACCINE ANTIGENS

C. rodentium infection in mice can be exploited to address questions relating to enteric infections and the potential of LT based mucosal adjuvants for eliciting protective immunity. For example: Can combinations of intimin and LT-based adjuvants elicit protection in mice against *C. rodentium* colonization and/or disease? The oral infectious dose of a *C. rodentium* strain expressing intimin from EPEC (strain DBS255(pCVD438)) in CH3/Hej is

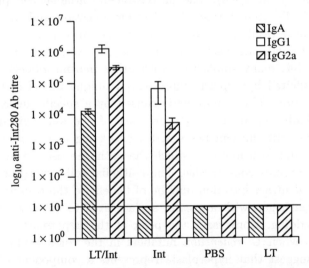

Fig. 1. Co-delivery of Int-280 with the mucosal adjuvant LT elicits a strong serum IgG and IgA response to Int-280. Groups of 5 C3H/Hej mice were intranasally immunized three times with 10 mg of Int-280 plus 1 mg of LT, 10 mg of Int-280 alone, 1 mg of LT alone or PBS alone. The data depicts the mean titre (plus standard error) of the antibody response to Int-280.

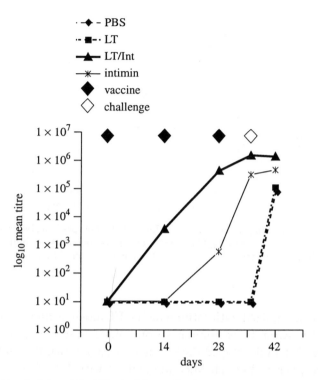

Fig. 2. Kinetic of the anti-Int280 total Ig antibody response in immunized mice pre- and post-challenge with DBS255(pCVD438). The data depict the mean serum Ig titre against Int-280.

approximately 10^5 bacteria. At this dose, infected mice become detectably colonized for approximately 24 days and develop visible hyperplastic colons by days 10–14. The development of hyperplasia is also associated with weight loss, although this is reversed when the numbers of *C. rodentium* in the colon begin to subside at around day 16. At higher infectious doses, the peak bacterial load in the colon is reached earlier (around day 7), as is the development of hyperplasia and weight loss. Whilst a higher infectious doses leads to earlier onset of disease, it also results in more rapid clearance of the infecting bacterium, which is usually absent from the colons of infected mice 16–18 days post-infection. C3H/Hej mice were intranasally immunized three times with a highly purified preparation of the C-terminal 280 amino acids of intimin (Int280) from EPEC together with LT. Control mice were

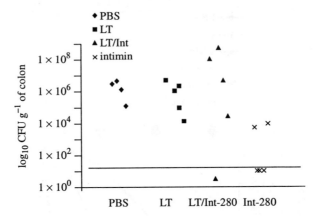

Fig. 3. Pre-existing immune responses to Int-280 do not prevent bacterial colonization by *C. rodentium* DBS255(pCVD438). The data depict the number of viable
DBS255(pCVD438) bacteria recovered from the colons of individual mice 14 days post-
challenge.

intranasally immunized with intimin alone, LT alone or phosphate buffered
saline (PBS). After three immunizations, mice that received Int280 ad-
mixed with LT had developed the strongest serum anti-Int280 IgG and IgA
antibody responses (Fig. 1). Mice immunized with Int280 alone mounted
anti-Int280 IgG responses, but not serum IgA responses (Fig. 1). Immunized
mice were orally challenged with DBS255(pCVD438) ten days after the last
immunization. On day 14 post-challenge, mice not previously immunized
with Int280 rapidly developed serum anti-Int280 antibodies (Fig. 2). The
extent of bacterial colonization and hyperplasia was measured 14 days after being challenged with DBS255(pCVD438). Microbiological analysis of
the colons of challenged mice suggested that the Int280-specific immune
responses elicited by vaccination were not sufficient to protect mice from
bacterial colonization nor the concomitant induction of colonic hyperplasia
(Fig. 3). Other EPEC-associated determinants are currently under investigation as vaccine antigens in this model of *C. rodentium* infection. In
addition, *C. rodentium* infection in other strains of mice has been evaluated with a view to investigating the immunological mechanisms through
which *C. rodentium* promotes mucosal T-cell unregulation and hyperplasia.
In summary, *C. rodentium* infection of mice represents a useful model in
which to further our understanding of EPEC immunobiology. Furthermore,

by using an 'expert' mucosal immunomodulator like *C. rodentium*, it is feasible to investigate mucosal T-cell regulation using a completely novel approach.

12. CONCLUDING REMARKS

We are in the middle of an intensive period of scientific investigations into the molecular basis of infectious diseases. As we begin to understand more about microbial pathogenicity, we appreciate the need to study pathogens *in vivo*, as they colonize, replicate and survive in host tissues. The design of novel and effective vaccines against mucosal pathogens will clearly benefit from detailed studies on the mechanisms of pathogenicity employed by the microbial pathogen and by monitoring the combined mucosal and systemic immune response associated with infection and recovery. In addition, many of the products of pathogenic bacteria might find specific use as mucosal vaccine components and/or more generic immuno-modulators. The importance of studies in the natural host or whole animal systems cannot be over-emphasized. In terms of immunity, there are usually extremely complex intercellular interactions occurring *in vivo* between different differentiated cell types and it is impossible to study many of these interactions *in vitro*. Indeed, the study of *in vitro* grown cells may provide misleading information on how regulatory networks are operating in real tissues. We are only just beginning to understand how the expression of immunity is regulated at mucosal surfaces and the use of bacteria and their products provides us with valuable tools for these ongoing investigations. As this work continues we can expect new and more effective vaccines to be designed and developed.

REFERENCES

Agren, L. C., Ekman, L., Lowenadler, B., Nedrud, J. G. & Lycke, N. Y. 1999 Adjuvanticity of the cholera toxin A1-based gene fusion protein, CTA1-DD, is critically dependent on the ADP-ribosyltransferase and Ig-binding activity. *J. Immunol.* **15**, 2432–2440.

Aizapurua, D. H. & Russell-Jones, A. 1988 Oral vaccination: identification of classes of proteins that provoke an immune response upon oral feeding. *J. Exp. Med.* **167**, 440–451.

Bajaj-Elliott, M., Poulsom, R., Pender, S. L., Wathen, N. C. & MacDonald, T. T. 1998 Interactions between stromal cell-derived keratinocyte growth factor and epithelial transforming growth factor in immune-mediated crypt cell hyperplasia. *J. Clin. Invest.* **102**, 1473–1480.

Bienenstock, J., McDermott, M., Befus, D. & O'Neil, L. M. 1978 A common mucosal immunological system involving the bronchus, breast and bowel. *Adv. Exp. Med. Biol.* **107**, 53–69.

Coster, T. S. (and 10 others) 1999 Vaccination against shigellosis with attenuated *Shigella flexneri* 2a strain SC602. *Infect. Immun.* **67**, 3437–3443.

Cropley, I., Douce, G., Roberts, M., Chatfield, S., Pizza, M., Rappuoli, R. & Dougan, G. 1995 Mucosal and systemic immunogenicity of a recombinant, non-ADP-ribosylating pertussis toxin: comparison of native and formaldehyde-treated proteins. *Vaccine* **13**, 1643–1648.

Donnenberg, M. S., Tacket, C. O., James, S. P., Losonsky, G., Nataro, J. P., Wasserman, S. S., Kaper, J. B. & Levine, M. M. 1993 Role of the *eaeA* gene in experimental enteropathogenic *Escherichia coli* infection. *J. Clin. Invest.* **92**, 1412–1417.

Douce, G., Turcotte, C., Cropley, I., Roberts, M., Pizza, M., Dominghini, M., Rappuoli, R. & Dougan, G. 1995 Mutants of *Escherichia coli* heat-labile toxin lacking ADP-ribosyltransferase activity act as non-toxic mucosal adjuvants. *Proc. Natl Acad. Sci. USA* **92**, 1644–1648.

Douce, G., Fontana, M., Pizza, M., Rappuoli, R. & Dougan, G. 1997 Mucosal immunogenicity and adjuvanticity of site-directed mutant derivatives of cholera toxin. *Infect. Immun.* **65**, 2821–2828.

Dougan, G. 1994 The molecular basis for the virulence of bacterial pathogens: implications for oral vaccine development. Colworth Lecture. *Microbiology* **140**, 215–224.

Elson, C. J. & Ealding, W. 1984 Generalised systemic and mucosal immunity in mice after mucosal stimulation with cholera toxin. *J. Immunol.* **132**, 2736–2743.

Fagundes Neto, U., Ferreira, V. D. C., Patricio, F. R., Mostaco, V. L. & Trabulsi, L. R. 1989 Protracted diarrhea: the importance of the enteropathogenic *Escherichia coli* (EPEC) strains and *Salmonella* in its genesis. *J. Pediatr. Gastroenterol. Nutr.* **8**, 207–211.

Finlay, B. B. & Falkow, S. 1997 Common themes in microbial pathogenicity revisited. *Microbiol. Mol. Biol. Rev.* **61**, 136–169.

Formal, S. B., Hale, T. L. & Kapfer, C. 1989 *Shigella* vaccines. *Rev. Infect. Dis.* **11** (Suppl. 3), S547–S551.

Frankel, G., Phillips, A. D., Novakova, M., Field, H., Candy, D. C., Schauer, D. B., Douce, G. & Dougan, G. 1996 Intimin from enteropathogenic *Escherichia coli* restores murine virulence to a *Citrobacter rodentium eaeA* mutant: induction of an immunoglobulin A response to intimin and EspB. *Infect. Immun.* **64**, 5315–5325.

Frankel, G., Phillips, A. D., Rosenshine, I., Dougan, G., Kaper, J. B. & Knutton, S. 1998 Enteropathogenic and enterohaemorrhagic *Escherichia coli*: more subversive elements. *Mol. Microbiol.* **30**, 911–921.

Garside, P., Mowat, A. M. & Khoruts, A. 1999 Oral tolerance in disease. *Gut* **44**, 137–142.

Giuliani, M. M., Del Giudice, G. D., Giannelli, V., Dougan, G., Douce, G., Rappuoli, R. & Pizza, M. 1998 Mucosal adjuvanticity of LT72, a novel mutant of *Escherichia coli* heat-labile enterotoxin with partial knock-out of ADP-ribosyltransferase activity. *J. Exp. Med.* **187**, 1123–1132.

Groux, H., O'Garra, A., Bigler, M., Rouleau, M., Antonenko, S., de Vries, J. E. & Roncarolo, M. G. 1997 A CD4^{+} T-cell subset inhibits antigen-specific T-cell responses and prevents colitis. *Nature* **16**, 737–742.

Hensel, M., Shea, J. E., Gleeson, C., Jones, M. D., Dalton, E. & Holden, D. W. 1995 Simultaneous identification of bacterial virulence genes by negative selection. *Science* **269**, 400–403.

Higgins, L. M., Frankel, G., Connerton, I., Goncalves, N. S., Dougan, G. & MacDonald, T. T. 1999a Role of bacterial intimin in colonic hyperplasia and inflammation. *Science* **285**, 588–591.

Hill, S. M., Phillips, A. D. & Walker-Smith, J. A. 1991 Enteropathogenic *Escherichia coli* and life threatening chronic diarrhoea. *Gut* **32**, 154–158.

Jackson, R. J., Staats, H. F., Xu-Amano, J., Takahashi, I., Kiyono, H., Hudson, M. E., Gilley, R. M., Chatfield, S. N. & McGhee, J. R. 1994 Oral vaccine models: multiple delivery systems employing tetanus toxoid. *Annls NY Acad. Sci.* **730**, 217–234.

Jerse, A. E., Yu, J., Tall, B. D. & Kaper, J. B. 1990 A genetic locus of enteropathogenic *Escherichia coli* necessary for the production of attaching and effacing lesions on tissue culture cells. *Proc. Natl Acad. Sci. USA* **87**, 7839–7843.

Kaper, J. B., Morris Jr, J. G. & Levine, M. M. 1995 Cholera. *Clin. Microbiol. Rev.* **8**, 48–86.

Klee, S. R., Tzschaschel, B. D., Falt, I., Karnell, A., Lindberg, A. A., Timmis, K. N. & Guzman, C. A. 1997 Construction and characterization of a live attenuated vaccine candidate against *Shigella dysenteriae* type I. *Infect. Immun.* **65**, 2112–2118.

Lagos, R., Fasano, A., Wasserman, S. S., Prado, V., San Martin, O., Abrego, P., Losonsky, G. A., Alegria, S. & Levine, M. M. 1999 Effect of small bowel bacterial overgrowth on the immunogenicity of single-dose live cholera vaccine CVD 103-Hg. *J. Infect. Dis.* **180**, 1709–1712.

Levine, M. M. & Dougan, G. 1998 Optimism over vaccines administered through mucosal surfaces. *Lancet* **351**, 1375–1376.

Levine, M. M., Woodrow, G. C., Kaper, J. B. & Cobon, G. S. 1997 *New generation vaccines*, 2nd edn. New York: Marcel Dekker, Inc.

Lycke, N. & Holmgren, J. 1986 Strong adjuvant properties of cholera toxin on gut mucosal immune responses to orally presented antigens. *Immunol.* **59**, 301–308.

McGhee, J. R., Mesteky, J., Dertbaugh, M. T., Eldridge, J. H. M. & Kiyono, H. 1992 The mucosal immune system: from fundamental concepts to vaccine development. *Vaccine* **10**, 75–81.

Nashar, T. O., Williams, N. A. & Hirst, T. R. 1998 Importance of receptor binding in the immunogenicity, adjuvanticity and therapeutic properties of cholera toxin and *Escherichia coli* heat-labile enterotoxin. *Med. Microbiol. Immun.* (Berlin) **187**, 3–10.

O'Callaghan, D., Maskell, D., Liew, F. Y., Easmon, C. S. F. & Dougan, G. 1988 Characterisation of aromatic-dependent and purine-dependent *Salmonella typhimurium*: studies on attenuation, persistence and ability to induce protective immunity in BALB/c mice. *Infect. Immun.* **56**, 419–423.

O'Hagan, D. T. 1990 Novel non-replicating antigen delivery systems. *Curr. Opin. Infect. Dis.* **3**, 393–401.

Pizza, M. G., Domenighini, M., Hol, W., Gianneli, V., Fontana, M. R., Giuliani, M., Magagnoli, C., Peppeloni, S., Manetti, R. & Rappuoli, R. 1994 Probing the structure–function relationship of *Escherichia coli* LT-A by site-directed mutagenesis. *Mol. Microbiol.* **14**, 51–61.

Rappuoli, R., Pizza, M., Douce, G. & Dougan, G. 1999 Structure and mucosal adjuvanticity of cholera and *Escherichia coli* heat-labile enterotoxins. *Immunol. Today* **20**, 493–500.

Roberts, M., Bacon, A., Rappuoli, R., Cropley, I., Douce, G., Dougan, G. & Chatfield, S. N. 1995 A mutant pertussis toxin molecule that lacks ADP-ribosyltransferease activity, PT-9K/129G, is an effective mucosal adjuvant for intranasally delivered proteins. *Infect. Immun.* **63**, 2100–2108.

Schauer, D. B. & Falkow, S. 1993a Attaching and effacing locus of a *Citrobacter freundii* biotype that causes transmissible murine colonic hyperplasia. *Infect. Immun.* **61**, 2486–2492.

Schauer, D. B. & Falkow, S. 1993b The *eae* gene of *Citrobacter freundii* biotype 4280 is necessary for colonization in transmissible murine colonic hyperplasia. *Infect. Immun.* **61**, 4654–4661.

Schauer, D. B., Zabel, B. A., Pedraza, I. F., O'Hara, C. M., Steigerwalt, A. G. & Brenner, D. J. 1995 Genetic and biochemical characterization of *Citrobacter rodentium* sp. *J. Clin. Microbiol.* **33**, 2064–2068.

Su-Arehawaratana, P. (and 14 others) 1992 Safety and immunogenicity of different immunization regimens of CVD 103-HgR live oral cholera vaccine in soldiers and civilians in Thailand. *J. Infect. Dis.* **165**, 1042–1048.

Tacket, C. O., Losonsky, G., Nataro, J. P., Cryz, S. J., Edelman, R., Kaper, J. B. & Levine, M. M. 1992 Onset and duration of protective immunity in challenged volunteers after vaccination with live oral cholera vaccine CVD 103-HgR. *J. Infect. Dis.* **166**, 837–841.

Tacket, C. O., Sztein, M. B., Losonsky, G., Wasserman, S. S., Nataro, J. P., Edelman, R., Pickard, D., Dougan, G., Chatfield, S. & Levine, M. M. 1997 Safety and immune response in humans of a live oral *Salmonella typhi* vaccine strain deleted in *htrA* and *aroC*, *aroD* and demonstration in mice of their use as vaccine vectors. *Infect. Immun.* **65**, 452–456.

Tacket C. O. (and 11 others) 1999 Randomized, double-blind, placebo-controlled, multicentered trial of the efficacy of a single dose of live oral cholera vaccine CVD 103-HgR in preventing cholera following challenge with *Vibrio cholerae* Ol El for inaba three months after vaccination. *Infect. Immun.* **67**, 6341–6345.

Tsolis, R. M., Adams, L. G., Ficht, T. A. & Baumler, A. J. 1999 Contribution of *Salmonella typhimurium* virulence factors to diarrheal disease in calves. *Infect. Immun.* **67**, 4879–4885.

INDEX